D1319764

Water in Polymers

Stanley P. Rowland, EDITOR

Southern Regional Research Center

Based on a symposium
cosponsored by the
ACS Macromolecular Secretariat and
the Division of Biological Chemistry
at the 178th Meeting of the
American Chemical Society,
Washington, D. C.,
September 10–13, 1979.

ACS SYMPOSIUM SERIES **127**

RELEASED

AMERICAN CHEMICAL SOCIETY
WASHINGTON, D. C.   1980

75281

QD
1
A512
v. 127

Library of Congress CIP Data

Water in polymers.
(ACS symposium series; 127)

Includes bibliographies and index.

1. Polymers and polymerization—Congresses. 2. Water—Congresses.
I. Rowland, Stanley Paul, 1916–   . II. American Chemical Society. Macromolecular Secretariat. III. American Chemical Society. Division of Biological Chemistry. IV. Series: American Chemical Society. ACS symposium series; 127.

QD380.W37   547.8'4   80–13860
ISBN 0–8412–0559–0   ACSMC8   127 1–597 1980

Copyright © 1980

American Chemical Society

All Rights Reserved. The appearance of the code at the bottom of the first page of each article in this volume indicates the copyright owner's consent that reprographic copies of the article may be made for personal or internal use or for the personal or internal use of specific clients. This consent is given on the condition, however, that the copier pay the stated per copy fee through the Copyright Clearance Center, Inc. for copying beyond that permitted by Sections 107 or 108 of the U.S. Copyright Law. This consent does not extend to copying or transmission by any means—graphic or electronic—for any other purpose, such as for general distribution, for advertising or promotional purposes, for creating new collective works, for resale, or for information storage and retrieval systems.

The citation of trade names and/or names of manufacturers in this publication is not to be construed as an endorsement or as approval by ACS of the commercial products or services referenced herein; nor should the mere reference herein to any drawing, specification, chemical process, or other data be regarded as a license or as a conveyance of any right or permission, to the holder, reader, or any other person or corporation, to manufacture, reproduce, use, or sell any patented invention or copyrighted work that may in any way be related thereto.

PRINTED IN THE UNITED STATES OF AMERICA

ACS Symposium Series

M. Joan Comstock, *Series Editor*

Advisory Board

David L. Allara

Kenneth B. Bischoff

Donald G. Crosby

Donald D. Dollberg

Robert E. Feeney

Jack Halpern

Brian M. Harney

Robert A. Hofstader

W. Jeffrey Howe

James D. Idol, Jr.

James P. Lodge

Leon Petrakis

F. Sherwood Rowland

Alan C. Sartorelli

Raymond B. Seymour

Gunter Zweig

FOREWORD

The ACS Symposium Series was founded in 1974 to provide
a medium for publishing symposia quickly in book form. The
format of the Series parallels that of the continuing Advances
in Chemistry Series except that in order to save time the
papers are not typeset but are reproduced as they are submitted by the authors in camera-ready form. Papers are reviewed under the supervision of the Editors with the assistance
of the Series Advisory Board and are selected to maintain the
integrity of the symposia; however, verbatim reproductions of
previously published papers are not accepted. Both reviews
and reports of research are acceptable since symposia may
embrace both types of presentation.

CONTENTS

POLYSACCHARIDE INTERACTIONS WITH WATER

PREMEATION, TRANSPORT, AND ION SELECTIVITY

SYNTHETIC POLYMERS: WATER INTERACTIONS

PREFACE

Since early in this century, water in polymers has been of interest to scientists concerned with living matter and natural polymers. However, the significance and role of water in synthetic polymers and commercial resins have become recognized only recently, with significant initiation of research in this area and in allied fields.

The symposium was timely, with objectives (a) to bring together scientists from different parts of the field to exchange and discuss information, (b) to teach those reseachers who are presently active on the boundary of the field, and (c) to bring before the entire group some of the latest thinking in connection with the structure and behavior of water itself.

Several divisional cochairmen collected and arranged the papers in the symposium. I should like to thank them: John A. Rupley and Irwin D. Kuntz, Jr. of Division of Biological Chemistry; Daniel F. Caulfield of Cellulose, Paper and Textile Division; Curt Thies of Division of Colloid and Surface Chemistry; Cornelis A. J. Hoeve of Division of Organic Coatings and Plastics Chemistry; and P. Anne Hiltner of Division of Polymer Chemistry.

My special thanks go to Martin Karplus and John Rupley for consultation and advice and to Frank Stillinger for obligingly and capably filling the void when an ill Henry S. Frank had to cancel his opening paper for the symposium.

I thank the Petroleum Research Fund of the ACS for financial assistance granted to this symposium.

The expected professional expertise of the participants in quality of research and presentations was appreciated gratefully. The same extends to the preparation of manuscripts for this monograph. The symposium attendees found the papers and program to be highly informative. I hope that the reader, too, will find the monograph instructive, informative, and useful for perspective to methodology, results, and active researchers.

The chapters in the monograph are grouped into eight sections to suggest to the reader certain close relationships among the chapters within each section. However, there are other relationships across sections, some of which are brought out in the Introduction.

Southern Regional Research Center
AR, SEA, USDA
New Orleans, Louisiana 70179

January 3, 1980

STANLEY P. ROWLAND

INTRODUCTION

STANLEY P. ROWLAND
Southern Regional Research Center, P.O. Box 19687, New Orleans, LA 70179

IRWIN D. KUNTZ, JR.
Department of Pharmaceutical Chemistry, University of California,
San Francisco, CA 94143

For many reasons there is considerable concern for the nature of interactions between water and polymers. However, the two major reasons are: the water-polymer interaction is essential to biological processes, and the water-polymer interaction is often beneficial or detrimental to performance of commercial polymers. These primary concerns are evident throughout the chapters of this monograph, but superimposed atop these is a search for an understanding of the way(s), the driving force(s), and the consequence(s) of the interactions of water with dissolved, swollen or rigid polymeric surfaces.

The initial section of the monograph deals with The Structure of Water as evidenced by its character in the supercooled state [1] (numbers in brackets refer to chapters listed in the Table of Contents), as affected by a dipeptide [2], and as affected by a variety of solutes and polymers [3]. One chapter is regrettably missing; Henry S. Frank, a pioneer of stature in studies of the structure of water, became ill before the symposium and was unable to attend or to complete his manuscript.

The second section, Perspective: Macro- and Microinteractions of Water and Polymers, consists of three chapters. It covers a brief historical introduction leading to an insight into selected aspects of current thought on interactions of water with model solutes and proteins [4], consideration of water-polymer and ice-polymer interfacial regions [5], and examination of stages in the process of protein hydration assessed by a variety of types of measurements [6].

The chapters in sections 1 and 2 provide perspective pertinent to chapters in subsequent sections.

Section 3.	Proteins: The Mobile Water Phase	4 Chapters
Section 4.	Proteins: Ordered Water	4 Chapters
Section 5.	Polysaccharide Interactions with Water	3 Chapters
Section 6.	Permeation, Transport, and Ion Selectivity	7 Chapters

0-8412-0559-0/80/47-127-001$05.00/0
© 1980 American Chemical Society

An Overview of Results

The Structure of Water. Pertinent to the bound, non-
freezing, or interface water in polymers, a subject of subsequent sections, are the studies of Angell et al. (see ref. in
[1]). He shows that pure water at normal pressures cannot be
supercooled below -40°C and that virtually all physical properties of water point to a "lambda anomaly" at -45°C [1].
    The fact that surfaces disrupt the natural order of a bulk
phase serves as basis for Stillinger's attention to supercooling of water in small droplets or clusters [1]. His statistical
mechanical approach to the structure of ice Ih and water as
hydrogen-bonded in ordered and disordered polygonal structures,
respectively, results in a qualitative estimate of the depression of temperature of maximum density. His approach also explains the behavior of supercooled water in terms of structural
fluctuations in the bonded bicyclic octameric water network that
represents ice Ih.
    By molecular dynamics simulation of a dilute aqueous solu-
tion of an alanine dipeptide, Karplus [2] estimates the influence
of the solute on the dynamic properties of water to be limited to
the first solvation layer. The different structures of water in
the vicinity of polar and nonpolar groups were resolved to bulk-
like hydrogen bonding involving a reduced number of neighbors in
the former case and the absence of hydrogen bonding together with
decreased water mobility in the latter case.
    Overtone infrared spectroscopy described by Luck [3] is an
effective means for determining quantitatively the concentrations
of water in nonbonded and hydrogen bonded OH groups. Interesting
results have been obtained for a variety of situations, including
salt solutions, water-organic solvent mixtures, interface effects,
organic molecule hydration, and diffusion in polymeric substrates.
From such studies, Luck classifies water structure as (a) first
shell water hydrate, (b) second shell, disturbed liquid-like
water, and (c) liquid-like water. For salt transport in membranes,
for diffusion of dyes in fibers, and for life in plant and animal
cells, water of types b and c are essential.

    Protein-Water Interactions. There appears a general agree-
ment that all thermodynamic measurements, such as heat capacity
results of Rupley et al. [6] and Hoeve [7] or the NMR data on
frozen materials of Bryant [8], indicate that 0.3-0.4 g H_2O/g
protein form an unfrozen boundary layer at subzero temperature.
The water is primarily associated with polar groups. It seems

appropriate to call this layer "bound" water because its properties are demonstrably different from either supercooled water or
ice at the same temperature. Is this layer a useful concept in
dilute solution? The thermodynamic measurements do not provide a
direct answer because one cannot use a nearby phase boundary to
make the sharp distinctions possible below the freezing point.
One source of information, hydrodynamic measurements, has recently
been reviewed by Squire and Himmel (1). Their analysis suggests
large amounts of water associated with globular proteins, but
even their restricted data set shows large variations between the
amount of hydration calculated from sedimentation or diffusion
results. Because the cube of the friction factor enters into the
calculation, highly accurate data are required to resolve this
matter. The issue of hydrophobic hydration is obscure. Karplus'
calculation [2] and much early work do suggest that water near
hydrophobic surfaces should have modified properties. To date,
the thermodynamic experiments at low temperature and the structural studies (see below) give little direct indication of how
many such water molecules are present and whether their properties
are different from either "bulk" or "bound" water molecules.

Several papers deal with the spatial arrangements of water
molecules near proteins. The neutron-scattering experiments of
Schoenborn [12] set a high standard for careful crystallographic
analysis. Refined x-ray studies are also in progress. Ordered
water molecules are found hydrogen-bonded to polar sites or to
each other in very short chains. Some dozens of such water molecules are observed at high occupancy. The number of water molecules placed by these methods is becoming smaller as the structures become more refined. Hermans [11] suggested that part of
the difficulty is the high salt concentration in many crystals.
We expect this question to be answered as resolution improves and
neutron scattering comes into wider use.
On the theoretical side, work discussed by Karplus [2] and
by Hermans [11] and previous studies of Hagler and Moult (2) and
Stillinger and Rahman (3) make it clear that a similar structural
ordering is found with standard potential functions. Although no
simulation of a protein solution with internal degrees of freedom
for the protein and a large number of water molecules has been
conducted, we may expect that similar structural sites will be
maintained in solution.
Two issues deserve comment. First, there is as yet no
evidence for clatherate water structures surrounding globular
proteins. Second, there is no sharp "boundary" on structural
features. The diffraction experiments encourage us to think of
a spectrum of structural sites characterized by an order parameter
ranging from highly ordered to completely random distribution over
two or three layers of water molecules.

It has been difficult to unravel the microscopic energies of interaction between water molecules and a protein surface by direct experiment. Thermodynamic methods on two component systems never permit a rigorous molecular dissection of this kind. Infrared and Raman spectroscopy, as described by Luck [3], have some promise.

The theoretical attack represented by Stillinger [1], Karplus [2], and Hermans [11] has become increasingly productive. In brief, water-water energies are relatively large for nonbonded interactions and water-protein interactions are roughly comparable in strength, making water-protein surfaces of the "soft-soft" kind in Adamson's nomenclature [5]. A clear hierarch is emerging, with interaction strengths decreasing in the progression: ion-ion > water-ion > water-polar = polar-polar = water-water > water-hydrophobic. The major issues are (a) how good are the current potential functions for quantitative studies, and (b) how should one treat entropy terms that are surely important, if not crucial, to our final understanding of these systems.

Several chapters deal with the dynamics of water molecules near protein surfaces. Recent dielectric experiments by Grigera and Berendsen (4) and Hoeve [7] suggest that low (1 MHz) and intermediate (100 MHz) dielectric dispersions result largely from protein-water interface effects, from protein rotation, and from sidechain reorientation. NMR experiments, represented here by Koenig [9] and Bryant [8], offer clear evidence that water molecule motion reflects both rigid rotation of the protein in solution and some other motion in the nanosecond region (-30°C to +30°C). Lillford's theory [10] of water-proton relaxation considers compartmentalization and relates complex relaxation to heterogeneous mass distribution down to ∿10μ dimensions. The hydrodynamics of coupling the macromolecular motions to the water relaxation processes are currently obscure but are under study. Hundreds, if not thousands, of water molecules per protein are implicated in the nanosecond motion. The details of the NMR theory are incomplete but pose a serious interpretive issue when taken with the neutron scattering results which report a much smaller number of water molecules strongly affected by protein surface. Of course, there is no fundamental reason that the dynamics (governed by activation barriers) and the static structure (controlled by the width and depth of the potential wells) must exactly correspond. Extension of molecular dynamics calculations to and beyond the nanosecond region may provide insight into these questions.

Polysaccharide-Water Interactions. These polymers exhibit a range of hydrophilicity comparable to that of proteins and interact with water in many ways similar to those already noted for

proteins. The subject is covered here by a diversity of
substrates, representing different compositions, solubilities,
and crystallinities.

Bound water, primarily observed as a nonfreezing component,
is reported in two chapters; but by comparison to the value
0.04-0.10 g H_2O/g cotton cellulose (5), the unfrozen boundary
layer found by Deodhar and Luner [16] for wood pulp is 0.4-0.6 g
H_2O/g and that measured by Ikada et al. [17] for mucopolysaccharides is 0.4-0.7 g/g. The former authors associate nonfreezing
water in wood pulps with pore size of the cellulose pulp,
estimating 40 Å as the largest pore that can carry 100% non-
freezing water. Actual pore measurements coupled with various
measurements of water-polymer interactions, such as those covered
by Rupley et al. [6] would be revealing and are in order on wood
pulps and cotton celluloses. The effect of salts in structuring
and destructuring water in the pores of wood pulps [16] (see
Luck [3]) seems complicated by the necessity for the salt to
penetrate small pores to be effective.

The salt-free crystalline polysaccharides reviewed by Bluhm
et al. [15] are stabilized in characteristic crystalline unit
cells by specific amounts of water. Two kinds of locations have
been proposed for the water molecules; one is unique, i.e., the
water lies clustered in an existing interstitial cavity between
double helices of B-starch. The other has water bound at specific sites within each unit cell. Additional water in this second
type expands one or more unit cell dimension. This almost continuous expansion of the unit cell with increasing content of
water may represent a more ordered aspect of the same interaction that occurs between water and accessible, disordered
surfaces of celluloses crystallites (and other imperfectly
crystalline polysaccharides).

At an opposite extreme from crystalline water-containing
polysaccharides are the mucopolysaccharides described by Ikada
et al. [17]. In addition to the nonfreezing water that characterizes these polymers in the solid state, one to three thermal
transitions measured on solutions suggest complex interactions
with water.

Permeation, Transport, and Ion Selectivity. Under this
heading, we bring together those compositions that have in common
a pore-like structure and that have been studied in this regard in
these chapters.

Tirrell et al. [21] measured nonfreezing water to the extent
of 0.36-0.74 g H_2O/g copolyoxamide membrane, noted that these
values are a direct function of surface area, and suggested that
bound water exists in bulk polymer and over greater distances
than a few molecular diameters on the polymer surfaces. The
model developed by Belfort and Sinai [19] for porous glass predicts that bulk water can form in pores above 27-38 Å (see

Deodhar et al. [16]), and, therefore, desalting cannot be expected
for pores larger than this range. Permeation measured by Kim
et al. [20] for hydrophilic solutes in aqueous media through
poly(hydroxyethyl methacrylate) (p-HEMA) is consistent with transport by bulk-like water. Permeation is reduced for crosslinked
p-HEMA or HEMA copolymers as a result of non-bulk-like water.
Southern and Thomas [22] showed that hydrophilic impurities in
rubber provide loci for water molecules and account for diffusion
of aqueous solutions through this hydrophobic substrate.

Breuer et al. [18] developed a mechanism for the swelling of
hair from consideration of the contribution that swelling makes
to the overall free energy changes accompanying the water absorption. The dominant thermodynamic driving force for water absorption is site binding, i.e., interaction with discrete polar side
chains and peptide bonds. But this force is supplemented with
capillary condensation and entropic gains from the mixing of
water with polymer chains. The mechanism is similar to that
described for cellulosic fibers (5). Breuer's mechanism with
minor modifications might be applied to other highly hydrophilic
polymers in this section. From studies of the selectivity of
porous synthetic ion exchange resins, Marinsky et al. [23] concluded that differences in excess free energy from ion-solvent
interactions contribute most importantly to differences in
affinity of ion pairs for the resin and, therefore, to ion
selectivity.

In studies of a completely different type of porous structure, Lipshitz and Etheredge [24] showed that articular cartilage
is anisotropic in flow of interstitial fluid and that its properties are a function of the impedence to flow during and following
compression.

Water Interactions with Bulk Polymers and Resins.
Starkweather [25] describes the water first absorbed in nylon 66
as most tightly bound and as nonfreezing. Bretz et al. [32]
refer to tightly bound water at the level of 0.02-0.03 g/g nylon
66, but it may approach 0.08 g/g [25]. The water that is absorbed
at low levels and that is uniformly distributed in such relatively
hydrophobic polymers seems to fall in this category. This behavior appears to be one of the ways in which water is absorbed near
polar groups in amorphous regions of partially crystalline
polymers or throughout amorphous polymers.

The spatial arrangement of absorbed water at low partial
pressure is proposed to be in hydrogen bonds between two amide
units in nylon 66 by Starkweather [25], at hydrogen bonding sites
(ester or carboxyl groups) in amorphous acrylic polymers by
Brown [26], at polar groups in epoxy resins by Moy and Karasz
[30], and in the ionic cluster phases of perfluorosulfonic acid
polymers by Pineri et al. [28,29]. Above a characteristic level
of water for each polymer, the water appears in a second form,
generally termed "clusters"; the terminology is used rather

broadly [26]. Clusters may consist of an average of three water molecules bonded at polar sites [25] or a separate phase of water in a less polar polymer [27].

In both nylon 66, studied by Starkweather [25], and high-performance epoxy resin, described by Moy and Karasz [30], the water is believed to hydrogen bond between polar sites in the polymers, acting as crosslinks at low temperatures but plasticizing at elevated temperatures because of the greater thermal mobility of water compared to segments of the polymer chain.

Water-induced plasticization of polymers by disruption of intermolecular hydrogen bonding between polymer chains that are generally considered to be hydrophobic is rather common. Moy and Karasz [30] show that the lowering of T_g for an epoxy-diamine resin is proportional to the amount of water in the system. Johnson et al. [27] report that polysulfone and poly(vinyl acetate) show enhanced low-temperature β-loss transitions in proportion to the unclustered water. Clustered water in poly(vinyl acetate) has no effect on T_g, although T_g shifts with increasing amount of unclustered water. Fuzek [31] found that water absorbed by synthetic fibers and silk at room temperature and 65% RH substantially lowers T_g's, the effect being reflected in several different fiber properties. Wet soaking has an additional effect. The action of water in poly(methyl methacrylate) is interpreted by Moore and Flick [33] as one of general plasticizing character. These effects are not limited to synthetic polymers and resins (note silk, above); Scandola and Pezzin [13] note the lowering of T_g of elastin by water and describe other aspects of the interaction that are closely related to the plasticization of poly(vinyl chloride) and polycarbonate.

The effect of water in polymers and resins is not simple and is not necessarily predictable on the basis of the foregoing comments. Some complexity is indicated by the necessity for a two-parameter sorption isotherm to correlate water sorption [26] and by the antiplasticizing effect of low levels of water at low temperature [25]. The epoxy composites of Illinger and Schneider [34] show large changes following water sorption, suggesting changes in structure of the substrate resulting from the sorption. A more subtle situation appears in the studies of Bretz et al. [32]; as water is imbibed by nylon 66, fatigue crack propagation decreases to about one-fifth and then increases about three fold. The level of water corresponding to maximum fracture energy is 1 H_2O/4 amide units, approximately half the value at which substantial clustering begins [25]. Comments in the foregoing sections also suggest some of the complexities that might be anticipated in more detailed studies.

A Final Word

In a symposium of this kind, which is oriented toward a phenomenon and whose ideal objective is a broad understanding of the nature and limitations of the phenomenon, it seems that more questions have been raised than answered. However, despite the diversity of polymeric substrates under examination and the wide spectrum of experimental and theoretical approaches, there do appear to be strong threads of commonality in general results and conclusions among many, if not all, of the chapters. We heartily recommend the reading of each and all of the following chapters with the thought that there is much to be learned from each and that one reader may, with additional research, put the phenomenon into perspective and quantitative array.

Literature Cited

1. Squire, P. G.; Himmel, M. E. Arch. Biochem. Biophys., 1979, 196, 165-177.

2. Hagler, A. T.; Moult, J. Nature, 1978, 272, 222-226.

3. Stillinger, F. H.; Rahman, A. J. Chem. Phys., 1978, 68, 666-670.

4. Grigera, J. R.; Berendsen, H. J. Biopolymers, 1979, 18, 47-57.

5. Rowland, S. P., in "Textile and Paper Chemistry and Technology"; J. C. Arthur, Jr., Ed., Symposium Series No. 49, American Chemical Society, Washington, D.C., 1977, pp. 20-45.

RECEIVED January 4, 1980.

THE STRUCTURE OF WATER

Thermal Properties of Water in Restrictive Geometries

F. H. STILLINGER

Bell Laboratories, Murray Hill, NJ 07974

Melting and freezing transitions are examples of cooperative phenomena. They are dramatic changes on account of their suddenness of occurrence, and because of the discontinuities that they cause in various intensive quantities (volume, heat capacity, viscosity, self-diffusion rate, etc.). These sharp changes come into existence only because of the presence of an enormous number of interacting molecular degrees of freedom. Normally melting and freezing are observed for matter in bulk, comprising roughly Avogadro's number (6.022×10^{23}) of molecules, and for all intents and purposes this can be reckoned as infinitely large.

However there are exceptional circumstances wherein melting and freezing are observed for much smaller aggregates of molecules. In the case of water it is possible to induce melting and freezing in sufficiently small systems that the phase transitions ought to exhibit modifications due to finite size effects. This might be expected to occur for fine droplets in aerosols, for some emulsions of water in oils, and for very small water clusters that can be made to form in polymeric solids (1,2). The phase transition modifications arise not only from the reduced number of molecular degrees of freedom _per se_, but also from the fact that a substantial portion of the material will be present in a boundary, or interfacial, region. One expects modifications in both equilibrium (transition "rounding") and kinetic (supercooling) behavior.

Droplets or clusters in the size range from microns downward are those expected to show measurable deviations from bulk equilibrium behavior. For the sake of quantitative orientation, Table I shows representative numbers of molecules contained in spherical droplets of water in this size range, and it also shows the fraction of those molecules within a typical molecular distance (5Å) of the surface.

The objective of this paper is to provide a theoretical framework for understanding the melting, freezing, and supercooling behavior of water in small dusters and droplets. Although attention will be focussed on spherical geometry, the basic ideas involved can be generalized to other shapes.

HYDROGEN BOND PATTERNS IN WATER

Both ice and liquid water consist of space-filling networks of hydrogen bonds. That of the former is regular and static; that of the latter is irregular and mobile. The hydrogen bonds in ice are arranged so that each molecule participates in exactly four bonds, and the bonds are spatially disposed so as to form polygons with only even numbers of sides (hexagons, octagons decagons, ...) (3). The random network in the liquid evidently contains polygons of both even and odd numbers of sides

0-8412-0559-0/80/47-127-011$05.00/0
© 1980 American Chemical Society

TABLE I. Contents of spherical water droplets.[a]

Cluster Diameter (microns)	3.00	1.00	0.30
Number of Water Molecules in Cluster	4.73×10^{11}	1.75×10^{10}	4.73×10^{8}
Number of Molecules Within 5Å of Surface	4.72×10^{8}	5.25×10^{7}	4.71×10^{6}
Fraction of Molecules Within 5Å of Surface	0.0010	0.0030	0.0100

Cluster Diameter (microns)	0.10	0.03	0.01
Number of Water Molecules in Cluster	1.75×10^{7}	4.73×10^{5}	1.75×10^{4}
Number of Molecules Within 5Å of Surface	5.20×10^{5}	4.57×10^{4}	4.74×10^{3}
Fraction of Molecules Within 5Å of Surface	0.0297	0.0966	0.2710

[a] The droplets are at 0 °C, and are presumed to possess the macroscopic liquid density up to the geometric surface.

intermixed (4); a variety of defect structures is present as well, including broken bonds and bifurcated hydrogen bonds. Thus the invariant fourfold coordination in ice is replaced upon melting by indefinite coordination that varies from one to five, and averages about 2.5 near the melting point (4). Needless to say, the hydrogen bonds that are present in liquid water are much more strained on the average than those in ice.

The famous density maximum in liquid water at 4 °C represents a balance point between two opposing tendencies. On the one hand there is the continuation of the process initiated at the melting point, namely conversion of bulky hydrogen bond structures to more compact forms (the negative volume of melting signifies this change). On the other hand there is present the natural thermal expansion of all liquids, which tends to accelerate in magnitude with increasing temperature. It is plausible to suppose that the first of these is associated with gradual disappearance in the liquid of unstrained polygonal and polyhedral bonding patterns, as they are replaced by highly strained and bond-broken arrangements that permit more efficient packing of molecules.

The volumetric properties of small water droplets or clusters can be measurably influenced by the significant fraction of material in the surface region (see Table I). The structural character of the hydrogen bond network in the bulk phase is certain to be modified in the interfacial region if the droplet has either a free surface or is in contact with hydrophobic material. Obviously more broken bonds (*i.e.*), unsatisfied bond positions) must be present. But additionally the organization of those bonds which are present, for example into polygons, will differ statistically from the situation deep beneath the surface. It is generally believed, on the basis of experimental observations, that the surface of ice roughly in the range $-10\,°C$ to $0\,°C$ is covered by a mobile "liquid-like" layer ($\underline{3}$), which ought then to contain odd-sided polygons and bifurcated hydrogen bonds. In a crude sense, surfaces act to disrupt the natural order present in the bulk phase.

This last point leads to an interesting prediction concerning the volumetric behavior of liquid water droplets or clusters. Since their surface regions ought to be more disordered at any given temperature than their interiors, the temperature of minimum volume (maximum mean density) should be displaced below $4\,°C$. The magnitude of this effect presumably would depend on the chemical characterer of the region outside the droplet or cluster, and is difficult to estimate with precision. However a crude estimate is possible by supposing the outer $5\mathring{A}$ behaves as a "normal" nonaqueous liquid with thermal expansion $10^{-3}/\,°C$, and the remainder as bulk water. Table II shows the resulting predictions for the temperatures of maximum mean density for the droplet sizes considered in Table I. The depressions are substantial for the smaller sizes.

TABLE II. Temperature of maximum mean density
for spherical droplets of water.

Diameter, Microns	T_{max}, $°C$
∞	3.98
3.00	3.92
1.00	3.80
0.30	3.36
0.10	2.13
0.03	-2.06
0.01	-12.88

STATISTICAL MECHANICAL THEORY

It is mandatory to consider the underlying statistical mechanical formalism in order to systematize the preceding ideas. For this purpose we will examine the partition function for an N-molecule water cluster. The isothermal-isobaric ensemble is appropriate ($\underline{5}$) since constant normal stress (denoted below by "pressure" p) seems relevant to most cases of interest. The partition function for this ensemble, $\Delta_N(\beta,p)$, depends upon the inverse temperature parameter $\beta=(k_B T)^{-1}$ and the external pressure p. We will presume that the droplet is constrained to spherical shape, though with variable radius. The connection to thermodynamics is established by the fact that

$$\ln \Delta_N = -\beta G, \tag{1}$$

where G is the Gibbs free energy.

An expression for Δ_N in terms of molecular quantities is the following:

$$\Delta_N(\beta,p) = \Delta_N^{(0)} < exp(-\beta\Theta)>; \tag{2}$$

here $\Delta_N^{(0)}$ is a normalizing factor that has no further importance in subsequent analysis, and Θ is the sum of the potential energy Φ for the water molecules and the pressure-volume product pV,

$$\Theta = \Phi + pV. \tag{3}$$

The Boltzmann factor average in Eq. (2) is an *a priori* average over all N molecule orientations and positions within the spherical cluster volume V, and over V with a suitable upper limit.

The evaluation of Δ_N in Eq. (2) would be a formidable task even if Φ were known in all detail (unfortunately it is not). The reason is that the required average entails a multiple integral of order 6N, since there are three translational and three rotational degrees of freedom per molecule. Table I shows that N is still huge even for very small clusters.

In spite of this difficulty we can understand the behavior of Δ_N to some extent by classifying molecular configurations according to a small number of independent parameters. We choose two such parameters, the quantity Θ already introduced and a structural order parameter Ψ. The latter is intended as a measure of the ice-like order present. Conceivably we could take Ψ to equal the number of hydrogen-bond hexagons present, however this would not distinguish ice *Ih* from the cubic modification *Ic* that is presumably unstable under conditions of interest here (6). Alternatively and more appropriately Ψ could represent the number of bonded bicyclic octamers (see Figure 1), that are present in ice Ih but *not* in ice Ic. For the remainder of this analysis we will use this second choice. By accounting for all the distinct (and overlapping) ways that octamers can be traced out on the perfect ice lattice, we find that the maximum value of Ψ is $N/2$, to within terms of order $N^{2/3}$.

Once the decision has been made to classify configurations according to their Θ and Ψ values, we can write the average in Eq. (2) in a formally simple manner:

$$< exp(-\beta\Theta)> = K^{-1} \int\int exp[w(\Theta,\Psi)-\beta\Theta]\,d\Theta\,d\Psi,$$
$$K = \int\int exp[w(\Theta,\Psi)]\,d\Theta\,d\Psi. \tag{4}$$

In this expression $exp(w)$ stands for the configuration space weight to be assigned to the given Θ, Ψ pair. K is a temperature-independent normalization factor.

The discontinuous liquid-solid transition automatically emerges from the first of Eqs. (4) in the large$-N$ limit. This can be demonstrated by using "peak integration" to evaluate that first integral. This approximation, which becomes increasingly accurate as N increases, exploits the fact that dominant contributions to that integral arise from regions at which the quantity

$$Y \equiv w(\Theta,\Psi) - \beta\Theta \tag{5}$$

is at a maximum in the two-dimensional Θ, Ψ space. By making a local Gaussian approximation for each such maximum we conclude that

$$< exp(-\beta\Theta)> = 2K^{-1}\sum_{j} (Y_{j11}Y_{j22})^{-1/2} exp(Y_j). \qquad (6)$$

Here j indexes the maxima, and Y_j stands for the value of Y at maximum j. The quantities Y_{j11} and Y_{j22} are the second partial derivatives of Y at j along the two principal directions of curvature there.

On physical grounds we expect only two relevant maxima to contribute substantially to Eq. (6), at least when the temperature is around 0 ℃ and pressure around 1 bar. These two correspond respectively to hexagonal ice and to liquid water. In this temperature range one should keep in mind that Θ is dominated strongly by the potential energy part Θ; pV is far smaller in magnitude. For the "ice" maximum Θ is low (many well-formed hydrogen bonds) while Ψ is high (many characteristic ice-like structures). By contrast the "liquid" maximum will display higher Θ (fewer and more strained hydrogen bonds) and lower Ψ (most bonding patterns not ice-like).

The suddenness of the first-order melting or freezing arises from the fact that the Y_j are of order N. It is clear that after exponentiation one of the terms in Eq. (6) will dominate the other almost totally, except in a narrow switch-over temperature range around the transition point. This switch-over range is expected to be proportional in asymptotic width to N^{-1}, and represents the extent of smearing of the transition due to finite system size. Latent heat is absorbed or discharged over this finite temperature interval if in fact nucleation can occur.

It will be profitable to examine a schematic diagram of the level curves of Y, for β chosen to make the two maxima equal in magnitude (in the infinite-N limit this will be precisely the thermodynamic transition point). Such a diagram appears in Figure 2. This Figure shows that between the two equal-altitude maxima there will be a saddle point along a ridge line. When N is reasonably large this ridge line will correspond to side-by-side coexistence of liquid-like and ice-like regions within the droplet, the relative proportions of which determine position along the ridge line. Since an interface must exist between the regions the ridge line will lie below the maxima by an amount of order $N^{2/3}$.

At the temperature for which Figure 2 is shown the two maxima are the positions of the Y_j that are to be inserted in Eq. (6). These may generically be designated by Y_L and Y_C, for "liquid" and "crystal", respectively. They are the positions of simultaneous contact for a horizontal plane resting atop the $Y(\Theta,\Psi)$ surface.

As temperature varies, Y_L and Y_C will shift from the locations of the maxima shown in Figure 2. The shifts can be identified with points of rolling contact for the initially horizontal plane as it is tipped so as to remain parallel to the Ψ axis. This rolling constraint arises from the fact that β multiplies only Θ in Y in the first integral of Eq. (4). The Θ-direction slope of the rolling plane is just $\Delta\beta$, the change in β from its initial value at the horizontal orientation.

Under the supposition that $Y(\Theta,\Psi)$ is continuous and at least twice differentiable in its two variables, the loci of rolling contact (one through each maximum in Figure 2) will be at least piecewise continuous curves. These curves trace out the dominant Θ,Ψ "structure" for the liquid and solid phases, and they are indicated in Figure 2 as full curves in the respective ranges of thermodynamic stability.

The relative positions of the two maxima on the Y surface will change with N, even after accounting for the fact that Ψ and Θ are extensive quantities in the large-N limit. The presence of the structure-disrupting surface has a greater influence on an ice phase than on a liquid phase. As N declines, Ψ/N decreases and Θ/N increases for both maxima, but the changes are larger for ice than for liquid. Further-

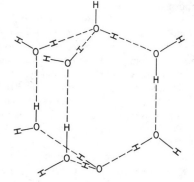

Figure 1. Bicyclic octamer structure, a pattern of hydrogen bonding that appears exclusively ice Ih. Several isomers are possible, depending on the disposition of hydrogens along the given linear hydrogen bonds.

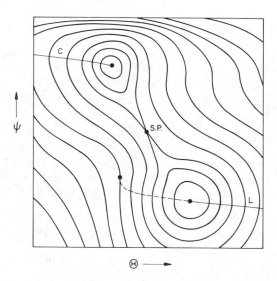

Figure 2. Level curves for Y at the melting temperature (maxima at equal altitudes). Curves C and L are loci of rolling contact for a plane parallel to the Ψ axis. The dotted portion of the L curve corresponds to supercooled liquid, and it ends at a point with vertical tangent.

more these shifts, if viewed at constant β, will be associated with a change of relative altitude of those maxima. This latter effect causes the previously mentioned broadening of the reversible thermodynamic transition with decreasing N to be accompanied by a shift to lower temperature.

SUPERCOOLING

Nucleating the crystal phase from liquid cooled just below the thermodynamic freezing point can kinetically be especially difficult in small clusters. This stems partly from the fact that impurity "seeds" are less likely to be present in any given cluster, and partly from the proportionality of homogeneous nucleation rates to the volume of the system. Consequently supercooling becomes the rule rather than the exception, and this fact has been useful in laboratory studies of supercooled water (7). It should be stressed that a range of N values evidently exists which on the one hand is small enough to prevent nucleation, while on the other hand is large enough so that the intensive properties observed for the supercooled water are essentially independent of system size.

Considering this situation it is thus useful in Figure 2 to continue the rolling loci through the respective maxima to map out metastable phase extensions. In principle this is just as possible for Y_C as for Y_L though melting of a crystal usually is immediately initiated at its already amorphous surface. [A water inclusion in the interior of an AgI crystal, or some other epitaxial structure-promoter, might provide an exception.] The metastable extension for the liquid is indicated in Figure 2 as a dashed curve.

The differential geometry of our constrained rolling contact demands that the contact loci cross the level curves only at points where the latter have tangents parallel to the Ψ axis. It is important to recognize that this requirement can lead to sudden disappearance of a locus. In mathematical terms the locus can suffer a "catastrophe", with an endpoint that itself has a vertical tangent. Such a catastrophic endpoint hence would manifest a diverging rate of change with temperature of the concentration of ice-like structural elements measured by Ψ.

The possible existence of an endpoint for the supercooled liquid locus is particularly interesting in view of the experiments of Angell and coworkers (7,8,9,10). They find that pure water at ordinary pressures (even very finely dispersed) cannot apparently be supercooled below about $-40\ ℃$, and that virtually all physical properties manifest an impending "lambda anomaly" at $T_s \cong -45\ ℃$. The most striking features of this anomaly are the apparent divergences to infinity of isothermal compressibility, constant-pressure heat capacity, thermal expansion, and viscosity. We now seem to have in hand a qualitative basis for explaining these observations.

The supercooled liquid catastrophe, if it exists, would necessarily be associated with diverging fluctuations in the structural order parameter Ψ. This stems from the fact that the Y surface develops a vanishing curvature in the Ψ direction as this endpoint is approached. Because the bicyclic octamer elements are bulky, fluctuations in their concentration amount to density fluctuations. Diverging density fluctuations then imply diverging isothermal compressibility. Furthermore the infinite slope of the metastable liquid locus at its endpoint implies the divergence of thermal expansion. Potential energy fluctuations remain essentially normal, so constant-volume heat capacity remains small. But the volumetric divergence creates an unbounded constant-pressure heat capacity.

Thus far we have seen that differential geometry of the $Y(\Theta,\Psi)$ surface *can* produce a metastable liquid catastrophe, not that it *must*. Demonstration of the latter will have to invoke the special character of the interactions in water, specifically the strong directionality of hydrogen bonding which produces tetrahedral coordination.

The key observation seems to be that the structure-measuring octamer units ought to experience a mean attraction for one another in the liquid medium. They are in fact geometrically suited to link up through directed hydrogen bonds, and indeed it is so that the extended lattice of ice Ih can be entirely broken into bicyclic octamer units (11). In addition an isolated octamer unit embedded in a locale of strained and broken bonds would be in great jeopardy, for those neighboring defects would exert torques and forces that would distort it severely and often past the point where it would continue to qualify for inclusion in Ψ. The net result is a statistical tendency of the surviving octameric units to aggregate, and the lower the temperature the greater the degree of aggregation.

This thermally-sensitive sorting out process will not segregate bicyclic octamers alone but will also include a few other types of well-bonded (and bulky!) species. Figure 3 shows two which qualify, each fused to an octamer without mutual distortion. The first includes pentagons; the second is a structure which appears in cubic ice. Because bonding arrangements such as these exist one must keep in mind that the strongly bonded aggregates are not automatically ice Ih fragments.

The aggregation or clumping of bulky and well-bonded structures is clearly going to be a cooperative process, and as such it will produce long-range coarsening of the texture of the fluid medium. This process is entirely analogous to that which creates long-range fluctuations at the critical point of a condensing gas. In the present case it will produce divergent Ψ fluctuations as already discussed above. On structural grounds it appears most natural for the aggregates to grow in a dendritic, rather than globular, fashion.

Just as in that analogous case of conventional critical fluctuations it is useful to introduce a correlation length ξ for the scale of inhomogeneity (12). In the present context ξ gives an average linear dimension for regions of anomalously high (or low) concentration of the structural units defining Ψ. As temperature declines for the supercooled liquid ξ increases, becoming infinite at T_s. Since these structural fluctuations are also density fluctuations it should be possible *in principle* to detect them and to measure their size ξ by light scattering or small-angle X-ray scattering experiments on strongly supercooled water. *In practice* these experiments would probably be very difficult.

The presence of strongly bonded regions, presumably composed of low-mobility molecules, obviously will inhibit hydrodynamic flow just as polymer dissolved in low-molecular-weight solvent does. Furthermore this effect will amplify as temperature declines. Thus it is not surprising that viscosity should experimentally diverge at T_s.

It is in the nature of random mixing statistics for the various strongly-bonded units in a low-density fluctuation region that several bicyclic octamers would occasionally link up to form a larger fragment of ice Ih. Such events become more and more likely in a given amount of supercooled liquid as temperature declines toward T_s. These multiple octamer structures are the most likely precursors of a nucleation event. Thus the supercooled liquid contains literally the seeds of its own destruction. This observation rationalizes the inability experimentally to supercool pure water below about $-40\,°C$ without spontaneous freezing. The only way to avoid this fate seems to be to use much smaller clusters and high cooling rates; perhaps condensing clusters within thin film polymeric media might be a suitable preparative technique which would permit the required subsequent rapid cooling.

The droplet size range for convenient observations of supercooling

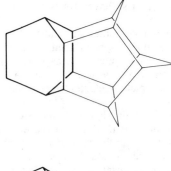

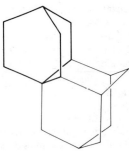

Figure 3. Bonding of bicycic octamer units (heavy lines) to unstrained structures that do not occur in ice Ih. Oxygen atoms occur at vertices, and hydrogen bonds between water molecules are illustrated only schematically by lines. In the lower part of the diagram the octamer is fused to a structure occurring in ice Ic.

anomalies was mentioned earlier. Unfortunately this range shrinks as the temperature of measurement declines toward T_s. Firstly, the rapid increase in homogeneous nucleation rate as temperature decreases places ever more stringent upper limits on droplet size. Secondly, the correlation length ξ must remain small compared to droplet diameter to maintain size-independence of intensive properties; but this causes the minimum permissible size to increase as T_s is approached from above. The situation seems inexorably to draw a veil of unobservability about the lambda anomaly at T_s, while still permitting the careful experimentalist to infer its properties by extrapolation.

CONCLUSIONS

It is appropriate to end this theoretical discourse with a statement about the future role of computer simulation studies, which have much to offer this field. Pure water has been the object of intense study by this approach and a lot has been learned (4, 13). More recently molecular dynamics computer simulation has turned to the modelling of alkanes and polymers (14, 15). It seems natural and inevitable to combine the two specialties to illuminate molecular motions and structure for water droplets in alkane or polymeric hosts.

Thus far the N values feasible in digital computer simulation has fallen below those shown in Table I. The largest number that has been used to date is 1728 (13). However rapid advances in computer technology, including especially parallel processing capabilities, are likely to increase this number by at least an order of magnitude. That increase would cause overlap with the smallest cluster size considered in Table I.

Specific polyhedral structures such as those in Figures 1 and 3 cannot be detected in the liquid phase by any known experimental method. However a digital computer can be instructed to identify these patterns in the course of a simulation study. The mean concentration of bicyclic octamer units could thus be determined at any given temperature, along with their tendency to aggregate in regions whose mean size ξ could also be determined.

Studies of this sort could provide vital quantitative underpinning for the qualitative ideas presented here. The resulting expansion of knowledge ought to go far toward completing our understanding of melting and freezing in water, of its supercooled state, and of its behavior in small clusters.

ABSTRACT

When water is finely dispersed as an aerosol, an emulsion, or as small clusters in polymeric host media, its thermal behavior can deviate significantly from that exhibited by bulk water. The reasons for these deviations are examined, and a statistical-mechanical approach for their study is proposed. A rough estimate is obtained for the depression of the temperature of maximum mean density for small spherical droplets. An explanation is advanced (in terms of specific structural fluctuations) for the singular behavior of strongly supercooled water that has been observed in emulsions near -40 ℃ by Angell and collaborators.

LITERATURE CITED

[1]   Bair, H. E.; Johnson, G. E., "Analytical Calorimetry, Vol. 4", Porter, R. S. and Johnson, J. F., Eds.; Plenum: New York, 1977; pp. 219-227.

[2]   Bair, H. E.; Johnson, G. E.; Merriweather, R. *J. Appl. Phys.* 1978, 49, 4976-4984.

[3]   Fletcher, N. H. "The Chemical Physics of Ice"; Cambridge U.P.: Cambridge, 1970.

[4]   Rahman, A.; Stillinger, F. H. *J. Am. Chem. Soc.* 1973, 95, 7943-7948.

[5]   Hill, T. L. "Statistical Mechanics"; McGraw-Hill: New York, 1956; pp. 66-68.

[6]   Eisenberg, D.; Kauzmann, W. "the Structure and Properties of Water"; Oxford U.P.: New York, 1969; pp. 90-91.

[7]   Rasmussen, D. H.; MacKenzie, A. P.; Angell, C. A.; Tucker, J. C. *Science* 1973, 181, 342-344.

[8]   Angell, C. A.; Shuppert, J.; Tucker, J. C. *J. Phys. Chem.* 1973, 77, 3092-3099.

[9]   Speedy, R. J.; Angell, C. A. *J. Chem. Phys.* 1976, 65 851-858.

[10] Angell, C. A. "Water, A Comprehensive Treatise, Vol. 7"; Franks, F., Ed.; Plenum: New York, 1980.

[11] Stillinger, F. H.; David, C. W. *J. Chem. Phys.* 1980, 00, 0000-0000.

[12] Stanley, H. E. "Introduction to Phase Transitions and Critical Phenomena"; Oxford U.P., New York, 1971.

[13] Geiger, A.; Stillinger, F. H.; Rahman, A. *J. Chem. Phys.* 1979, 70, 4185-4193.

[14] Weber, T. A. *J. Chem. Phys.* 1979, 70, 4277-4284.

[15] Helfand, E.; Wassermann, Z. R.; Weber, T. A. *J. Chem. Phys.* 1979, 70, 2016-2017.

RECEIVED January 3, 1980.

Solvation: A Molecular Dynamics Study of a Dipeptide in Water

MARTIN KARPLUS

Department of Chemistry, Harvard University, Cambridge, MA 02138

PETER J. ROSSKY

Department of Chemistry, University of Texas, Austin, TX 78712

Water has an essential role in living systems and is ultimately involved in the structure and function of biological polymers such as proteins. However, in this contribution we shall focus primarily not on what the water does for the biopolymer but rather on the effects that the biopolymer has on the water that interacts with it. Of interest are alterations in the structural, energetic, and dynamic properties of the water molecules. Studies of the rotational mobility of water molecules at protein surfaces have been interpreted by dividing the solvent molecules into three groups (1). The most rapidly reorienting group has a characteristic rotational reorientation time (τ_r) of not more than about 10^{-11}s. The next most rapid group exhibits a rotational reorientation time of about 10^{-9}s and has been tentatively identified as the water molecules that are strongly associated with ionic residues. The third group exhibits a τ_r of about 10^{-6}s; these solvent molecules are considered to be essentially irrotationally bound to the macromolecules; an example might be the four waters in the interior of the bovine pancreatic inhibitor. The population exhibiting the fastest times is expected to include molecules which form hydrogen bonds to the peptide backbone and those which are influenced by the presence of nonpolar groups. It is this group which forms the major part of the solvation shell and, therefore, is likely to play the dominant role in the solvent effect on protein properties. Because of the difficulties involved in studies of protein solutions per se, it is of particular interest to investigate systems of small molecules that incorporate functional groups present in proteins.

In this contribution we describe the results of a molecular dynamics simulation of such a molecule, the alanine dipeptide in aqueous solution. In such a simulation, one treats a sample of molecules with fixed volume and an energy and density corresponding to the system of interest. Given the internal and interaction potentials for the molecules in the box, and certain initial conditions for the coordinates and momenta of each particle, one solves the classical equations of motion for all of the particles

0-8412-0559-0/80/47-127-023$05.00/0
© 1980 American Chemical Society

to obtain the phase space trajectory of the entire system over a
period of time. An initial integration period during which
certain properties (e.g., individual particle velocities) are adjusted is used to obtain a system that is equilibrated. After
the equilibration period, integration is continued for a length
of time sufficient to yield time averages that approximate equilibrium averages. In addition, the time evolution of the particle trajectories can be used to determine time-dependent properties.

We examine structural and dynamic aspects of both the dipeptide solute and the aqueous solvent. For the dipeptide, primary
emphasis is placed on the internal motions. The size and dynamical character of fluctuations relative to the average structure
are investigated in vacuum and in the presence of solvent. The
dipeptide vibrational degrees of freedom have frequencies varying
from approximately 50 (dihedral angle torsions) to 3500 cm^{-1}
(bond stretching), corresponding to characteristic times in the
range of 7×10^{-13} to 1×10^{-14}s. For such a range in characteristic times, a significant variation in solvent effects (e.g.,
damping of fluctuations) is expected.

The structural and dynamic properties of the aqueous solvent
in the region immediately surrounding the dipeptide solute are of
special interest. The principal questions which we address are:
First, how is the dynamic behavior of the solvent altered by the
proximity of the solute? Second, what is the range of influence
of the solute; that is, are the effective solvent-solute interactions of sufficiently short range that it is reasonable to regard the water molecules in contact with the polar (peptide)
groups as qualitatively different from those in contact with the
nonpolar (methyl) substituents? Finally, we investigate the
structural origins of observed differences between the dynamic
properties of the bulk solvent and that in contact with the solute.

Model

The details of the model used to simulate the dipeptide solution have been presented previously (2); a brief review of the
interactions present in the system and the methods used to carry
out the simulation is given here. The alanine "dipeptide" solute
($CH_3C'ONHCHCH_3C'ONHCH_3$), shown in Figure 1, is a neutral molecule
terminated by methyl groups, rather than by the carboxylic acid
and amino groups of an amino acid. This is an appropriate choice
to obtain a system that models an amino acid as part of a polypeptide chain. The structure shown in Figure 1 is the equatorial
C_7 conformation (C_7^{eq}) that is the global minimum in the dipeptide potential surface in vacuum and is believed to be the favored
conformation in both aqueous and nonaqueous solutions.

The internal degrees of freedom of the dipeptide are governed
by a molecular mechanics force field (2) that includes terms

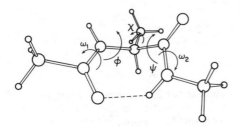

Figure 1.   *Alanine dipeptide in the equatorial C₇ conformation, (φ,ψ) ≃ (−60°, 60°). The structure is (left to right) CH₃-CONHCHCH₃CONHCH₃; the dashed line represents the internal hydrogen bond.*

corresponding to harmonic bonds, anharmonic bond angles, dihedral angle torsions, and nonbonded Lennard–Jones and electrostatic interactions. No dipeptide degrees of freedom are constrained in the simulation. The water molecules are modeled by a modification of the ST2 model of Stillinger and Rahman (3). The model consists of four point charges placed within a single Lennard–Jones sphere centered at the oxygen atom; two positive charges are located at the hydrogen atom positions, and two negative charges are located at positions representing the lone-pair orbitals. The only modification made in the ST2 model is to allow internal flexibility in the water molecules. The intermolecular interactions among the waters are given by pair-wise potentials. For the two molecules W_1 and W_2, we have

$$V_{W_1 W_2} = 4\varepsilon_W \left\{ \left(\frac{\sigma_W}{r_{O_1 O_2}} \right)^{12} - \left(\frac{\sigma_W}{r_{O_1 O_2}} \right)^6 \right\}$$

$$+ \sum_{i,j=1}^{4} \frac{q_i^{W_1} q_j^{W_2}}{r_{ij}} S(r_{O_1 O_2}) \qquad (1)$$

where σ_W and ε_W are the parameters characterizing the Lennard–Jones interaction, q_i^W is the ith charge in water molecule W, $r_{O_1 O_2}$ is the intermolecular oxygen-oxygen distance, $S(r)$ is a switching function (3). The interactions between each water molecule and the dipeptide are given by a sum of Lennard–Jones and electrostatic terms of the form

$$V_{WD} = \sum_{\substack{\text{dipeptide} \\ \text{atoms, } \lambda}} \left[4\sqrt{\varepsilon_W \varepsilon_\lambda} \left\{ \left(\frac{\bar{\sigma}_\lambda}{r_{O\lambda}} \right)^{12} - \left(\frac{\bar{\sigma}_\lambda}{r_{O\lambda}} \right)^6 \right\} + \sum_{j=1}^{4} \frac{q_j^W q_\lambda}{r_{j\lambda}} \right] \quad (2)$$

where $\bar{\sigma}_\lambda = (\sigma_W + \sigma_\lambda)/2$ and $r_{O\lambda}$ is the water oxygen-dipeptide atom distance. For the chosen values (2) of dipeptide-atom Lennard–Jones parameters, σ_λ and ε_λ, and charges, q_λ, the water molecules associated with the solute peptide groups have reasonable energies and geometries. In particular, the optimal association energies (kcal/mol) for the four types of hydrogen bonds in the system in order of increasing strength (the water HOH bond angle is fixed at the tetrahedral angle) are NH $\ldots$ H_2O (−6.0) < H_2O $\ldots$ H_2O (−6.8) < C = O $\ldots$ H_2O (−7.4) < N−H $\ldots$ O = C (−8.1). The values of the association energies are adjusted to the water-water interaction energy given by the ST2 model so that the hydrogen bond strengths are all similar, in accord with available calculations and data.

The simulation is carried out on a sample consisting of one dipeptide and 195 water molecules in a cubic box with an edge length of 18.2194 Å; the density of 1.004 g/cm^3 is in accord with experiment. The dipeptide solute is surrounded by approximately two molecular layers of water at all points. After the equilibration, the simulation analyzed in the current work corresponds to 4000 time steps of 3.67 x 10^{-16}s, or 1.5 ps on a molecular time scale. The mean solvent kinetic temperature is 303°K and that of the dipeptide is 298°K.

Solute Properties

During the simulation, the solute remains in the vicinity of the C_7^{eq} minimum. This is not to be interpreted as implying that the C_7^{eq} is the most stable solution structure, because there is a very small probability of observing a large conformational change in such a short time.

To determine the effect of the water on the dipeptide in solution, we did a corresponding simulation of the dipeptide dynamics in the absence of solvent. Neither the average structure nor the magnitude of the local fluctuations of the dipeptide is strongly affected by the solvent environment. The results are summarized in Table I, where we give results for typical bonds, bond angles, and dihedral angles (see Figure 1). With the exception of the fluctuations in the dihedral angles ψ and χ the observed differences are within the statistical error of the calculation.

Correlation Functions. We next consider dynamic correlations of the solute fluctuations. The time correlation function for solute structural fluctuations is defined as

$$C_A(t) = \frac{<\Delta A(\tau)\Delta A(t + \tau)>\tau}{<\Delta A^2>} = \frac{<\Delta(0)\Delta A(t)>}{<\Delta A^2>} \qquad (3)$$

where A is a particular structural parameter (e.g., bond length, dihedral angle), $\Delta A(t) = A(t) - <A>$, $<\Delta A^2> = <(A - <A>)^2>$, and the brackets indicate an average over the simulation; in the numerator of eq 3, the average includes all values, τ, in the simulation.

The unperturbed harmonic vibration of an isolated bond of length b would lead to a correlation function, $C_b(t)$, which oscillates without decay for all times. However, shifts in either the frequency or phase of the oscillation results in an eventual decay of $C_b(t)$ to zero after a time when the phase of $\Delta b(t)$ is, on the average, completely random with respect to that of $\Delta b(0)$. Even in an isolated molecule, such a decay can occur due to coupling of the motion of various degrees of freedom with different frequencies, although over long times the correlation function cannot remain zero since there is no dissipation. The change in

Table I

Average Solute Structure

A^a	$\langle A \rangle$		$\langle \Delta A^2 \rangle^{1/2}$	
	vacuum	solution	vacuum	solution
Bonds[b]				
$C'_L - O_L$	1.235	1.237	0.023	0.028
$N_L - H_L$	0.994	0.997	0.018	0.012
$C_\alpha - C_\beta$	1.544	1.542	0.041	0.035
$N_R - C_R$	1.461	1.459	0.034	0.032
Bond Angles[b]				
$C_L - C'_L - O_L$	122.31	121.89	3.30	3.23
$N_L - C_\alpha - C_R$	114.73	114.46	4.25	3.93
$N_L - C_\alpha - C_\beta$	107.71	108.15	3.51	3.85
$C'_R - N_R - H_R$	120.97	120.68	3.97	4.29
Dihedral Angles[b,c]				
ϕ	−67.21	−63.96	9.67	7.83
ψ	63.45	59.33	11.53	22.57
ω_1	−179.25	−179.67	8.74	9.67
ω_2	−179.68	178.08	14.72	12.39
χ	−59.10	−62.88	9.96	32.25

[a]All structural parameters, A, are defined in Figure 1; $\langle A \rangle$ = mean value, $\langle \Delta A^2 \rangle^{1/2} = \langle (A - \langle A \rangle)^2 \rangle^{1/2}$. [b]Bonds in Ångstroms, bond angles and dihedral angles in degrees. [c]The vacuum minimum occurs at $\phi = 66.2°$, $\psi = -65.3°$, $\omega_1 = 179.2°$, $\omega_2 = 179.9°$.

the correlation function in solution is determined by the effectiveness of solvent-solute collisions in dissipating the solute's dynamical information. In water, these "collisions" can involve the repulsive forces characteristic of hard sphere-like systems and the strong hydrogen bonding forces; weak attractive van der Waals interactions are expected to have only a small effect. Collisions with solvent are more likely to affect the solute motion if the latter is associated with a small characteristic force constant or if the mass of the solute structural component involved is small; also, damping is generally more effective for motions involving structural components of increased spatial dimensions. In the current study, a comparison of different dipeptide structural motions shows the expected qualitative differences in behavior.

We illustrate the results by presenting the correlation functions for (see Figure 1) a typical bond ($C_\alpha - C_\beta$) in Figure 2, a typical bond angle ($N_L - C_\alpha - C_R'$) in Figure 3, and the dihedral angle χ in Figure 4. In each figure, we show the result obtained in solution at the top and that obtained in vacuum at the bottom. The spectral density as a function of frequency ω corresponding to $C_A(t)$ is

$$\widetilde{C}_A(\omega) = \int_o^\infty dt \, \cos(\omega t) C_A(t) \qquad (4)$$

The limited knowledge of $C_A(t)$ forces us to truncate the time integral at t_{max}, rather than at infinite time; the calculated spectral density function is shown in an inset in each case (the amplitudes are in arbitrary units). Negative values of $\widetilde{C}_A(\omega)$ result from the finite upper limit on the integration.

On the picosecond time scale considered, no significant damping is seen in the oscillatory correlation functions describing the high-frequency ($\omega \gtrsim 300$ cm^{-1}) bond-length stretching and bond-angle bending modes. It is clear that for these high-frequency motions, the behavior of the time correlation functions is very similar in solution as compared to vacuum, and that in both environments there is no evidence for a significant zero-frequency component in the spectral density. The latter is expected if the correlation function contains a decaying component.

Clear evidence of solvent damping is found for the torsional angle χ (Fig. 4) for which the vacuum motion can be seen to involve principally a single frequency. During the current simulation, the motion involves only libration and not overall reorientation of the methyl group. By comparison with the result in the absence of solvent, it can be seen that the solvent is effective in damping the oscillatory motion of the methyl group. The behavior is manifest by the appearance of a low-frequency component in the spectral density, $\widetilde{C}_\chi(\omega)$. The short-time behavior of the solution correlation function ($t \lesssim 0.1$ ps) is roughly consistent with underdamped motion calculated from a Langevin equation,

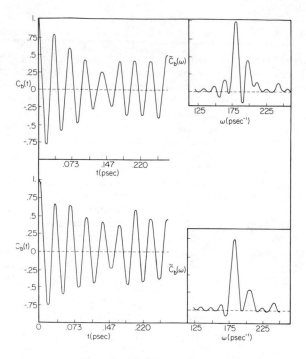

Figure 2.   Time correlation function and spectral density for fluctuations in the bond length, C_α—C_β (Figure 1). Top: in solution; bottom: under vacuum.

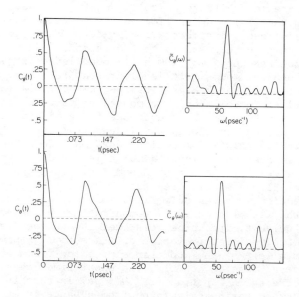

Figure 3.   Time correlation function and spectral density for fluctuations in the bond angle, N_L—C_α—C_R' (Figure 1); Top: in solution; bottom: under vacuum.

$$I \frac{d^2 \chi}{dt^2} + I\omega_0^2 \chi + f \frac{d\chi}{dt} = F_R(t) \tag{5}$$

where I is the moment of inertia of the methyl group (3 amu $\overset{\circ}{A}^2$),
ω_0 is the harmonic vacuum frequency (35 ps^{-1}; see Fig. 4), f is a
frictional coefficient, and $F_R(t)$ is a white noise random force.
The time correlation function calculated for motion governed by
the Langevin equation is shown as a dotted line at the top in
Figure 4. It is in reasonable agreement for short times with the
correlation function $C_\chi(t)$ found in solution when the characteristic time, 2I/f, is chosen to be 0.05 ps; the corresponding spectral density is shown dotted in the inset. It is evident that
for times greater than about 0.15 ps the correlation functions are
not in agreement. This discrepancy is reflected in the spectral
density; the Langevin equation predicts only a single peak in
$\tilde{C}(\omega)$ ($\omega \sim 25$ ps^{-1}) and not two. We note that the initial decay
of $C_\chi(t)$ corresponds to an apparent solvent drag which is much
smaller than hydrodynamic estimates that assume stick boundary
conditions. For example, treating the methyl group as a sphere of
radius a = 2.5 $\overset{\circ}{A}$ (the van der Waals radius) to obtain the frictional coefficient, f, f = $8\pi\eta a^3$ (η is the shear viscosity, 0.01 P)
we find 2I/f equal to 0.0003 ps, i.e., about 150 times smaller
than that observed. In this sense, the observed drag is nearer to
the hydrodynamic slip boundary condition limit; the exact slip
limit for a sphere corresponds to f = 0 and an infinite relaxation
time. The relatively long relaxation time is consistent with the
results of experimental studies of the rotational motion of small
nonassociated molecules (4).

Structure and Dynamics of the Solvent

     The structural and dynamic properties of the water in the
immediate vicinity of the dipeptide, the so-called solvation
"shell", can best be characterized by dividing the water molecules included in the simulation into the groups according to
whether they are near solute polar groups (polar), near nonpolar
groups (nonpolar), or outside of the first solvation layer (bulk).
This division is shown schematically in Figure 5. The central
blank area immediately surrounding the solute corresponds to the
region from which the centers (i.e., the oxygen atoms) of the
solvent molecules are excluded by the solute. The outer square
corresponds to the wall of the box which encloses the centers of
all 195 water molecules. Individual water molecules are classified into groups according to the average distance between their
oxygen atom and each of the amide H and carbonyl O atoms of the
two peptide links and three methyl carbon atoms; the average is
taken over the time of the stimulation. Each of the 195 water
molecules is assigned to one of the three solvent classes; polar
if the mean distance to any of the four polar atoms is less than

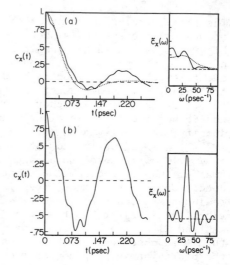

Figure 4.   Time correlation function and spectral density for fluctuations in the dihedral angle, χ (Figure 1). Top: in solution; bottom: under vacuum. The functions shown dotted at the top are obtained from a Langevin equation (see text).

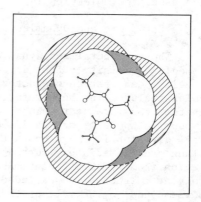

Figure 5.   Schematic of the solvation regions defined in the text; "polar," dotted; "nonpolar," cross-hatched; "bulk," within outer square and outside shaded regions.

4 Å, nonpolar if the mean distance is greater than 4 Å to a polar
atom but less than 5 Å to a methyl carbon, and "bulk" if neither
preceding criterion is met. The division into solvent groups
results in the assignment of 14 molecules as "polar", 20 as
"nonpolar", and the remaining 161 molecules as "bulk". Although
only a fraction of the molecules in the polar group participates
in strong bonds with the dipeptide, their proximity to the polar
atoms leads to a significant interaction energy with the peptide
groups.

One quantity of interest that is obtained from the simulation
is the pair radial distribution function, $g(r)$. In Figure 6, we
show the water molecule oxygen–oxygen pair distribution function,
$g_{OO}(r)$, obtained from the current simulation including all solvent
pairs. The results is the same as that obtained from simulations
of bulk water (3), within the statistical accuracy of the calculation. We note that the function $g_{OO}(r)$ is characterized by narrow
peaks and troughs, a result of the hydrogen-bonded structure. The
first peak occurs at 2.85 Å corresponding to the energy minimum of
the O—H...O hydrogen bond. The average distribution of water
oxygen atoms around the methyl group carbon atoms, $g_{OO}(r)$, is
shown in Figure 7; the result is the average over the three solute
methyl groups. In contrast to Figure 6, the first peak is broad.
The center of the peak occurs at about 3.7 Å, comparable to the
average water molecule, methyl group van der Waals contact distance of about 3.8 Å. Since water molecules can make contact with
the methyl groups only within a restricted solid angle around the
carbon atom (due to the presence of the remainder of the solute
attached to the methyl group), the height of the first peak in
$g_{OC}(r)$ is reduced relative to the value that would be obtained if
the group were completely exposed to solvent (e.g., as in a
methane molecule in solution). The breadth and radial position of
the first peak are comparable to that found in studies on argon-
like systems (5); that is, the first peak occurs at nearly the van
der Waals contact distance and is relatively broader than that in
$g_{OO}(r)$ for water, particularly on the larger r side of the peak.
In argon, as for $g_{OC}(r)$, the structure is determined by the repulsive core of the Lennard-Jones spheres, rather than by the strong
attractive hydrogen bond forces characterizing pure water.

The translational and rotational mobility of the solvent
molecules can be characterized by time correlation functions. The
time correlation functions for rotational motion are

$$C_\ell(t) = \lim_{t' \to \infty} \frac{1}{t'} \int_0^{t'} d\tau \, P_\ell(\hat{\mu}(\tau) \cdot \hat{\mu}(t + \tau))$$

$$= \langle P_\ell(\hat{\mu}(0) \cdot \hat{\mu}(t)) \rangle \equiv \langle P_\ell(\cos \theta(t)) \rangle \quad (6)$$

(where $P_\ell(x)$ is the Legendre polynomial of order ℓ) measure the average reorientation rate of the molecular dipole direction, given at time t by the unit vector, $\hat{\mu}(t)$. As in Equation 3, the averages in Equation 6 are evaluated using all pairs of configurations separated by a time t during the simulation of finite length, t'. In particular, we have

$$C_1(t) = \langle \cos \theta \rangle \qquad (7)$$

and

$$C_2(t) = \langle (3 \cos^2 \theta - 1)/2 \rangle \qquad (8)$$

These correlation functions decay to zero as the molecular orientation becomes randomized with respect to its initial value; $C_2(t)$ typically decays more quickly than $C_1(t)$. The calculated results for $C_1(t)$ (Equation 7) are shown in Figure 8 for the different groups of water molecules. The initial rapid decay of the correlation function during the first 0.05 ps corresponds to overall molecular oscillation (libration) with a loss of phase memory, but without significant net reorientation. In this short time period, the three groups appear to behave similarly. However, for longer times, there are differences in the decay rates of $C_1(t)$. The molecules in the polar class reorient at a rate similar to that exhibited by the bulk class, while those in the nonpolar class reorient more slowly. We can obtain an easily comparable measure of the decay rates for each solvent class from the computed functions by carrying out a least-squares fit to a single exponential; for this fit, we consider the period from 0.25 to 0.6 ps. The narrow solid line drawn through each curve in Figure 8 corresponds to such an exponential fit. The relaxation times obtained in this way, denoted τ_1, are shown in Table II, which also includes values for τ_2 and for the translational diffusion constant D of the various groups of water molecules. From Table II, it is clear that, both for translational and rotational motion, the water molecules in the neighborhood of the methyl groups (nonpolar) are less free (small D, larger τ_1 and τ_2) than the water molecules associated with C=O or N–H groups (polar) or the "bulk" water. Experimental estimates for reorientation rates, primarily from NMR, suggest a factor of 2 to 3 for water molecules in the neighborhood of nonpolar solutes (6).

To determine the origin of the differences among the groups of water molecules, we examine certain time-averaged structural and energetic properties. We consider first some average properties related to the bonding energetics. Of interest are the strengths of the hydrogen bonds involving the different solvent species, the total interaction energies in the presence and absence of the solute, and the number of hydrogen bonds formed by each molecule. Figure 9 shows the calculated distributions of water-water pair interaction energies. In the figure, the

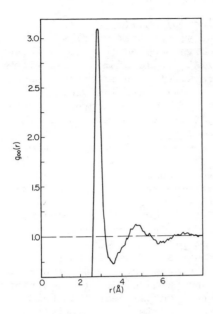

Figure 6. Water oxygen–oxygen pair correlation function computed, including all water molecular pairs

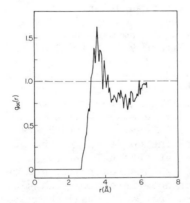

Figure 7. Water oxygen–methyl group carbon pair distribution, averaged over the three solute methyl groups

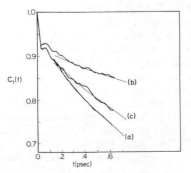

Figure 8. Rotational reorientation of water molecular dipole direction for l = 1 (Equation 7); (a) "bulk"; (b) "nonpolar"; (c) "polar," as defined in text.

relative origins of the curves labeled (b) and (c) are shifted
upward by 0.008 and 0.004, respectively; the particular values
given on the ordinate refer to the curve labeled (a). Each curve
gives the probability $P(\epsilon)$ of observing a pair of molecules with
interaction energy, ϵ, when all pairs within the potential range
(8 Å) of each other are included; the curves are individually
normalized such that their integral is unity. The three separate
distributions correspond to (a) all distinct molecular pairs in
the system; (b) all pairs which include at least one molecule in
the nonpolar class; and (c) all pairs which include at least one
molecule in the polar class. The peak at $\epsilon = 0$ includes the relatively large number of molecular pairs which are well separated
in space and therefore have very small average interaction energies; the number of such pairs is finite owing to the finite potential range. The general shape of these curves, including the
appearance of a local maximum in the negative energy region, is
that expected from studies of pure water (3).

To define hydrogen-bonded water pairs, we use an energy criterion (-3 kcal/mol) in accord with the shape of the distribution
in Figure 9; this value corresponds approximately to the position
of the local minimum that separates the central peak from the
bonding region. However, the precise value is not essential to
the interpretation of the results of the simulation.

The strength of intermolecular bonding for each type of
solvent is reflected by characteristics of the peak occurring
near $\epsilon = -5$ kcal/mol in the curves of Figure 9. It is clear that
there are only small differences, if any, in the position of the
peak; that of the polar group is at a slightly more positive energy, and that of the nonpolar group at a slightly more negative
energy than that of the total system. An alternative comparison
is obtained from the calculated mean pair energy, $\bar{\epsilon}$, for bonded
pairs ($\epsilon \leq -3$ kcal/mol). The relative values of $\bar{\epsilon}$, which are
given in Table III, are in accord with the positions of the peaks
in the figure. The shift is much smaller than $k_B T$ (~ 0.6 kcal/
mol), indicating that changes in hydrogen bond energies, per se,
cannot account for the observed differences in dynamic behavior.

It is important, however, to note that very small changes in
the mean bond energy can have significant thermodynamic effects.
The bonding region for the nonpolar group (curve (b)) includes
contributions from 40 distinct molecular pairs in a typical configuration. Consequently, a shift of only -0.05 kcal/mol in the
mean bond energy (Table III) contribute an enthalpy change of -2
kcal/mol. This result suggests that enthalpies of solution are
sensitive to small changes in the average bond energies. The total energy of solution from the gas phase obtained from the calculation is -6.72 kcal/mol. This can be thought of as resulting
from a cancellation between the change in "nonpolar" water-water
bonding (-3.4 kcal/mol), "polar" water-water bonding (+19.32 kcal/
mol), and water-solute interaction (-22.64 kcal/mol). Of the
latter contribution, -15.82 kcal/mol arises from the "polar" water

Table II

Characteristics of Translational and Rotational Mobility

H_2O class[a]	D, $10^{-5}cm^2/s$	τ_1, ps	τ_2, ps
total	3.24	3.4	1.4
bulk	3.45	2.7	1.1
polar	2.8	3.7	1.8
nonpolar	0.68	8.6	3.1

[a]As defined in the text.

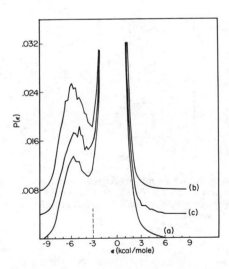

Figure 9. Normalized distributions of pair interaction energies among water molecules; a hydrogen bond is defined by $\epsilon \leq -3$ kcal/mol, indicated by the vertical mark on the abscissa: (a) all pairs in the system; (b) one of the pair in the "nonpolar" class; (c) one of the pair in the "polar" class. Each curve is integrally normalized to unity.

Table III

Water–water Hydrogen Bond Energies[a]

class[b]	mean bond energy, $\bar{\epsilon}$
total	−5.25
nonpolar	−5.35
polar	−5.21

[a]Calculated from results shown in Figure 23; a hydrogen bond is defined as $\epsilon \leq -3$ kcal/mol. [b]As defined in the text.

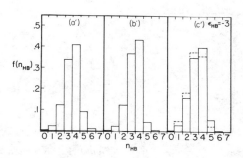

Figure 10.   Fraction of water molecules participating in n_{HB} hydrogen bonds, according to the bond definition, ϵ_{HB}. Distribution are: (a') "bulk"; (b') "nonpolar"; (c') "polar". Dashed lines include only water–water bonds; solid lines include also water–dipeptide bonds.

molecules, -3.60 kcal/mol from the "nonpolar" molecules, and 3.22 kcal/mol from the "bulk". Hence, although the long-range interactions are individually small (-0.02 kcal/mol for each "bulk" molecule), they are not insignificant in the total.

We now examine the number of hydrogen bonds formed by a typical water molecule in each of the three groups. Histograms presenting the average fraction, $f(n_{HB})$, of water molecules which participate in n_{HB} hydrogen bonds are shown in Figure 10; the energy criterion is $\epsilon_{HB} = -3$ kcal/mol. It is clear that the distributions are very similar. Thus, a typical water molecule participates in roughly the same number of hydrogen bonds in any of the three environments. For the polar group, we see that the bonds to the dipeptide contribute significantly; they tend to shift the peak in the distribution to higher values of n_{HB} and make the result more similar to that in the bulk than if these bonds are excluded. The average number of hydrogen bonds is 3.45 for bulk, 3.35 for nonpolar, and 3.28 for polar water.

From the distribution functions, the number of molecules in the first water shell around each water molecule is found to be 5.75 (bulk), 4.95 (polar), and 4.70 (nonpolar); that is, the water molecules in the bulk have roughly one more nearest-neighbor water molecule than those in the first solvation shell of the dipeptide. Since the average number of hydrogen bonds formed by molecules in the nonpolar group is essentially equal to that in the bulk, there must be bonding among a significant higher fraction of nearest-neighbor pairs in the former than in the latter. For the solvent molecules near a polar group the average number of neighbors capable of hydrogen bonding is not reduced, since a hydrogen bond to the solute can take the place of that to a water molecule.

The formation of the same number of hydrogen bonds by the water molecules at the surface of a nonpolar solute as by those in bulk water entails significant restrictions on the orientation of the former. From the schematic representation shown in Figure 11, we see that the maximum number of favorable water-water interactions can occur if none of the hydrogen atoms or lone-pair orbitals of the solvent molecule is directed toward the nonpolar group. If θ is the angle between an O-H bond direction (or O-lone pair direction) and the axis defined by the methyl carbon and solvent oxygen, the ideal orientation corresponds to $\theta = 0$. This molecular orientation is typical of crystalline clathrate hydrate compounds. In the simplest picture, optimum solvent hydrogen bonding can be achieved as long as one of the four bonding directions in the water molecule points away from the nonpolar surface. Consequently, we consider the four charges in the model water molecules as equivalent and compute a distribution for their orientations. To examine the calculated distribution of orientations, the angle θ is redefined as the angle formed by any one of the four charges of the solvent molecule, the solvent-center of mass, and the methyl carbon atom. The lower curve in Figure 12 shows the calculated

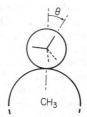

Figure 11.   Schematic of water molecule orientation   near   a   nonpolar   (—CH₃) group.

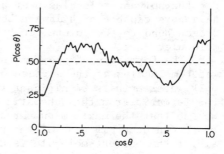

Figure 12.   Distribution of orientations of water molecules near methyl groups (lower curve); θ as shown in Figure 11. All four solvent charges are included in the distribution, as described in the text.

distribution of charge directions, averaged over the "nonpolar" solvent neighbors of the three methyl groups. The expected orientational bias of charges away from the nonpolar group is seen; that is, the distribution peaks around $\theta = 0$ ($\cos\theta = 1$) and has its minimum at $\theta = 180°$ ($\cos\theta = -1$). The probability of orientations with one charge directed away from the methyl group (i.e., at $\theta = 0$) is approximately three times that found with one toward it (i.e., at $\theta = 180°$). There is a broad secondary peak in the region $-0.1 > \cos\theta > -0.8$. This corresponds to the maximum at $\theta = 0$ and is expected from the three other charges at $\cos\theta = -1/3$ for four tetrahedrally arranged charges; there is a corresponding minimum in the neighborhood of $\cos\theta = 1/3$. It is evident from the width of these maxima and minima that there is a significant dispersion in the orientations found in solution; contributions from nonoptimal orientations (e.g., that involving the two OH bonds bridging the methyl group) are clearly important.

In summary, the present study of the kinetic and structural properties of the aqueous solvent surrounding an alanine dipeptide has provided a consistent, qualitative picture of solvation. The significant influence of the solute on the dynamic properties of the water molecules is limited to a first solvation layer. Further, the influence of individual functional groups is localized. The solvent "structure" which is induced in the vicinity of nonpolar groups, and the concomitant configurational confinement of water molecules, is a result of the maintenance of bulk-like intermolecular hydrogen bonding within the constraint of a reduced number of neighbors capable of participating in the bonds. The nonpolar groups are incapable of forming hydrogen bonds and it is this property which distinguishes them from the polar groups. The decreased mobility of the solvent near nonpolar groups arises primarily from configurational (entropic) barriers rather than energetic barriers. Although certain geometrical aspects of the system are "clathrate-like", the term is misleading in its implications with respect to the number and strength of intermolecular bonds. The bulk-like dynamics observed for solvent near the solute polar groups is consistent with the interpretation that the polar groups interact with neighboring water molecules in approximately the same way as do other water molecules.

Additional details pertinent to this report are discussed elsewhere (7).

Abstract

The characteristics of water in the neighborhood of mixed-functional solutes containing both polar and nonpolar groups will be reviewed and illustrated by the detailed results obtained from a molecular dynamics simulation of a dilute aqueous solution of an alanine dipeptide. Both the solvent effect on the dipeptide and that of the dipeptide on the solvent will be described.

Literature Cited

1. Cooke, R.,; Kuntz, I.D. Annu. Rev. Biophys. Bioeng., 1974, 3, 95.
2. Rossky, P.J.; Karplus, M.; Rahman, A. Biopolymers, 1979, 18, 825.
3. Stillinger, F.H.; Rahman, A. J. Chem. Phys., 1974, 60, 1545.
4. Berne, B.J.; Pecora, R. "Dynamic Light Scattering", Wiley, New York, 1976.
5. Verlet, L. Phys. Rev., 1967, 159, 98; ibid, 1968, 165, 201; Rahman, A.; Stillinger, F.H. ibid, 1964, 136, A405.
6. Hertz, H.G. Prog. Nucl. Magn. Reson. Spectrosc., 1967, 3, 159.
7. Rossky, P.J.; Karplus, M. J. Am. Chem. Soc., 1979, 101, 1913.

RECEIVED January 4, 1980.

The Structure of Aqueous Systems and the Influence of Electrolytes

WERNER A. P. LUCK

Institut Physikalische Chemie, Universität Marburg, D-3550 Marburg, Federal Republic of Germany

One of the most important questions in applied chemistry is the structure of water in solutions and biologic cells or membranes. Progress in modern experimental techniques, such as interpretation of spectra, opens opportunities for useful information and models.

1. Structure of Pure Water.

1a. The Observation Method. Properties of nonpolar liquids can be described surprisingly well by a hole model (1). In the case of water the partition of molecular orientation raises a further complication. The angle dependent H-bond interaction (2,3) determines about 2/3 of the intermolecular energy of 11.6 kcal/mol in ice. Among all others, water molecules stand out because of their high OH group content of 110 mole/l at room T and an equal content of lone pair electrons, Θ. Therefore the discussion of the H-bond content, as a dominant factor, leads to a good approximation of the structure of water (4-12). Because of the dominating role of the fundamental I.R. spectroscopy many overlooked that the overtone spectroscopy is a very useful tool to determine the content of H-bonds quantitatively(13,14). Solutions of molecules with OH or NH groups show, contrary to all others, an anomalous strong concentration dependence of the OH or NH overtone bands (13,14). The sharp band ν_F, which appears at low concentrations, is replaced at higher concentrations by a broad frequency-shifted band ν_b. The ν_F-band has been established as vibration of non-H-bonded, so called "free", OH groups c_F. The second band ν_b, with a larger half width $\Delta\nu_{1/2}$ but similar area $\int \varepsilon d\nu$ of the extinction coefficients ε has been associated with H-bonded species (13,14). Its frequency shift $\Delta\nu$ is proportional to the H-bond energy ΔH_H (Badger-Bauer rule). Some critics of the overtone method ignored the fact, that H-bonds change

0-8412-0559-0/80/47-127-043$07.25/0
© 1980 American Chemical Society

- in contrast to overtones - the area $\int \epsilon d\nu$ of fundamentals by a factor of 20 or more and therefore the fundamental vibrations make the view "opaque" for determination of the "free" OH.

The determination of the concentration C_B of H-bonds with I.R. overtone bands is complicated: i.e., overlapping with other bands and simultaneous one-quantum-absorption by 2 H-bonded molecules (15). Therefore, the more exact method is: to determine $C_F = 0_F C_0$ (C_0: concentration by weight; 0_F: fraction of "free" OH), the content of "free" OH by the sharp ν_F band of undisturbed molecules and to calculate $C_B = (1-0_F) C_0$.

With this precise technique the spectra of OH containing solutions could be analyzed quantitatively as chemical equilibrium:

$$OH_{free} \quad + \quad \theta_{free} \rightleftharpoons OH_{bond} \tag{1}$$

Equilibrium constants determined by this technique are among the most accurate ones in physical chemistry (2).

With increasing temperature T in the spectra of pure alcohols appears more and more the same band of "free" OH (5). For pure alcohols the extinction coefficient, ϵ_{max}, of the "free" OH is smaller and the corresponding $\Delta\nu_{1/2}$ is larger than in solutions, so as to make the area $\int \epsilon d\nu$ the same. The abbreviation "free" does not mean "gas-like" but just not H-bonded; it includes other interactions like the dispersion type. This appearance of "free" OH in alcohols can be measured quantitatively by the ϵ_{max} of overtone maxima or by $\int \epsilon d\nu$. In water too this band of "free" OH can be observed (with highest accuracy with HOD) (Figure 1 and 2) (4-12).

The water overtone bands show an isosbestic point below 150°C (6,7,8,16,17), for example in the first overtone of HOD the region 7100 cm^{-1} $<\nu_F$ <6800 cm^{-1} belongs to free or weakly bonded OH and 6800 cm^{-1} $< \nu_b$ <6200 cm^{-1} to H-bonds-bands. This is an indication of changing equilibrium between two different types of molecules with different absorption properties, and leads to an approximate model of water (4). It describes water as a chemical equilibrium between OH and θ (see equation 1).

The T-dependence of the equilibrium constant K of equation 1 demonstrates for T <200°C the H-bond energy of water: $\Delta H_H =$ - 3.7 kcal/mol (8,15) in water and about - 4 kcal/mol in alcohols.

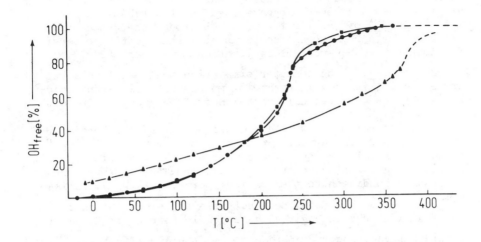

Figure 1. Fundamental parameter of hydrogen-bonded liquids: the spectroscopically determined content of nonhydrogen-bonded OH groups in (▲) liquid water, (●) methanol, and (■) ethanol (liquid–vapor equilibrium) (4).

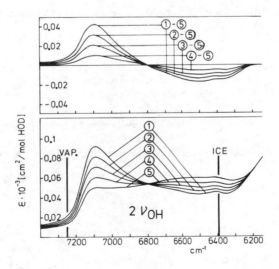

Figure 2. Bottom: first overtone of liquid HOD at: (1) 90°C, (2) 70°C, (3) 50°C, (4) 30°C, and (5) 10°C. Top: difference spectra (90°–10°C, etc.) (15). The isosbestic point confirms a model of two different OH groups.

1b. The Two-State Model. The computer simulation of water
at room T too "leads to an unambigious division of pairs into hydrogen bonded and non-hydrogen bonded" states (18) with an oscillation (19,20) of the first type of OH around the H-bond angle
ß = o and a "non-hydrogen bonded" state;"these entities persist
far longer than H_2O molecules vibrational periods" (21). - The
conclusion (22): "the computer simulation would exclude a two
state model", is only relevant for models with different molecules but not with two different kinds of OH groups, like our model assumes.

This approximate two state model of water with bonded OH
(which may have an angular partition around ß = o) and "free" OH
is basis for calculations of heat enthalpies, heat of vaporization,
specific heat, surface energy, density and T-dependences under
saturation condition to the critical point T_c (4,23). Unlike
earlier models this model does not need adjusted constants. As
one example equation 2 gives the specific heat C_σ of H_2O at the
liquid saturation curve instead $c_{v,id}$, the specific heat at ideal
gas state:

$$C_\sigma = C_{v,id} + \frac{dO_F}{dT} \left[2\Delta H_H + RT \right] + (2 + O_{\ddot{F}}) \frac{3}{2} R \quad T<200°C \quad (2)$$

Equation 2 established the good agreement between the simple model of equation 1 with experiments (4,23).

The model of two different OH groups is a simplified one. A
discussion whether water is a continuum or a system of two species
is misdirected. A model has to fit experimental data and to predict unknown properties of nature in the simplest way. This is
fulfilled with our water model. The continuum theory may be correct as object in view of future research but has been ineffi-
cient to date for describing properties of water quantitatively.

2. Structure of Aqueous Electrolyte Solutions.

Solution structure is a complex field. We can only consider selected points. The selection will be based on new ideas based on
the water spectra of aqueous systems and their consequences.

2a. Understanding of the Hofmeister Ion Series. On the basis
of the Hofmeister or lyotropic ion series colloid chemists have
known that aqueous electrolytes cannot be explained completely by
the simple discussion of ion charges. The author has interpreted
this ion series as one in which the water structure undergoes
change (8-12, 24-27). Added salts change in first approximation
the water overtone spectra as a T-change. Therefore we have undertaken quantitative measurements of the so called structure

temperature T_{str} (28) as the temperature T, which pure water would have, possessing the same extinction coefficient ε in the region of free or weak H-bonded OH.

The structure temperature T_{str} of electrolytes with 1 M anion concentration and solution temperature T = 20°C gave the series: (determined at 1,156 nm (24)) $Ba(ClO_4)_2 > NaClO_4 > NaClO_3 >$ KSCN = $KNO_3 > NaI > KBr> KCl > NaCl > BaCl_2 > MgCl_2 > Na_2SO_4 \approx$ 20°C $> (NH_4)_2SO_4 > Na_2CO_3 > MgSO_4$.

Results of such studies are as follows:
1. This spectroscopic series is very similar to the Hofmeister series.
2. It is necessary to study salt concentrations higher than 0.2 molar to recognize spectral changes.
3. There are two salt groups: a) salts with $T_{str} > T$ (solution) called structure breakers and b) salts with $T_{str} < T$ (solution) called structure makers. The boundary between "makers" and "breakers" depends in part on the solution T (24).
4. The position in this series is determined mainly by the anions; cations play a secondary order role.

Commentary:
1. The Hofmeister ion series gives the order: in which changes occur in water structure. The positions in this series denote specific effects, one parameter being the ion size (9).
2. Dilute concentrations induce only small structural change of the 110 mole OH/l and of the about 100 mole H-bonds/l at room T. In the medium concentration region of 0.2 m < C < 2-3m, the range of 1 or more solute molecules per "cluster" (average extension of H-bonded molecules), a change of water structure becomes important.
3a. Figure 3 gives one example of the parallelism between T_{str} and salt effects for solutions or colloids. There is plotted the turbidity point T_k of aqueous solutions of i-C_8H_{17} -⬡ - -$(OCH_2CH_2)_9OH$ (abbreviation PIOP-9) in presence of salts (C_{anion} = 1 mol/l) against T_{str}. The usefullness of the heuristic nomenclature T_{str} is demonstrated in figure 4. In this plot of the transition temperature T_{trans} of ribonuclease the spectroscopic T_{str} -values of KSCN and $(NH_4)_2SO_4$ solutions are shown by arrows at the same salt concentration and measured transition T. Figure 4 demonstrates that T_{str} under these conditions is near the transition T without salts. This may mean that the averaged H-bond system of water in presence of salts is at T_{trans} similar to that in water at T = 61°C, or that the H-bond structure of water at

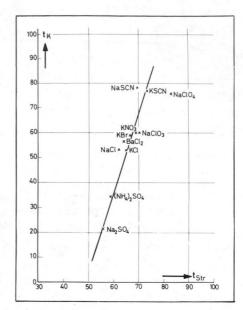

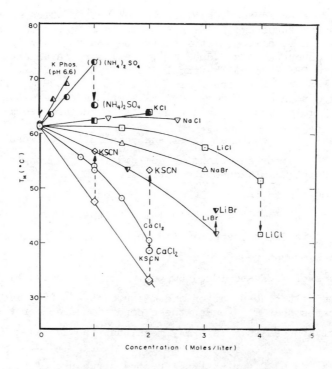

Figure 3. Relationship between the turbidity temperature, T_K, and the spectroscopically determined structure temperature, T_{str} (at 65°C). Indicated is the ion influence of the water structure ([anion] = 1M).

Figure 4. Ionic dependence of transition T, T_M, of ribonuclease (29): (---) Structure T, determined spectroscopically at the same conditions as related T_M values. Indicated is the cause of the change in T_M by the ion influence of the water structure.

$T = 61^\circ C$ is similar to that at T_{trans} in presence of different ions.

3b. T_K of PIOP-9 without salts is $64^\circ C$. $T_{str} > T$ (solution) means a higher content of "free" OH, or of weakly H-bonded OH, and causes salt-in effects; the opposite holds for salts with $T_{str} <$ T (solution). Salt-in or salt-out effects with other solutes (9,12,26) demonstrate that the boundary between salt-in or salt-out effect depends also on the organic solute. There seems to be a rivalry between H-bonds of water-water, water-ions and water-organic solute, which depends on bond strength to the organic solute and whether it is a H-bond acceptor or donor.

4. The contributions of anions to position in the T_{str}-series seems to be primarily an effect of our method; OH vibrations are more sensitive to the direct anion interaction than the cation interaction. But additional experiments demonstrate that the important role of anions is valid for other properties too.

1. Example: The turbidity point T_K of PIOP-9 is more sensitive to added anions than to cations (10,24,26,30). The following anions induce a T_K region influenced by cations:

Slight cation influence on T_K

anion	ClO_4^-	SCN^-	I^-	NO_3^-	Cl^-	CN^-
T_K ($^\circ$C)	80-75	78-74	78-69	68-60	69-63	54-50

anion	SO_4^{--}	CrO_4^{--}
T_K ($^\circ$C)	35-22	35-32

The turbidity point is nearly insensitive on the PIOP-9 concentration indicating that the H-bonds of OH to the ether oxygen play the dominant role in the solubility process of PIOP-9. The greater influence of anions on T_K of PIOP-9 than cations make ion adsorption improbable but the influence of ions on the OH groups determines the mechanism of change of T_K. Otherwise, one should expect that the cations play the dominant role interacting with the lone pair electrons of the ether groups.

2. Example: The partial molar volume of water V_1 depends more strongly on the type of added anion than cation. V_1 is equal for added NaCl or $MgCl_2$, if we plot V_1 as function of the Cl^- concentration (9). In mixtures of structure breaking and making ions of medium concentrations both effects are additive (9). This may be a common property of electrolyte mixtures.

2b. Salt-Out Effects and Spectral-Changes Effects. H-bonds
in liquid water are cooperative by virtue of angle dependence;
the probability to induce a second orientation defect is larger
near the first defect with unfavourable angles. As a consequence
of orientation defects, the "free" OH may not be distributed statistically. Defects may be more or less accumulated. This is
similar to the idealized cluster model, in which the H-bond defects could be considered as fissure planes between ordered areas
of very short life times ($\approx 10^{-11}$ sec.). By knowledge of the content of "free" OH we may estimate in an idealized approximation
N, i.e. the number of water molecules H-bonded together (8-12).
At room T the average is $N \approx 300$. From T_{str} we could estimate the
change in $(N-N_e)$ as a result of added ions as function of the solution T (8-10). The extension N of H-bonded systems increases
(for structure makers) or decreases for structure breakers (8).
The most common salt NaCl has only a small effect on O_F or T_{str};
on the base of this observation we predicted that salt-out effects
of NaCl should be smaller in comparison with strong structure
makers and its T-dependence should be small. Our measurements of
the partition coefficient of p-cresol between cyclohexane and water solutions confirmed this (9). The salt-out effect by Na_2SO_4
decreases with T but our method has been insensitive to such a
change for NaCl. Assuming in a rough model that the reduced p-
cresol content in the salt solution is caused by the ion-hydration
shells, which would not solvate organic solutes, we could estimate
the size of this hypothetical hydration number H.N. This H.N. of
Na_2SO_4 estimated by partition effects is close to the spectroscopically estimated change of the extension of the H-bonded water
systems (8).

2c. "Structure temperature" as Nomenclature. Doubts about
the term structure temperature are caused by careless generalisa-
tion of it. The author proposed to determine T_{str} - which has
been offered qualitatively by Bernal and Fowler (28) - quantitatively in the spectral region of free or weak bonded OH only for
the purpose of analysing and predicting salt effects on other solutes. Other solutes will mainly act with weak H-bonded water
groups; therefore this nomenclature seems to be reasonable. But
this nomenclature as "terminus technicus" is an abbreviation
based on complex relations and has a different meaning than the
simple words. Philip and Joliceur (31) found a linear relation
between T_{str} and the heat of transfer ΔH^o of the ions from H_2O to
D_2O. ΔH^o depends upon interaction energy of ion-water. This relationship implies that T_{str} is a measure of the average H-bond
strength.

2d. Gel-Formation. The usefulness of T_{str} may be demonstrated with other systems, too. The dye-pseudoisocyanine forms
aqueous gels in concentrations of dye/H_2O = 1/1000 (32). The
melting point of this gel in D_2O is about 3.8° higher than in H_2O,
a difference similar to that of the melting points of D_2O/
H_2O. This could be described heuristicly: D_2O has an H-bond
strength at T like H_2O at T-3.8°; thus its "T_{str}" is T-3.8°. This
relation can be applied (10) too for the melting temperature of
RNA-ase⁻ (E.Coli) (33,34); or some bacteria are deactivated in
D_2O at high T in a manner similar to that in H_2O some degress
cooler (10,35). The anomalous gel-formation of water with very
small contents of solutes like isocyanine or SiO_2 (36) may be
described as fixation of the flickering mechanism of the H-bond
system of water, which is assumed in time intervals of the relaxation time of water of about 10^{-11} sec. Solutes with H-bond
acceptors weaker than water may be solved in water initially near
the free or weakly bonded OH (37), which may start the flickering
meachanism in pure water. If they have interactions with solutes
the flickering mechanism may be reduced. This picture may lead
to the concept of organic hydrates which have a longer life time
than the water H-bonds (see section 3e.).

2e. Second Approximation of Salt Effects on Water. We could
show that T_{str} of electrolyte solutions determined in the frequency region of H-bonds gives detailed changes of the ion series
and of T_{str} (38). Choppin and Buijs (39) demonstrated that there
is a parallelism between the increase of the extinction coefficients in the region "free" OH and a decrease in the absorption
region of H-bonded groups. Joliceur and al. (16) confirmed with
difference spectra our results of the T_{str} series. In addition
they found detailed differences in the change of water spectra by
ions in the region of H-bond bands. Figure 5 shows Joliceurs'
difference spectra. We have added lines for spectra interpretation based on our spectra analysis. The double line at 960 nm
gives the region of free OH absorption, the line about 1030 nm is
the position of ice like linear H-bonds, and the middle line about
990 nm gives the position of a third middle band. This middle
band is necessary for the computer analysis of the water spectra
(7). We assume this as the centre of an angle partition of a
third type of interactions of OH groups by medium strong H-bonds.
In this position two OH groups may lie antiparallel in form of a
cyclic dimeric ring with ß $\approx$ 110° (3). The H-bond interactions of
this angle unfavoured interaction may be only 40 % per OH bond,

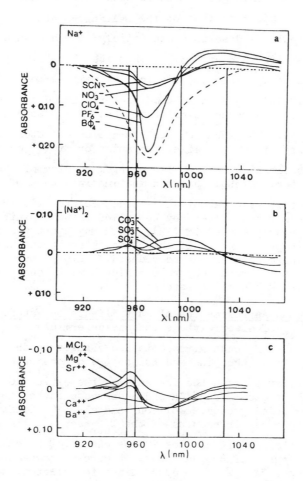

Figure 5. Difference spectra for pure H_2O/ion solutions (16). Vertical lines indicate maximum frequencies of (left to right): "free" OH; angle-unfavored hydrogen bonds; and linear (ice-like) hydrogen bonds. Different ions influence the three types of OH differently.

which means because two OH are engaged about 80 % of the energy
of one linear H-bond. By smaller O...O distances the dispersion
energy in this configuration seems to be larger (about 1 kcal/mol
higher than in the linear position). Thus, this configuration
may play an intermediate metastable role in water. Figure 5 a
shows the structure breaking effects of some large anions. Figure

5b, a loss of free OH and of the antiparallel positions $H-O...H$ with $H...O-H$
and an increase of strong H-bonds by structure maker ions and
Figure 5 c demonstrates the structure maker effect for some cations in the free OH region and an increase of intensity in the
region of medium H-bond interaction and partially in the region
of strong H-bonds (note: by the Badger-Bauer-rule $\Delta\nu$ is proportional to the interactions energy).

2f. Most Anions Weak H-bond Acceptors? The structure maker
effect could be understood easily by an orientation of water dipoles in the Coulomb fields of ions. Until recently the cause of
the structure breaker effect of a large group of ions was not discussed with exception of the indication of greater ion radius.
In crystalline hydrates distances and orientations of H_2O are in
optimum position for H-bonds or interactions with ions (40). The
author has stressed that the $\Delta\nu$ of H_2O or D_2O in hydrates is mainly smaller than $\Delta\nu$ of ice (12,41). If we could apply the Badger-
Bauer-rule to water-ion interactions too, this would generate the
question "can the interactions OH...anions be smaller than the
interaction of OH...OH?" The consequence of this assertion seems
reasonable; the fairly high hydration energies of ions are given
per mole of ion pairs and would need a number of water molecules.
In that case the hydration energy of ions per mol OH may be much
smaller. From this point of view it is easy to understand why
salts have a finite solubility in spite of large hydration energy.
A fairly well-soluble salt like NaCl needs a minimum of 9 molecules of water or 18 OH groups to be solvated. Therefore, the
interaction energy per mole OH may be by about 18 times smaller
than the so-called hydration energies which are given in tables
(or about 36 times smaller if calculated per mol sum of OH plus
lone pair electrons). One difference between the water-water
interaction and water-anion interaction is the saturation effect
by one H-bond and the possibility of higher coordination numbers
of anions. There are some indications that the cation-water interaction may be substantially larger than the anion-water interaction, which can be observed mainly with the vibration spectra.
In addition the anion-cation interaction should have an appreciable effect on the so-called hydration energy, according to the
assumptions of the Debye-Hückel theory.
     But we could deduce from the small $\Delta\nu$ for ions that in agree-
ment with the observed "structure breaker effect", there are some

anions which have   weaker interaction energies per OH group compared with the linear H-bond energy in water. As a consequence
we would need more than 1 H_2O molecule in the ion hydration sphere
and there may be some flickering aggregates consisting of anion...
...(H_2O)...cation.

The simplified equilibrium of pure water given in equation
(1) may now be enlarged to a coupled equilibrium:

$$OH_{free} + \theta_{free} \rightleftharpoons OH_b \qquad K = \frac{\left[OH_b\right]}{\left[OH_F\right]\cdot\left[\theta_F\right]} \qquad (1a)$$

$$x\ OH_{free} + A^- \rightleftharpoons OH_xA^- \qquad K_A = \frac{\left[OH_xA^-\right]}{\left[OH_F\right]^x\ \left[A^-\right]} \qquad (3)$$

$$y\ \theta_{free} + K^+ \rightleftharpoons \theta_yK^+ \qquad K_K = \frac{\left[\theta_yK^+\right]}{\left[\theta\right]^y\ \left[K^+\right]} \qquad (4)$$

$$\frac{K}{(K_A)^{1/x}\ (K_K)^{1/y}} = \frac{\left[OH_b\right]}{\left[OH_xA^-\right]\left[\theta_yK^+\right]}\ \left[K^+\right]^{1/y}\ \left[A^-\right]^{1/x} \qquad (5)$$

Every water...ion interaction (equations 3 and 4) can be described with its own coupled equilibrium of different hydrations
steps as shown in equation 6:

$$OH_{free} + OH_{x-1}A^- \rightleftharpoons OH_xA^- \qquad (6)$$

The interactions of equations 3,4 and 6 may be smaller or bigger
than in 1. The structure of electrolyte solutions depends therefore in Hofmeister ion series on the specific ion-water interaction, on the coordination numbers and on the concentration if
the hydration spheres are large enough to disturb each other.
Exceptions to the rule $\Delta\nu$(electrolytes) $<\Delta\nu$ (ice) are:
1.) solutions of CsF (55), which have high solubility in water.
Both factors could be related to high interactions of water and
CsF; 2.) Solid $Al(Cl)_3 \cdot 6H_2O$, which shows hydrolysis based on
strong interactions; 3.) Solutions of strong acids and bases, in
which both groups induce broad bands with high intensities at
the long wave length side of the ice fundamental or overtone bands
(26,42). For H-bond bands of solutions or liquids $\Delta\nu_{1/2}$ is pro-

portional to $\Delta\nu$ (41). These broad bands of strong bases and acids may indicate parts of the solution with stronger H-bond interactions as $H_2O...H_2O$ (41). This would agree with the high solubility of acid or base ions. The spectra of aqueous solutions of strong acids and bases reproduce the abnormal properties of such solutions compared with salt-ions.

2g. Hydration Numbers. In special cases spectra provide information on hydration numbers. $NaClO_4$ or $NaBF_4$ induce strong intensive water bands in the ν-region indicating very weak H-bonds. By detailed analysis of the coupling between the two OH stretching frequencies, the symmetric ν_1 and the asymetric ν_3 (43,44) and comparison with HOD spectra, it is possible to isolate the spectra of two complexes: 1.) An asymetric ClO_4^- ...DOD.. ...DOD (1:1), 2.) A symmetric ClO_4^- ...DOD ... ClO_4^- (2:1). - By an analysis of the area of the different bands we can estimate the content of different H-bonded D_2O (upper part of Figure 6). In the lower part of Figure 6 the concentration of D_2O is calculated per ClO_4^- , the hydration number HN. At low concentration in the 1:1 complexes HN is nearly 4, decreasing with C (ClO_4^-). The 2:1 complexes, which correspond to the crystalline monohydrate of $NaClO_4$, have HN about 1 such as in the crystalline state. The decreasing HN of the 1:1 type is related to a decrease of the water content per volume unit with increase of salt concentration. Both effects may depend to a type of coupling between oriented hydration spheres of cations and anions. The quotient of contents of 1:1 to 2:1 complexes depends on C (ClO_4^-), the content of 2:1 increasing with C. This leads to the assumption of a 2:1 complex ClO_4^-...DOD...ClO_4^- and the exclusion of a symmetric complex involving one ClO_4^- symmetrically positioned between two OD groups of one D_2O.

Therefore we could assume that the OD axis is not far from the $O...ClO_4^-$ axis similar to predictions from computer simulations (45) for anion...HOH.

The effect of correlations between hydration spheres of anions and cations at higher concentrations has to be assumed at other solutions too. 1 mole/l salts reduce the D_2O content of 55 mole/l at $25°C$ to the values: KF:53; LiCl:51.5; CsCl:47.7; and Bu_4NBr:26.3. With this reduction of water concentration by ions, the H-bond equilibria of equations 1-5 are also changed.

This reduction may be related to the volume of the ions, but the structure maker Na_2SO_4 does not change the volume concentration of water considerably. Electrostriction effects of the ion fields may compensate partially the Na_2SO_4 ion-volume in the solution.

The various equilibria of equations 1-5 generate a lot of parameters depending on the position in the ion series. The difference spectra (difference from pure water) of Figure 7 (55), give some indications. The ions (exception F^-)induce excess intensity in the region of weak H-bonds (1940 nm) and reduce the intensity in the regions of strong H-bonds (2050 nm) and of the "free OH" (1900 nm).

The reduction of "free OH" by most ions would mean that weak interacting ions fill free OH of pure water; the strong interaction F^- ions may be engaged with weak H-bonded H_2O of pure water, too.

The relation between different hydration spheres can be observed by the concentration dependence of salting-out effects of organic solutes (9,10,12,26). There may be, in addition, rivalry between the hydration of ions and of the organic solutes. Figure 8 shows that the turbidity point of PIOP-9 increases until KI concentration reach about 2 mol/l and then decreases above 2 m. As different from the case of pure water, T_K depends for high concentration of structure breaking ions on the concentration of the organic solute, too. The T maximum with NaI is near 4 m; with LiI > 3 m, the Li-maximum is not easily observable because $T_K > 100^\circ C$ (26).

Because of the many parameters it is not easy to calculate properties of aqueous solutions with the simple model of section 1. For instance, the known negative partial molar specific heats of ions (46,47) may be related to a reduction of the intermolecular part of the specific heat of water, the part $(dO_F/dT) \Delta H_H$ in equation 2. This part gives the energy of the change of H-bonds per degree. dO_F/dT is given by the slope of $O_F = f(T)$ (see Figure 1). With additions of KF, LiCl or CsCl (15 water: 1 ion-pair) the slope of O_F is decreased by 20 % by CsCl and 30 % by LiCl or KF. Bu_4NBr, which has a large positive partial molar specific heat, increases dO_F/dT by 30 %. Quantitative calculations require knowledge of interaction energies ΔH_H in presence of ions.

3. Mixtures: Water/Organic Liquids/Electrolytes.

3a. Solubility and Acceptor Strength of the Organic Molecules

Weak acceptors for H-bonds will interact mainly with the "free" OH in water. This could be demonstrated with solutions of NH_3/H_2O

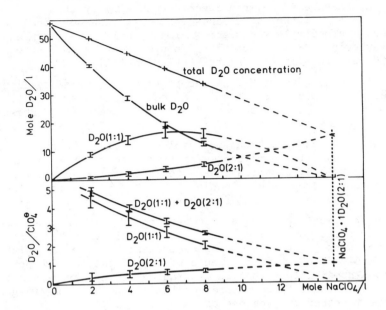

Figure 6. Top: concentration of total D_2O, liquid-like D_2O in 1:1 or 2:1 complexes with ClO_4^- as a function of $NaClO_4$ concentration at 20°C. Bottom: hydration numbers of ClO_4^- in 1:1 and 2:1 complexes and the mean values of both at 20°C.

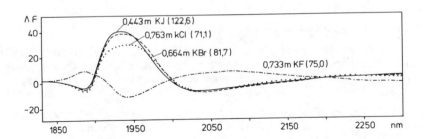

Figure 7. Difference spectra for pure water/saturated electrolyte solutions at 23°C. KI, KCl, and KBr reduce the content of free OH (1890 nm) and the content of strongly hydrogen-bonded OH (about 2050 nm) and increase the content of weakly bonded OH (1940 nm). KF has the opposite effect (55).

spectroscopically (37). Acceptor strength similar to that of water or stronger could disturb the H-bond system of pure water.
The position of hydrophilic molecules in the series of H-bond
acceptors can be determined spectroscopically by the Badger-Bauer
rule from $\Delta\nu$ of the water bands (12,43,44). We obtained the
series: acetonitrile < dioxane<< dimethylsulfoxide < pyridine <
triethylamine.A comparison with ice spectra gives acceptor strength
of H_2O between pyridine and triethylamine (12).

Molecules without H-bond acceptors are solvated by dispersion
forces. The solubility of small molecules increases with its critical T, which is a measure of dispersion forces (1). Starting
with molecules of the size of XCH_3 (X:F, Cl, Br) the solubility
in water decreases with the molarvolume or the size of the molecules (10,48). Large molecules disturbe the water structure.
There are some discussions that water has a gas-hydrate like structure around hydrophobic molecules (9-12,48). This pentagondodeca-
hedral arrangements of H_2O form bigger holes than ice, all OH or
lone pair electrons being at the surface of the pentagondodecahe-
dral, or point to outside, none to the interior. This arrangement needs no "free" OH. - The same process, a structure with a
minimum of "free" OH is the cause of the so called "hydrophobic
bond", a nomenclature that is misleading. Only the tendency of
the water H-bonds to be closed constrains the hydrophobic groups
to come together in presence of water.

A mixture of organic solutes in water can change the hydrophobic properties of solutes. For example the turbidity point
of PIOP-9 can be increased or decreased by adding organic molecules. A rise of T_K was observed for 0.5 mol/l additives:

$C_4H_8(OH)_2$ > $H_2HCON(CH_3)_2$ > $NH_2COOC_2H_5$ > $i-C_3H_7OH$ > NHC_3H_7OH >
C_2H_5OH > tetrahydrofurane; a reduction of T_K was observed for
glucose < $H_2C=CH-C\equiv N$ < $i-C_4H_9OH$ < paraldehyde.

The first group of organic molecules may screen hydrophobic
groups, the second one may screen hydrophilic groups. Joliceur
et al. (16) have observed a change in the water structure by
different organic solutes and describe it by T_{str} too. In agreement with the PIOP-9 method they found two groups of organic solutes: 1. T_{str} > T (solution) 2. T_{str} < T (solution).

3b. Micelle Formation and Electrolytes. Molecules with more
or less separated hydrophobic and hydrophilic groups form
"clusters" of their hydrophobic groups in the shape of micelle
nuclei. Above the critical concentration (CMC) all added monomers have to form micelles as a result of the equilibrium, n
monomers = micellle. CMC depends on the HHB-value, the balance
between hydrophilicity and hydrophobicity. HHB is sensitive to
added salts parallel to the Hofmeister ion series. For instance,

the equilibrium constants K_{mi} of micelle formation of $i-C_8H_{17}-$
- ⟨◯⟩ - $(OCH_2CH_2)_{40}$-OH (PIOP-40) determined by a U.V. method
(27,49) are reduced by 0.5 mol/l salts in the series: LiCl <
NaCl < BaCl$_2$ < NaHCO$_3$ < (NH$_4$) SO$_4$< Na$_2$(SO$_4$) (0.4 m) < Na$_2$CO$_3$
(0.3 m) (at 20°C). The maximum change of K_{mi} is by a factor of
0.01.

Molecules which are weakly soluble in water, can be solvated
in the hydrophobic micelle nucleus. This is an important technical process for instance for detergents. As levelling agent in
textile chemistry micelles act as buffers for dyestuffs, the dyes
are solubilized in the micelle nucleus and this equilibrium holds
the "free" dye concentration in the water phase more or less constant during the dyeing process (49,50).

Electrolytes in the dyeing bath have two effects: 1) salt-
out effect of dye to the fibre (51), and 2)"salt-out effect" of
the levelling agent increasing the micelle concentration (27).
The salt out effect on dyes favours the wash-fastness of tex-
tiles: the partition coefficient of the dye between fibre and
washing bath is increased by structure making salts (52).

3c. Interface Effects and Electrolytes.
The surface tension
of water is increased by ions in the series (26):

$$SO_4^{--} > Cl^- > Br^- > NO_3^- > I^- > SCN^-$$

$$Li^+ > Na^+ > K^+$$

This effect is caused by two factors: 1.) the change of the water
structure by change of the H-bond system as a real effect on surface energy (1) and 2.) the change of the number of water molecules per cm^2 of surface (1,4,53) by ions. Consequently a change
of the surface area-pressure diagrams of monolayers of $C_{18}H_{37}-$
$(OCH_2CH_2)_3OH$ (C18-03) by ion additions to the water subphase has
been observed (54). The change of the water structure by ions in
the series of Hofmeister alter the interaction energy between
the water subphase and the ethylenoxide groups of (C18-03) and
the required force to get the same monolayer area varies similarly
to a T-change (54).

3d. Ion Solubility in Mixtures: Water Organic Solvents.
Solubility of salts in hydrophilic solvents is smaller than in
water; only some perchlorates or Li salts have comparable solubilities (Figure 9 gives the reciprocal ion solubility expressed
in mole solvent (alcohol or acetone) per mole salt on a log scale
(55). Data collected by Barthel (56) demonstrate that at 25°C the
partial molar volume of ions in methanol at infinite dilution is

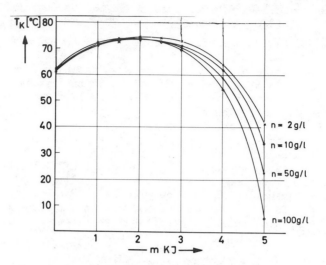

Figure 8. Rivalry of ion and ether hydrates of PIOP-9 at high ion contents. Above 2m KI, the tubidity point, T_K, depends on the PIOP concentration (ng/L), and above 4m, the structure-breaking effect of K disappears.

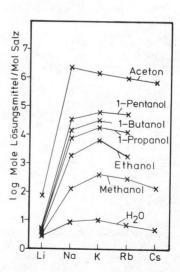

Figure 9. Reciprocal solubility of chlorides in water and alcohols on a logarithmic scale, demonstrating the important role of the network water structure on ion solubilities (55)

about 25 $\pm 4 (cm^3)$ smaller than in H_2O and about 4 ± 3 (cm^3) bigger in formamide without indicating systematic variations from the given average values. In ternary mixtures: water/acetone or ethanol/salts, the reciprocal salt solubility expressed in H_2O present per ion pair (Figure 10) is nearly constant until solvent concentration reaches about 250 H_2O per solvent molecule (the size of the water "clusters") (55). Above this solvent concentration the solubility of salt decreases sharply. In mixtures of water/acetone/K_2CrO_4 two phases appear above 0.03 mole acetone per mol water; the second phase has 0.24 acetone per water and the salt content is reduced by a factor of 5.5. These types of experiments indicates that ions need a group of water molecules to be dissolved. With higher concentrations of solvent molecules the content of such groups of H_2O decreases and, therefore, the salt solubility decreases.

The relationship drawn in Figure 10 has some similarities to properties of aqueous salt solutions: 1) the behaviour changes above concentrations higher than 1 solute per about 300 H_2O, 2) the solubility of ions requires a group of H_2O molecules, and 3) there is a rivalry between ion and solvent hydration (see 2 phases acetone/H_2O).

3e. Organic Hydration. The foregoing relationship and others suggest the existence of hydrates of organic molecules. For instance above the turbidity point T_K of PIOP-9 or similar compounds the organic phase has a fairly high water content (26), depending on the difference $(T-T_K)$, starting with about 20 H_2O per ether group and decreasing till 2 H_2O. Additions of ions to the water phase decreases this water content to a limiting value of 2 H_2O per ether oxygen, also. This mixture of PIOP-9 with 2 H_2O per ether group has a sharp viscosity maximum, a maximum of the velocity of sound, and a special x-ray structure (10). This structure has been described as meander structure of the ethylenoxide groups or as a helix. Models of this hydrate show that there are water positions possible with all H-bond angles ß = o. H-bonds bridges of 2 H_2O from one ether oxygen to the next to nearest neighbors may be amplified by cooperative mechanism of the 2 H_2O.

In this case the dihydrate may be of special stability. Warner (57) has discussed possible relations between hydration of biopolymers and ice-like distances of the H-bond acceptor groups of these polymers (57).

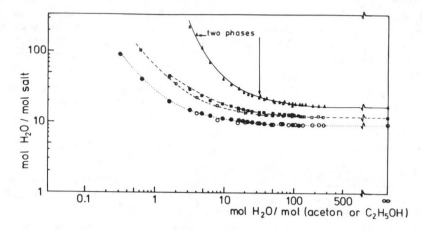

Figure 10.   Reciprocal salt solubilities in the ternary systems: water/acetone or ethanol/salts at 23°C as a function of the solvent mixture ratio: (○, ●) NaCl, (□, ■) KCl, (△, ▲) K₂CrO₄; open symbols represent acetone systems, closed symbols represent ethanol systems.

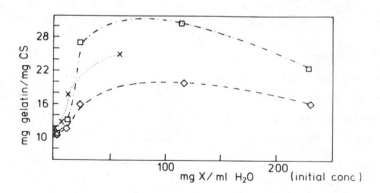

Figure 11.   Ratio of gelatin and chondrotine sulfate (CS) in the coacervates depends on additives (X): (□) NaCl; (×) NaHPO₄; (◇) sodium cyclopentanecarboxylate. The coacervate was prepared by mixing 5 mL gelatin (1%) and 3 mL CS (1%) at 40°C and cooling to 20°C (55).

3f. Coacervates. In a two phase system, water/organic solvent, the phase enriched with the solvent is called coacervate (Bungenberg de Jong, 58). The organic phase above the turbidity point of PIOP-9 behaves like a coacervate. Its composition can be changed strongly by ion additives (26,10,55,59).

The following table demonstrates in the second line the influence of different ions on the water content of the organic phase (mole H_2O/ether group) and in the third line the influence on the PIOP-9 content of the water phase (g PIOP-9/1) at $80^{\circ}C$.

salt	NaSCN	NaClO$_4$	NaI	Without salt	NaNO$_3$
H$_2$O content	18	14.3	11.5	4.8	4.4
PIOP-9 "	27	2.2		0.84	

salt	NaCl	ZnCl$_2$	NaBr	CaCl$_2$	LiCl	NaOH
H$_2$O content	4	4	3.9	3.8	3.7	3.3
PIOP-9 "	0.5	0.39	0.26	0.49	0.67	0.82

salt	KCl	MgCl$_2$	NH$_4$Cl	AlCl$_3$	NaF	Na$_2$CO$_3$
H$_2$O content	3.2	3	3	2.8	2.6	2.2
PIOP-9 "			0.46	0.49	0.34	0.42

salt	NaHCO$_3$	CuSO$_4$	MgSO$_4$	ZnSO$_4$	Na$_2$SO$_4$
H$_2$O content	2.2	2	1.8	1.7	1.6
PIOP-9 "	0.38				0.13

A second experiment of this type (59) establishes the strong influence of salts on coacervates and demonstrates the importance of this observation for biological systems. Figure 11 shows the composition of a coacervate in the system water/gelatine/chondrotinesulfate/X. Different additions of X change the ratio gelatine /chondrotinesulfate depending on the concentration and type of additive (59). The two main components are important components of cartilage. Cartilage and collagen behave similarly to coacervates (59), which can depend on the type and concentration of ions in the aqueous phase. These could encourage researchers of environmental protection institutions to prove influences of ions on the human body. Compare the difference by a factor 10^4 of ions on gelation of sols (60), or the shift of the melting point of gelatine (Δ T till 35°) by different ions (61), and further examples (10).

Already Ostwald (62) distinguished primary hydration spheres around organic solutes and a second diffuse hydrate sphere. Bungenberg de Jong assumed coacervate formation as combination of the primary hydration spheres and separation from the diffuse one. This process is similar to ion pair hydrate formation (63,64,65), where the primary hydration sphere is connected (Figure 12). From our model experiments with PIOP-9 we would assume as preliminary step of coacervation the connection of the second hydration shells (Figure 12).

The coacervate formation of PIOP-9 at higher T (at low T only one phase exists)belongs to a group of processes, which is favoured at high T (polymerisation of tabacco mosaic protein or hemoglobin of anaemia sickle cells, division of fertilized eggs, precipitation of poly-L-proline above 25°C etc.) (10,66). Lauffer (66) calls these effects "entropy driven processes" and assumes that the entropy increase of water is its cause. The increase in hydrophobic interaction also favours these changes.

4. Aqueous Systems.

We give three different examples for water properties in complicated systems of the solid type.

4a. Polyamide Fibres. The water content of polyamide fibres is important for dye diffusion processes (67,68,69). Acid dyes have a very low diffusion in polyamides at ambient humidity and room temperature; the diffusion can be accelerated by increasing the relative humidity. For instance the diffusion rate of a Perliton dye at 60°C contacted with liquid water is similar to that in air at 150°C at low humidity (68).

The water diffusion coefficient D in polyamide itself increases with water concentration. D of H_2O (80°C) in 6-polyamide is reduced at 25 % relative humidity to 18 % of its value at 100 % relative humidity (68). This may be important for insects in arid areas. Drying at the surface may reduce the water diffusion from inside.

Salts change the water contents of polymers and therewith the diffusion processes. In 6-polyamide at 70°C we found an increase of water uptake to 132 % with 1 m $NaClO_4$ in the water phase and a decrease to 82 % with 1 m Na_2SO_4 (10).

An important and often neglected factor is the pretreatment of polymers. For instance, the partition coefficient of strong acids between 6-polyamide/water is about 200 at p_H = 4-5. To reduce the acid content of a 6-polyamide fibre equilibrated with HCl of p_H = 4 to 50 %, it is necessary to treat it about 200 times in equilibrium with pure water at volume ratios 1:1 (52).

4b. Water in Biologic Systems. In recent years there have
been discussions on the important role of different types of water molecules in biological cells, differentiated as "bonded" and
"unbonded" (10,70-76). To prove this assumption we applied our
I.R. overtone method to water in cartilage or collagen (55,77).

In these probes there is a frequency shift of the water bands
at low humidities (Figure 13). In the region to about 50 % relative humidity the first hydration shell of the biopolymers is
formed. In this region of the formation of the first hydration
shell the polymer I.R. band changes with water additions, indicating small change of the polymer properties. After filling the
first hydration shell the water uptake increases rapidly; the water frequency changes in the direction of weaker H-bonds, but the
polymer band is now constant (Figure 13). Such experiments conducted with glucose, amylose, amylopectin, glycogen and dextrine,
establish two different types of water: 1) first hydration shell
(relative humidities < 50 %) water is a little stronger bonded than
in the liquid state, 2) secondary hydration (50 % < relative humidities < 98 %) with little disturbed liquid-like water structure
and 3) at high water contents, liquid-like water. These three
types were confirmed with difference spectra in the three humidity
regions (55). The differences are very similar in each state but
quite different from to another humidity region.

The water amount of the third type can be influenced by ions.
The water uptake of freshly prepared bovine nasal cartilage is
different if contacted with liquid water or solutions (115 % pure
water, 106 % at 0.08 NaCl).

The existence of different types of water, called bonded and
non bonded, has been established by the spectroscopic method. The
first hydration water was found to be a little more strongly bonded
than in the liquid state. Instead of "bonded water" we call it
hydrate water. The excess water is liquid-like with small disturbance.

On the basis of the spectra it would be better to call the second type liquid-like water rather than unbonded, because it is
characterized by normal H-bonding and the bond-differences to liquid water are only small. The biological consequences of the two
types of water are unexpectedly large compared with the small spectral differences. Seed does not grow at low humidity: food stability can be realized by small reduction of the water activity with
sugar, salt etc.; microorganism do not grow on food as a result of
small reductions of the relative humidity. These changes seems to
depend upon liquid-like water. The differences between the two or
three different water states could be interpreted too with the
structure temperature. In the middle part of Figure 13 is given
on the right scale the wavelength of the water maxima in tendon or
gelatine at different relative humidities and on the left scale the
structure temperature T_{str}, which pure water should have indicating
the same wavelength maxima. T_{str} may give a more reasonable scale

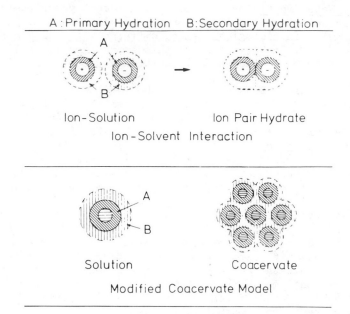

A : Primary Hydration   B : Secondary Hydration

Ion-Solution   Ion Pair Hydrate

Ion-Solvent Interaction

Solution   Coacervate

Modified Coacervate Model

Figure 12.   Similarity between the model of ion-pair hydration and hydrates in a coacervate

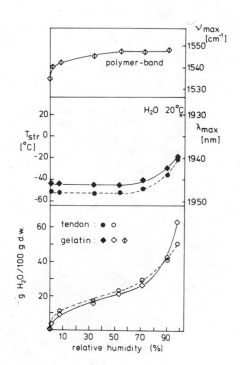

Figure 13.   Hydration of tendon or gelatin. Top: change of a polymer IR band during formation of the first hydration shell. Middle: during the formation of the first hydration shell, the water IR overtone band indicates slightly stronger hydrogen bonds, similar to liquid-like water (T_{str}: —45° to —50°C). Bottom: desorption isotherm (55). Symbols: ($\bullet$, $\bigcirc$) tendon, ($\diamond$, $\diamondsuit$, $\diamondsuit$) gelatin.

to understand that small changes of the position of the water I.R.
maxima are accompanied with big biologic effects.

There are a lot of salt effects on biological samples, which
can be ordered by the Hofmeister ion series (10). All these effects
indicate changes of the water structure and its importance for
living cells.

4c. Mechanism of Desalination Membranes. The decrease of the
water content of PIOP-9 coacervates with T seems to offer a method
for desalination as follows: 1) add PIOP-9 to sea-water, 2) warm
up to the turbidity point, 3) separate the aqueous phase, 4) heat
the organic phase, and 5) separate the water phase that has
appeared. This experiment was carried out but in the new water
phase the ion concentration was similar to the original. In this
coacervate these may be two different types of H_2O: the primary hydration phase and a more diffuse secondary water phase, which is
liquid-like and solvates ions. This leads to the postulate that
desalting materials should have hydration water but no liquid-like
water phase. Uptake of water that is liquid-like   could dissolve ions should be suppressed by steric hindrance (78,79). Such
experiments done with copolymers of ethylene/propyleneoxide derivatives (reduced H_2O affinity) gave an ion reduction in step 5
(78,79,80).

Spectra of water in desalination membranes (cellulose acetate,
polyimide, porous glass) show in contrast to biological studies a
smaller $\Delta\nu$ and indicate a weaker H-bonded system in these membranes compared with liquid water (78,79).

The spectroscopic overtone results indicate a water structure
in membranes different from liquid water. The observed weaker H-
bond system would favour a quicker flux. The water uptake in membranes gives an estimate of the size of "pores" between the polymer or glass chains, calculated in units of H_2O size: 1) cellulose
acetate, about 4, 2) examined glass membranes 8, and 3) technical
glass fibres about 5. This could mean that in these layers there
are not sufficient water for the ion solubility mechanism. Experiments by Pusch (81) established this concept; ion rejection decreases in the series: Na_2SO_4 > $CaCl_2$ > NaF > NaCl. Na_2SO_4 needs
more water molecules to be dissolved in the membrane pores than
NaCl. Disturbance of the water structure by the formation of hydrates with membrane material further reduce the efficiency of the
membrane water to solvate ions.

On the basis of the membrane research we prefer the following
nomenclature of water types in aqueous systems: 1) first shell of
water hydrate, 2) second shell of disturbed liquid-like water and,
3) liquid-like water. The salt transport in membranes, the diffusion of dyes in fibres or the biochemistry in living cells need
the existence of water of types 2 or 3.

Other methods to study different water types see references (10,73,74).

List of Symbols

θ : lone pair electrons

C_F : concentration of non H-bonded OH groups

O_F : fraction of non H-bonded OH groups

dO_F/dT : slope of O_F in Fig. 1

C_o : concentration by weight

C_B : concentration of H-bonded OH groups

OH_F : non-bonded or "free" OH

OH_b : H-bonded OH

ΔH_H : H-bond energy

ν_F : frequency of the band maximum of non H-bonded OH groups

ν_b : frequency of the band maximum of H-bonded OH groups

$\Delta\nu$: frequency shift between: "free OH" and "H-bonded OH"

$\Delta\nu_{1/2}$: half width of bands

ε_{max} : extinction coefficient at band maximum

T : Temperature

T_c : critical temperature

T_{str}, t_{str} : structure temperature of a solution, T of pure water with similar content of non H-bonded OH

T_K, t_k : lowest temperature of two phase formation

$T_{trans}=T_M$: transition $-T$ of ribonuclease

C_σ : specific heat in liquid state at saturation line

$C_{v,id}$: specific heat at constant volume of ideal gas state

R : Gasconstant

V_1 : partial molar volume of water in electrolyte solutions

β : H-bond angle, $\beta=0$ if angle between axis OH and lone pair electrons is zero

A^- : anion

K^+ : cation

HN : hydration number

$OH_x A^-$: hydrated anion with HN = x

$\theta_y K^+$: hydrated cation with HN = y

PIOP-9   : p-iso-octylphenol with 9 ethylenoxide groups

CS   : chondroitinesulfat

d.w.   : dry weight

I.R.   : Infra Red

Literature Cited

1. Luck, W.A.P. Angew.Chem., 1979, 91, 408;
   Angew.Chem.Int.Ed., 1979, 18,350.
2. Luck, W.A.P. Naturwissenschaften, 1965, 52, 25, 49;
   1967, 54, 601.
3. Luck, W.A.P., in Schuster-Zundel-Sandorfy (Ed.), "The Hydrogen Bond", Verlag North Holland Publ.: 1976, p.527.
4. Luck, W.A.P. Angew.Chem., 1979, in press.
5. Luck, W.A.P.; Ditter, W. Ber.Bunsenges.Phys.Chem., 1968, 72,365.
6. Luck, W.A.P. Ber.Bunsenges.Phys.Chem., 1963, 67, 186;
   1965, 69, 626.
7. Luck, W.A.P.; Ditter, W. Z.f.Naturforschung, 1969, 24b, 482.
8. Luck, W.A.P., "Structure of Water and Aqueous Solutions"; Verlag Chemie/Physik Verlag; Weinheim, 1974, p. 222, 248.
9. Luck, W.A.P., in Schuster-Zundel-Sandorfy (Ed.), "The Hydrogen Bond"; Verlag North Holland Publ.: 1976, p. 1369.
10. Luck, W.A.P. Topics in Current Chemistry, 1976, 64, 113.
11. Luck, W.A.P. ; Schiöberg, D. Advan.Mol.Relaxation Processes, 1979, 14, 277.
12. Luck, W.A.P. Progr.Colloid & Polymer Sci., 1978, 65, 6.
13. Mecke, R. Wissenschaftliche Veröffentlichungen 1937-1960, Festschrift zum 65. Geburtstag, Freiburg 1960.
14. Luck, W.A.P., in Felix Franks (Ed.), "The Hydrogen Bond in Water:" "A comprehensive Treatise", Plenum Publishing Corp.: New York, 1973, p. 225.
15. Schiöberg, D.; Buanam-Om, C; Luck, W.A.P.; Spectroscopy Letters, 1979, 12, 83.
16. Paquette, J.; Joliceur, C. J.Sol.Chem., 1977, 6, 403.
17. Worley, J.D.; Klotz, I.M. J.chem.Physics, 1966, 45, 2868.
18. Stillinger, F.H.; Rahman, A. J.Chem.Phys., 1972, 67, 1281.
19. Geiger, A.; Rahman, A.; Stillinger, F.H. J.Chem.Phys., 1979, 70, 263.
20. Geiger, A., Karlsruhe 1979, Film.
21. Rahman, A.; Stillinger, F.H. J.Chem.Phys., 1971, 55, 3336.
22. Stillinger, F.H.; Rahman, A. J.Chem.Phys., 1974, 60, 1545.
23. Luck, W.A.P. Discuss.Faraday Soc., 1967, 43, 115.
24. Luck, W.A.P. Ber.Bunsenges.Phys.Chem., 1965, 69, 69.
25. Luck, W.A.P. Proc. of the 4th Internat. Symposium on Fresh Water from the Sea, Heidelberg, 1973, p. 531.
26. Luck, W.A.P. Fortschr.chem.Forschung, 1964, 4, 653.
27. Luck, W.A.P. III.Internat.Kongr.f.Grenzflächenaktive Stoffe, 1960, 264, Köln.
28. Bernal, J.D.; Fowler, R.H. J.Chem.Phys., 1933, 1, 515.
29. Von Hippel, P.H.; Wong, K.Y. J.Biol.Chem., 1965, 240, 3909.
30. Luck, W.A.P. "Ullmanns Enzyklopädie der techn.Chemie", Verlag Chemie; Weinheim, 3.ed., Vol. 18, 1967, p. 401.
31. Philip, P.R.; Joliceur, C. J.Phys.Chem., 1973, 77, 3076.
32. Luck, W.A.P. Naturwissenschaften, 1976, 63, 39.
33. Hermans, J. Biochim.Biophys.Acta, 1959, 36, 534.

34. Lewin, S. Biochem.J., 1966, 99, 1P; Arch.Biochem. Biochem. Biophys., 1966, 115, 62.
35. Hübner, G.; Jung, K.; Winkler, E. "Die Rolle des Wassers in biologischen Systemen", Berlin: Akademie Verlag und Braunschweig: Vieweg 1970.
36. Gmelins "Handbuch der anorganischen Chemie" Nr. 15, Silicium B. Verlag Chemie, Weinheim, S. 414, 439, 447, 448, 453, 503.
37. Luck, W.A.P. J.Chem.Phys., 1970, 74, 3687.
38. Luck, W.A.P.; Zukovskij, A.P. Ed. A.J.Sidorovo "Molecular Physics and Biophysics of Water Systems", Leningrad University, 1974, p. 131.
39. Choppin, G.R.; Buijs, K. J.Chem.Phys., 1963, 39, 2042.
40. Falk, M., Knop, O. Ed. F. Franks "Water, A Comprehensive Treatise" Plenum Press, New York-London, 1973, p. 55.
41. Buanam-Om, C.; Luck, W.A.P.; Schiöberg, D. Z.Phys.Chem., 1979, in press.
42. Zundel, G. "Hydration and Intermolecular Interaction", Academic Press, New York, 1969.
43. Schiöberg, D.; Luck, W.A.P. Spectroscopy Letters, 1977, 10(8), 613.
44. Schiöberg, D.; Luck, W.A.P. J.Chem.Soc.Faraday Trans. 1979, 75, 762.
45. Palinkas, G.; Riede, W.O.; Heinzinger, K. Ztschr.Naturforsch., 1977, 82a, 1137.
46. Wicke, E. Angew.Chem.Int.Ed. 1966, 5, 106.
47. Rüterjans, H. et al. J.Phys.Chem., 1969, 73, 986.
48. Franks, F.; Reid, D.S. "Water A Comprehensive Treatise", Plenum Press, New York-London, 1973, p. 336.
49. Luck, W.A.P. Angew.Chem., 1960, 72, 57.
50. Luck, W.A.P. J.Soc.Dyers and Coulourists, 1958, 74, 221.
51. Luck, W.A.P. Melliand, 1960, 41, 315.
52. Luck, W.A.P. Chimia, 1966, 20, 270.
53. Luck, W.A.P. "Physikalische Chemie und Anwendungstechnik der grenzflächenaktiven Stoffe, Kongreßband VI. Internationaler Kongreß für grenzflächenaktive Stoffe in Zürich 1972", Carl Hanser Verlag, München, 1973, p. 83.
54. Luck, W.A.P.; Shah, S.S. Progr.Colloid & Polymer Sci., 1978, 65, 53.
55. Kleeberg, H. Thesis, Marburg, in preparation.
56. Barthel, J. "Ionen in nichtwäßrigen Lösungen", Dr.Dietrich Steinkopff Verlag, Darmstadt, 1976.
57. Warner, D.J. Ann.New York Acad.Sci., 1965, 125, 605; Nature, 1962, 196, 1055.
58. Bungenberg de Jong, H.G.; Kruyt, H.R. Koll.-Z., 1930, 50, 39.
59. Kleeberg, H.; Luck, W.A.P. Poster "Is cartillage a coacervate?" 6th Colloquium of the Federation of European Connective Tissue Clubs, August, 28.-30. 1978.
60. Kruyt, H.R. "Colloid Science", Vol.I, p.335, Elsevier, Amsterdam 1952.

61. Riese, H.C.A.; Vienograd, S.R. J.phys.chem., 1956, 60, 1299.
62. Ostwald, Wo.; Hertel, R.H. Koll.-Z., 1929, 47, 258, 357.
63. Franks, H.S.; Wen, W.V. Disc.Farad.Soc., 1957, 24, 133.
64. Lilley, T.H. in F.Franks (Ed.) "Water a comprehensive Treatise", Plenum Press, New York, 1973, Vol.3, p. 266.
65. Gurney, R.W. "Ionic Processes in Solution" McGraw Hill, New York, 1953.
66. Lauffer, M.A. Entropy-Driven-Processes in Biology, Berlin-Heidelberg-New York, Springer 1975.
67. Luck, W.A.P. 100 Jahre BASF, 1965, 259.
68. Luck, W.A.P. in H.Freund (Ed.) "Handbuch der Mikroskopie in der Technik", Umschau-Verlag, Frankfurt, Bd.VI, Teil I, S. 345, 1972.
69. Luck, W.A.P. Ber.Bunsenges.Phys.Chem., 1965, 69, 255.
70. Hauser, H. Lipids, in F.Franks (Ed.) "Water a comprehensive Treatise", Plenum Press, New York, 1975, p. 209.
71. F.Franks (Ed.) "Water a comprehensive Treatise", Plenum Press, New York, 1972-1975, Volumes 4. and 5.
72. International Meeting organized by CNRS Roscoff June 1975, Paris, Editions Centre National Recherche Scientifique, Nr. 246, 1975.
73. Hazlewood, C.F. Annals New York Acad.Sci., Vol.204, 1973.
74. Hazlewood, C.F. in W.Drost-Hansen a.Clegg, J. (Ed.) "Cell-Associated Water", Academic Press New York, 1979, p. 165.
75. Garlid, K.D. in W.Drost-Hansen a. Clegg, J. (Ed.) "Cell-Associated Water", Academic Press, New York, 1979, p. 293.
76. Berendsen, H.J.C. in F.Franks (Ed.) "Water a Comprehensive Treatise" Vol. 5, p. 293, 1975.
77. Kleeberg, H.; Luck, W.A.P. Naturwissenschaften, 1977, 64, 223.
78. Luck, W.A.P.; Schiöberg, D.; Siemann, U. Ber.Bunsenges.Phys. Chem., 1979, 83, in press.
79. Luck, W.A.P.; Schiöberg, D.; Siemann, U. Trans.Far.Soc., in press.
80. Luck, W.A.P. DBP Offenlegungsschrift, Nr.2151/207.
81. Pusch, W. in Luck, W.A.P.   (Ed.) Structure of Water and Aqueous Solutions, Chemie Verlag, Weinheim, 1974, p. 551.

RECEIVED January 4, 1980.

PERSPECTIVE:
MACRO- AND MICROINTERACTIONS OF
WATER AND POLYMERS

Water and Proteins: Some History and Current Perspectives

JOHN T. EDSALL

Biological Laboratories, Harvard University, Cambridge, MA 02138

Water is alone among the traditional four elements - earth, air, fire, and water - in being, not indeed an element, but a pure chemical compound. As Hall has noted: "Water appeared repeatedly as a life-giver in the creation myths of Greece, Egypt, Babylon, the Near East, and Persia" (1). Thales of Miletos, who lived about 600B.C., was credited by Aristotle and others with the view that life-as-soul (psyche) was immanent in water, and that water was essential for life-as-action (zoe).(1) Whatever the real views of the historical Thales may have been, the recognition of the deep significance of water for life is very old indeed.

More than 2000 years after Thales, Whewell (2) pointed out the unusual thermal properties of water, and the geological and biological significance of the fact that water is denser than ice, and becomes still denser as the temperature rises from 0 to 4°C. For Whewell's natural theology, these facts and others were manifestations of Divine Wisdom in the construction of the universe. Some 80 years after Whewell, L.J. Henderson (3) pointed out the unique position of water among all known liquids in possessing maximal, or nearly maximal, values for a wide range of thermal, electrical, and other properties, which rendered it a uniquely suitable medium for the properties of life as we know it. Henderson abjured any invocation of theological arguments like those advanced by Whewell, but did state that the remarkable properties of water, carbon dioxide, and other compounds of carbon, hydrogen, and oxygen, appeared to imply the existence of a "teleological order" in the properties of matter, favorable to the appearance of life in evolving worlds.

Twenty years after Henderson, Bernal and Fowler (4) published what probably still remains the most important single paper on water that has yet appeared. By 1933 the structure of ordinary hexagonal ice was known from x-ray diffraction, the shape and dimensions of the water molecule were known from spectroscopic studies, and its general electronic configuration was largely understood. Bernal and Fowler showed how water, at temperatures not too far above the freezing point, could be

0-8412-0559-0/80/47-127-075$05.00/0
© 1980 American Chemical Society

regarded as consisting of a sort of broken-down ice structure
for short-range interactions. They could explain in qualitative,
and often in semiquantitative terms, how the properties of
the molecule determined that remarkable array of physical
properties that had so impressed Henderson (whom they did not
mention). The approximately tetrahedral arrangement of the
charge distribution in the triangular molecule, with two
positive centers of charge (hydrogen bond donors) and two
negative centers (hydrogen bond acceptors) went far to explain
the high dielectric constant of water, its high surface
tension, its high heat of vaporization high melting point, and
many other properties. These brief remarks of course can give
no adequate indication of the range of new interpretations
offered in this remarkable paper. Of course some of the
suggested proposals of Bernal and Fowler were wrong, and were
soon rejected; but the general outlook their work conveyed
has been fundamental for later research.

In the last 20 years the study of water has undergone an
explosive development, involving the publication of thousands
of papers and several books, most notably the comprehensive
treatise in six volumes, edited by Felix Franks (5). Dr. H.
A. McKenzie of Canberra and I have embarked on a review of the
present state of our knowledge of water, especially in its
relation to proteins. The first part of this study has been
published (6); we are still at work on the second and final
part. We are not reporting on original research of our own,
and I have indeed retired from the laboratory several years
ago; but we hope to provide some perspective on aspects of
aqueous protein systems that have long preoccupied us. Here
I will briefly discuss only two points: (1) an aspect of hydrophobic interactions that has only recently become apparent, and
is still not widely noted, and (2) a few aspects of the
location and mobility of water molecules in protein crystals,
and (by inference) in solutions.

Effects of Pressure on Partial Volumes of Hydrophobic Solutes, at High Dilution

Ever since Walter Kauzmann's searching analysis of hydrophobic interactions (7), 20 years ago, the subject has been
a major concern of protein chemists and others. Such interactions have been invoked to explain an immense range of
phenomena. The interpretations were not always convincing, but
the remarkable behavior of dilute solutions of hydrophobic
compounds in water is compellingly clear. Here I aim only to
note one aspect of these phenomena, which I think deserves
more attention that it has received. I will consider only
the properties of such systems at infinite dilution of solute.
The already well-known features of these water-solute interactions are several. When a hydrophobic compound, or a compound

containing hydrophobic groups, is transferred to water from
a hydrophobic solvent, there is: (1) a large increase in chemical
potential of the solute, $\Delta \mu$, (hence low solubility in water).
In a homologous series each additional CH_2 group increases $\Delta \mu$
by about 3.3 kJ mol^{-1} (0.8 kcal mol^{-1}); (2) an enthalpy
change, varying rapidly with temperature; the transfer process
is exothermic at low, endothermic at higher temperatures;
(3) it follows from (2) that there is a large increase in heat
capacity. In a homologous series, the increment, per CH_2
group, of partial molar heat capacity in water, is near 90 J K^{-1}
mol^{-1} (22 cal K^{-1} mol^{-1}); for pure organic liquids it is
about one third as great; (4) there is a large negative
contribution to the entropy of mixing, as compared with that
for a hypothetical ideal solution; (5) there is generally a
shrinkage of volume, i.e. the partial molal volume of the
solute is less, by 5-20 cm^3 mol^{-1}, in dilute aqueous solution
than in a nonpolar organic solvent. These relations hold for
the noble gases dissolved in water, as well as for hydrocarbons
and compounds containing hydrocarbon residues. (6,7,8,9).

In the light of work from several laboratories, one may
add to this list a striking temperature dependence of the
partial molal compressibilities of hydrophobic compounds in
water. The partial molal isothermal compressibility K_2^0 of the
solute (component 2, in a two component system) is defined as
the limiting pressure coefficient of the partial molal volume,
with negative sign:

$$K_{2,T}^0 = -\left(\frac{\partial V_2}{\partial p}\right)_T \quad (c_2 \rightarrow 0, \quad p \rightarrow 1 \text{ bar}) \quad\quad (1)$$

Here V_2 is the partial molal volume of the solute. It is
difficult to get accurate measurements of the isothermal
compressibility at low pressures, and most recent measurements
have been done under adiabatic (isentropic) conditions by
measuring ultrasonic velocities in the medium, giving the
isentropic coefficient:

$$K_{2,S}^0 = -\left(\frac{\partial V_2}{\partial p}\right)_S \quad (C_2 \rightarrow 0, \quad p \rightarrow 1 \text{ bar}) \quad\quad (2)$$

Larionov (10) was the first, or one of the first, to measure
values of $K_{2,S}^0$ for certain alcohols in water; Alexander and
Hill (11) showed that, for 1-propanol, there was almost no
difference between the isothermal and isentropic coefficients,
over the temperature range from 0 to 40°C, and little difference
between them at higher temperatures up to 90°C. The striking
fact was that (in units of 10^{-4} cm^3 mol^{-1} bar^{-1}) the value of
$K_{2,T}^0$ was -30 at 0°C, -2 at 20°, + 15 at 40°. It became steadily

more positive, up to +56 at $90°$. Several further investigations have confirmed and greatly extended these observations. Some recent data for $K^{\varnothing}_{2,S}$ on primary alcohols, from two different laboratories - one in Norway, one in Japan (12, 13) are shown in Table I.

TABLE I

Partial Molar Isentropic Compressibilities ($K^{\varnothing}_{2,S}$) of Primary Alcohols in Water from 5 to 45°C at infinite dilution

$$K^{\varnothing}_{2,S} \times 10^4 \ (cm^3 \ mol^{-1} \ bar^{-1})$$

Alcohol	$5°$(13)	$25°$(13)	$25°$(12)	$35°$(12)	$45°$(13)
Methanol	7.3	12.6			15.9
Ethanol	-0.8	10.0			16.5
Propanol	-10.8	6.3	5.8	11.6	16.9
Butanol	-19.0	4.6	4.5	12.2	19.0
Pentanol	-27.1	2.6	2.3	12.4	21.3
Hexanol			0.5	13.3	
$\Delta K^0_{2,S}/\Delta CH_2$	-8.6	-2.5	-1.8	+0.6	+1.3

Data from Høiland and Vikingstad (12) and Nakajima, Komatsu and Nakagawa (13). For data on other homologous series, see Harada et al (14), and Cabani, Conti and Matteoli (15).

These data, which are typical of other homologous series studied, show at least two points. (1) In all cases the value of $K^{\varnothing}_{2,S}$ becomes steadily more positive with rising temperature. (2) At low temperature the increment in this quantity per CH_2 group is negative (average -8.6×10^{-4}) at $5°C$ but becomes positive around $25-30°$, and increasingly positive above that. Where the two laboratories made measurements under corresponding conditions, as with propanol, butanol and pentanol at $25°C$, the agreement is good.

The work of Høiland (12, 16) has shown a striking difference between the $K^{\varnothing}_{2,S}$ values in water and in an organic solvent, propylene carbonate, and some of the data from his investigations at $25°C$ are shown in Table II. The numerical values of $K^{\varnothing}_{2,S}$ in propylene carbonate are in all cases far larger and more positive than in water. The increment per CH_2 group (in units of $10^{-4} \ cm^3 \ mol^{-1} \ bar^{-1}$) is 10.4 in propylene carbonate, and -1.8 in water. Presumably the data in propylene carbonate are fairly typical for "normal" systems that do not show the special effects found in aqueous solutions.

TABLE II

Apparent Molal Isentropic Compressibilities ($K_{2,S}^0$) of Primary Alcohols in Propylene Carbonate (PC) and Water at 25°C

Alcohol	$K_{2,S}^0 \times 10^4$ (cm^3 mol^{-1} bar^{-1})		
	In PC	In Water	PC-Water
Propanol	50.5	5.8	44.7
Butanol	60.6	4.5	56.1
Pentanol	71.1	2.3	68.8
Hexanol	81.8	0.5	81.3
$K_{2,S}^0/$ CH$_2$	10.4	-1.8	12.2

Data in propylene carbonate from Høiland (16)
Data in water from Høiland and Vikingstad (12)

We must remember the minus sign in equations (1) and (2). A negative value of K_2^0 means that the partial molal volume of the solute __increases__ with increasing pressure. It has long been known that simple electrolytes behave this way - see for instance Harned and Owen (17, p. 710) or Friedman and Krishnan (18, pp 74-75). Of course no total system can respond to an increase of pressure by a reversible expansion; but the partial volume of a component can do so. For sodium chloride at 25°C and infinite dilution, the value of K_2^0 (either isothermal or isentropic) is near -50×10^{-4} cm^3 mol^{-1} bar^{-1}. The value for sodium sulfate is -154×10^{-4}. These values are numerically larger than, but of the same order of magnitude as, the negative value for (say) pentanol at 5°.

In the case of ionic solutes, these negative values clearly reflect the presence of a hydration shell around the ion, which is strongly resistant to compression; indeed the data suggest that increase of pressure expands the hydration shell, probably by increasing the number of water molecules that are encompassed within its domain. This description is, of course, quite lacking in precision; it is merely qualitative and suggestive. We may expect, in view of the great intensity of the electric field at the surface of an ion, that the values of K_2^0 for an electrolyte would change little with temperature; a moderate increase of thermal energy of the surrounding water molecules would count for little in the presence of such an intense field. Strangely, however, I have found in the literature no measurements of partial or apparent compressibilities of electrolytes in water, except at 25°; I have consulted several experts in the field, none of whom could give a reference to such data. More experimental information is urgently needed; the inference drawn above, that K_2^0 for electrolytes changes little with temperature, must remain tentative until we have more evidence.

On this assumption the partial molar compressibility of substances with hydrophobic groups provides an interesting contrast. Near $0^{\circ}C$ there appears to be a hydration region close to the non-polar surface, surely quite different in organization from the hydration shell of an ion, but with a greater degree of order than the bulk water further out in the solution. This concept of course is not new; the data on entropy of solution have obviously pointed to the same sort of conclusion. This local order, such as it is, can be readily disrupted by rise of temperature, so that the K_2^0 value becomes more positive as the temperature rises. However even at higher temperatures much of the order still remains, as shown by the fact that the K_2^0 values for such solutes in aqueous solution (see Table II) are still much lower than in non-aqueous solvents. The fact that, for 1-propanol in water, K_2^0 continues to rise steadily over the whole range from 0 to $90^{\circ}C$ (11) also suggests that some of the order in the water, induced by neighboring hydrophobic groups, still persists. More experiments, with a wider range of solutes, and over a considerable temperature range, are of course needed to test these inferences from the present available data. The temperature sensitivity of the K_2^0 data of course recalls the high values of the partial molal heat capacities of non-polar substances in water, reflecting the temperature sensitivity of the enthalpies of solution.

Such compressibility studies as those reported in (10,11,12, 13, 14) provide us with one additional approach to exploring the properties of hydrophobic systems in water, in addition to the five mentioned above. Although there is still no adequate theory of hydrophobic interactions, there are promising beginnings (9). Recent molecular dynamics calculations by Rossky and Karplus (19) on a dipeptide in water, and by Geiger, Rahman and Stillinger (20) have made impressive advances in portraying changes in the distribution, orientation, and hydrogen bonding in the immediate neighborhood of hydrocarbon residues, with some reminiscence of a fluctuating structure reminiscent of the order in clathrates (see also 21, 22). Geiger et al remark that "the dynamical data show that translational and rotational motions of solvation-sheeth water molecules are preceptibly slower (by at least 20 percent) than those in pure bulk water" (20, p. 263:abstract). Such changes are not dramatic, but they may be crucially important.

Location of Water in Protein Crystals by X-ray or Neutron Diffraction

These diffraction methods furnish the only way yet available to see the preferred locations of water molecules in relation to the surface and interior of a protein. They are also full of pitfalls and the data are subject to misinterpretation, even when the work is carefully done. Finney (23) has recently given a comprehensive and illuminating review of the whole field, emphasizing the factors affecting the stability or instability of native protein structure, and the role of water in stabilization of the structure. Here I offer a few comments.

Protein crystals commonly contain 40-50 per cent water by weight, sometimes more, and there is much evidence to indicate that the conformation of the protein in the crystal is in general extremely close to that in solution. Hence the crystal studies provide relevant evidence regarding water-protein interactions in dilute solutions. The diffraction patterns obtained require exposures of many hours; hence the structures calculated from them represent averaged positions for the atoms and molecules in the structure. Water molecules, in bulk water, at room temperature, have rotational relaxation times of the order of 10^{-11} s (10ps). The linear diffusion coefficient of water is near 2×10^{-5} cm^2 C^{-1}, so the root mean square linear displacement of a water molecule in 10ps would be of the order of 4-5 Å.

The NMR work of Seymour Koenig, presented at this symposium, would indicate that some water molecules in the neighborhood of a protein surface, may be retarded in their rotary motion by a factor of the order of 100 as compared with bulk water, with relaxation times in the nanosecond range. Native protein molecules also are far from static; they are continually undergoing rapid small internal motions and fluctuations; the evidence for this, from many different lines of experimental work, has recently been admirably summarized by Gurd and Rothgeb (24). Given this evidence, it is remarkable that diffraction measurements on protein crystals can specify the locations of substantial numbers of water molecules in the neighborhood of the protein surface. To obtain reliable evidence requires not only data of high resolution but also systematic refinement of the data for the native protein (25) after an approximate structure has been obtained with the aid of crystals containing isomorphous replacements with heavy atoms. The work that most closely meets these rigorous requirements is that from L.H. Jensen's laboratory on the small protein rubredoxin from Clostridium pasteurianum (Wattenpaugh et al 26, 27). The resolution achieved is 1.2 Å, and the structure is extensively refined. The protein contains 54 amino acid residues, and contains one iron atom, linked to the sulfhydryls of the four cysteine residues at positions 6,9,39, and 42. The water molecules are recognized as small, more or less spherical, regions of electron density, not forming a part of the continuity of the peptide chain. The observed electron density of the water sites, which specifies the occupancy of the site, varies greatly from one site to another. Watenpaugh et al reported 127 water sites, with occupancies ranging from near unity to 0.3; sites of lower occupancy were deleted from the model. In

retrospect, the authors concluded, it would have been better
to set the lower limit of occupancy at 0.5; this would have
reduced the number of reported water molecules to about 90.

Several of the water molecules of high occupancy are
involved in as many as five hydrogen bonds, to O and N atoms
of the protein and to other water molecules; this reflects the
fact that the structure is a dynamic one, and that the bonding
shifts from instant to instant. The occupancy figures represent
an average over time; they indicate the fraction of the time
that the site is occupied, but give no indication as to the
rapidity of exchange of water molecules in and out of the site.

Of the residues in the main protein chain, 12 of the peptide
NH groups are hydrogen-bonded to water, whereas 23 of the
peptide C=O groups are so bonded. (This tendency for more
C=O than N-H groups to be bonded to water has been found also in
other proteins). Nearly all the other peptide groups are
hydrogen-bonded to other atoms in the protein. Thus water
molecules serve to provide hydrogen bonds to peptide groups
that would otherwise lack such bonds. This represents a large
part of their role in stabilization of the protein structure -
a point emphasized by Finney (23).

The distribution of water molecules around the protein
surface in rubredoxin shows two distinct maxima, a major one
at 2.5-3.0Å, and a minor one at 4-4.5 Å. At further distances
the water merges into the continuum of bulk water. The well
defined water molecules are generally distributed over the
surface of the protein, except that the regions of the molecule
where hydrophobic side chains are concentrated are almost
devoid of identifiable water (see Stereo Figures 6,7,8, and 11
of reference 27). This does not,of course, mean that water
molecules are lacking in this region, but their high mobility
prevents them from being identified in definite sites. Even
around the charged and polar groups, where the x-ray work
identifies water in definite sites, the mobility, as judged
from NMR relaxation studies, is still high, with relaxation
times of the order of nanoseconds. In the past, many of us
have tended to think of water of hydration around a protein
as being far less mobile than this, but these intuitive judg-
ments were apparently illusory.

Some proteins, of course, contain internal water, and such
water molecules undoubtedly have far longer residence times
than those outside the protein surface. The smallest such
molecule is the pancreatic trypsin inhibitor, with only 58
residues in a tightly compact structure with four internal
water molecules (28). These form essentially an integral
part of the protein, bridging, through hydrogen bonds, regions
that would otherwise have unsatisfied bonds. The serine
proteases - chymotrypsin (29), trypsin (30), and elastase (31),
all have substantial numbers of internal water molecules - 24
to 25 for trypsin and elastase - and they occur, to a large

extent, in corresponding positions in these closely related structures. In hemoglobin (32) there are internal water molecules in the interfaces between the subunits - some 15 in the $\alpha_1 \beta_1$ interface, 4 in the $\alpha_1 \beta_2$ interface - which furnish hydrogen bonds that obviously serve to stabilize the binding between the subunits.

The neutron diffraction studies on myoglobin by Benno Schoenborn, reported elsewhere in this symposium, sound an important note of warning. Whereas in 1971 he had reported 106 water molecules per myoglobin molecule by neutron diffraction (33) and Takano in refined x-ray diffraction studies had found about 80 (34), Schoenborn in very careful work has now reduced his estimate to 42. It is known that false peaks of density can often be produced in calculations from x-ray or neutron diffraction. How many alleged water molecules in other structures may there be, that will not withstand closer scrutiny in future? This is a disturbing question that can only be answered by further research.

Abstract

This paper presents a very brief history of some past thought and research concerning water, a discussion of the unusual features of the partial molal compressibilities of hydrophobic substances in water, and a brief discussion of the evidence from x-ray and neutron diffraction for the locations of water molecules in protein crystals.

Acknowledgments

My primary debt is to H.A. McKenzie, with whom I have considered the problems of water in proteins for several years, (6,35). I have also profited from valuable discussions and correspondence with Walter Kauzmann, Seymour H. Koenig, I.D. Kuntz and L.H. Jensen; but none of them is to be held responsible for the views I have expressed. I am indebted to the National Scinece Foundation for continuing support (SOC7912543).

Literature Cited

1. Hall, Thomas S. "History of General Physiology"; University of Chicago Press, Chicago and London 1975; Vol. I, pp. 22-24.
2. Whewell, William "Astronomy and General Physics Considered with Reference to Natural Theology"; London, 1833.
3. Henderson, Lawrence J. "The Fitness of the Environment"; Macmillan, New York 1913; reissued by Beacon Press, Boston 1958; 317pp.
4. Bernal, J.D., and Fowler, R.H. J. Chem. Physics, 1933, 1, 515-548.
5. Franks, F., Ed. "Water: a Comprehensive Treatise"; Plenum Press, New York and London. Six volumes, 1972-1979.
6. Edsall, J.T., and McKenzie, H.A. Advances in Biophysics, 1978, 10, 137-207.
7. Kauzmann, W. Advances in Protein Chemistry, 1959, 14, 1-63.
8. Tanford, C. "The Hydrophobic Effect"; John Wiley and Sons, New York, 1973; 200pp.
9. Chan, Y.C., Mitchell, D.J., Ninham, B.W., and Pailthorpe, B.A. "Solvent Structure and Hydrophobic Solutions" in "Water: a Comprehensive Treatise"; (Franks, F. ed) Plenum Press, New York and London, 1979; Vol. 6, pp. 239-278.
10. Larionov, H.I. Russian J. Phys. Chem. 1953, 27, 1002-1012. (in Russian)
11. Alexander, D.M., and Hill, D.J.T. Austral. J. Chem. 1965, 18, 605-608.
12. Høiland, H., and Vikingstad, E. Acta Chem. Scand. 1976, A30, 692-696.
13. Nakajima, T., Komatsu, T., and Nakagawa, T. Bull. Chem. Soc. Japan 1975, 48, 788-790.
14. Harada, S., Nakajima, T., Komatsu, H., and Nakagawa, T. J. Solution Chem. 1978, 7, 463-474.
15. Cabani, S., Conti, G., and Matteoli, E. J. Solution Chem. 1979, 8, 11-23.
16. Høiland H. J.Solution Chem. 1977, 6, 291-297.
17. Harned, H.S., and Owen, B.B. "The Physical Chemistry of Electrolytic Solutions"; Reinhold, New York 1958; xxxiii + 803pp.
18. Friedman, H.L., and Krishnan, C.V. "Thermodynamics of Ionic Hydration" in "Water: a Comprehensive Treatise" (Franks, F. ed) Plenum Press, New York and London, Vol 3; pp. 1-143.
19. Rossky, P.J., and Karplus, M. J. Amer. Chem. Soc. 1979, 101, 1913-1937.
20. Geiger, A., Rahman, A., and Stillinger, F.H. J. Chem. Phys. 1979, 70, 263-276.
21. Owicki, J.C., and Scheraga, H.A. J. Amer. Chem. Soc. 1977, 99, 7413-7418.

22. Swaminathan, S., Harrison, S.W., and Beveridge, D.L. J. Amer. Chem. Soc. 1978, 100, 5705-5712.
23. Finney, J.L. "The Organization and Function of Water in Protein Crystals" in "Water: a Comprehensive Treatise" (Franks, F. ed) Plenum Press, New York and London 1979; Vol. 6, 47-122.
24. Gurd, F.R.N., and Rothgeb, T.M. "Motions in Proteins" Advances in Protein Chem. 1979, 33, 73-165.
25. Jensen, L.H. Ann. Rev. Biophys. Bioeng. 1974, 3, 81-93.
26. Watenpaugh, K.D., Margulis, T.N., Sieker, L.C., and Jensen, L.H. J. Mol. Biol. 1978, 122, 175-190.
27. Watenpaugh, K.D., Sieker, L.C., and Jensen, L.H. J. Mol. Biol. 1979, 131, 509-522.
28. Deisenhofer, J., and Steigemann, W. Acta Crystallogr. 1975, Section B, 31, 238-250.
29. Birktoft, J.J., and Blow, D.M. J. Mol. Biol. 1972, 68, 187-240.
30. Bode, W., and Schwager, P. J. Mol. Biol. 1975, 98, 693-717.
31. Sawyer, L., Shotton, D.M., Campbell, J.W., Wendell, P.L., Muirhead, H., Watson, H.C., Diamond, R., and Ladner, R.C. J. Mol. Biol. 1978, 118, 137-208.
32. Ladner, R.C., Heidner, E.J., and Perutz, M.F. J. Mol. Biol. 1977, 114, 385-414.
33. Schoenborn, B. Cold Spring Harbor Symp. Quant. Biol. 1972, 36, 569-574.
34. Takano, T. J. Mol. Biol. 1977, 110, 533-568.
35. Edsall, J.T., and McKenzie, H.A. "First Franco-Japanese Seminar, Nagoya, Nov. 3, 1978"; (Pullman, B., and Yagi, K. ed). Paper presented by H.A. McKenzie (in press).

RECEIVED January 4, 1980.

The Water–Polymer Interface

ARTHUR W. ADAMSON

Department of Chemistry, University of Southern California, Los Angeles, CA 90007

We present here some aspects of the surface chemistry and some explanatory models for water–polymer and related interfaces. The term "polymer" will be taken to mean an essentially organic material, of sufficiently high molecular weight and (or) sufficiently cross-linked that a stiff (as opposed to fluid) phase is involved. The material is insoluble in water, so that the term "water–polymer" interface refers to what is macroscopically an ordinary phase boundary. Typical polymers in the present context will be polytetrafluorethylene (PTFE), and polyethylene (PE). Solutions of macromolecules are thus not considered, nor is the related topic of so-called hydrophobic bonding, although some of what is discussed here is relevent to that subject.

The phenomenological surface chemistry of the water–polymer interface owes its modern development to the studies of Zisman and co-workers, (1) beginning in the 1950's. The useful and widely used critical surface tension quantity, γ_c, allowed practicing surface chemists to estimate contact angle on a polymer surface, given the liquid surface tension. The empirical observation was that $\cos \theta$, where θ is the solid–liquid–vapor contact angle, is linear in γ_L, the surface tension of the non–wetting liquid; the intercept at $\cos \theta = 1$ being defined as γ_c. Each type of polymer can be characterized by a γ_c value; further, γ_c is an approximately constitutive quantity in the sense of being additive in functional group contributions.

Although Zisman has appeared never to attribute a specific fundamental meaning to his γ_c, others have considered γ_c to be the surface tension that the solid would have were its cohesive forces the same as those acting across the solid–liquid interface. Thus for a hydrocarbon solid, which has been supposed to interact only through dispersion forces both cohesively and across the solid–liquid interface, γ_c should be the actual surface tension of the solid–vapor interface. This is essentially what is observed; γ_c for PE is approximately that measured for a high molecular weight liquid hydrocarbon. On the other hand, if water is considered to interact with a hydrocarbon only through dispersion

0-8412-0559-0/80/47-127-087$05.75/0
© 1980 American Chemical Society

forces (again, a common supposition), its γ_c relative to such interactions should be much lower than the actual surface tension of water. Figure 1 shows a Zisman plot for various liquids on ice (2); the line drawn indicates a γ_c value around 28 erg cm^{-2} at $-5°$C, or much less that the about 110 erg cm^{-2} estimated for the ice-vapor interface. (3) This γ_c can, however, be thought of as the contribution of dispersion forces to the surface tension of ice, or, alternatively and very qualitatively as the surface tension of a hypothetical solid having the structure of ice but whose cohesive forces were dispersion only in nature.

Qualitative ideas such as the above were made manageable by Good and co-workers, (4) with a key assumption, namely that for substances 1 and 2, 1-2 interactions could be interpolated from 1-1 and 2-2 interactions by a geometric mean law. On then equating energy and free energy of interaction, that is, on neglecting the entropy of bringing two phases into contact, due to structural changes, the equation

$$\gamma_{12} = \gamma_1 + \gamma_2 - 2\phi \sqrt{\gamma_1 \gamma_2} \tag{1}$$

was obtained. Here, ϕ is a term which arises from the different molar volumes of the two substances; empirical ϕ values were usually within 10% to 20% of unity. The model is illustrated in Figure 2. Fowkes (5) introduced the important qualification that for the cross, 1-2, term, only that contribution to γ_1 and γ_2 corresponding to the nature of the 1-2 interaction should be used. The square root term becomes, for the general case, $\sqrt{\gamma_1' \gamma_2'}$. Thus for a water (W)-Hydrocarbon (H) interfaces, across which only dispersion interactions were though to be important, one would write $\gamma_W' = \gamma_W^d$, where γ_W^d is the dispersion component of the water surface tension. However, since hydrocarbons interact by dispersion forces only, $\gamma_H' = \gamma_H$. For the reasons of tractability, the molar volume correction term, ϕ, was equated to unity. Thus

$$\gamma_{W-H} = \gamma_W + \gamma_H - 2 \sqrt{\gamma_W^d \gamma_H} \tag{2}$$

Although the operational origins are different, the explanatory concepts are similar enough that γ_c and γ^d have often been equated if the interface is one across which only dispersion forces were thought to be important. A variety of extensions and amendments to the above general approach have followed. Equation 2 has, for example, been extended to include "polar" interfacial interactions by adding a $\sqrt{\gamma_1^p \gamma_2^p}$ term. (6, 7) Thus for a water-polymer (P) interface, if the polymer contains oxygen, the added term would be $\sqrt{\gamma_W^p \gamma_p^p}$. These are semi-empirical extensions; the geometric mean rule should not generally be valid for dipole-polarization or dipole-dipole interactions.

Both Equations 1 and 2 can be applied to the Young equation for contact angle,

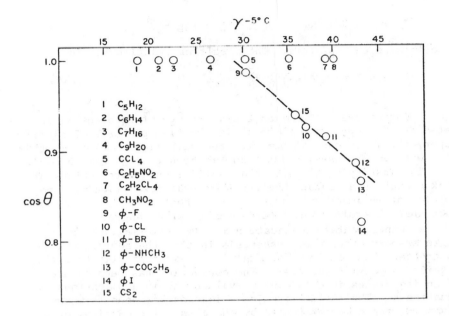

Figure 1.   Contact angles for various liquids on ice at $-5°C$

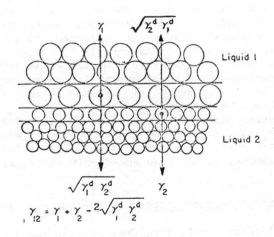

$$\gamma_{12} = \gamma_1 + \gamma_2 - 2\sqrt{\gamma_1^d \gamma_2^d}$$

Journal of Physical Chemistry

Figure 2.   *Good–Fowkes model for interfacial interactions (5)*

$$\gamma_L \cos\theta = \gamma_{SV} - \gamma_{SL} = \gamma_S - \gamma_{SL} - \pi^\circ \qquad (3)$$

to eliminate the $_{SL}$ term. The resulting relationship is

$$\cos\theta = -1 + 2\sqrt{\frac{\gamma_w' \gamma_p'}{\gamma_w}} - \frac{\pi^\circ}{\gamma_w} \qquad (4)$$

The γ_S introduced by Equation 2 and the γ_{SV} of Equation 3 are related by $\gamma_S = \gamma_{SV} + \pi^\circ$, where π° is the film pressure of adsorbed
vapor that is in equilibrium with the bulk liquid phase, in this
case water. The usual assumption has been that π° can be neglected. The justifying claim is that solid-liquid interactions are
weak enough in a contact angle system that adsorption of vapor
should not be important, especially on that majority of the surface that determines the experimental contact angle.

An experimental complication is the following. Contact
angle systems often show hysteresis in that the advancing angle
is greater than the receding angle; this is typically true for
liquids on polymer surfaces. The convention has been to use
advancing angles in obtaining γ_c values and in applications of
Equation 4. The qualitative justification has been that the
advancing angle is controlled by the less polar portions of the
surface, if it is heterogeneous, while the receding angle is
highly conditioned by polar sites. These last are often adventitious and hence not representative of the majority surface.

To summarize, the widely accepted convention is to treat the
solid-liquid interface as geometrically flat, with interactions
that obey the geometric mean law. The energy and free energies
of interaction are taken to be proportional; structural perturbations in the interfacial regions are neglected, as in the film
pressure of adsorbed vapor. If hysteresis is present, the advancing contact angle is taken to be the theoretically relevent
one. While departures from these assumptions have been considered, it usually is as a counterpoint, with no serious challenge
to the main picture being intended.

In the present paper, evidence is presented which indicates
that all of the above assumptions are seriously in error for the
water-polymer interface. The error is sufficient that a new
modelistic framework, rather than amendments to the existing one,
is needed.

Nature of Water-Hydrocarbon Interactions - Surface Restructuring

At the molecular level, a polymer surface consists of units
such as $-CH_2-$, $-CHO-$, $-CO-$, $-CF_2-$, aromatic rings, etc. Although
cross-linking may make the polymer macroscopically quite rigid,
at the molecular level, chain segments should have some mobility
since they interact with geometrically adjacent segments by relatively weak forces of the van der Waals type. At the molecular

level, a polymer surface should thus be much like that of a
molecular solid. Polyethylene surface should resemble that of a
crystalline long chain alkane, for example.

With this point in mind, it seems reasonable that the adsorption of small molecules on molecular solids should give information relevent to the interaction potential between such molecules
and a polymer surface. An important question is whether the
interaction can be attributed to dispersion forces only, or
whether other types of van der waals interactions are important.
We believe this last to be the case. In a series of studies on
the adsorption at 77 K of N_2, CO, and Ar on powdered molecular
solids such as ice, benzene, CO_2, CH_3OH, I_2, NH_3, etc., the
experimental heats of adsorption (in the small multilayer region
of the adsorption isotherms) did not order as would be expected
were they determined by a disperison potential function. (8) The
results raise serious questions as to the validity of the assumption that disperison forces are dominant in physical adsorption,
or at least, as to whether the conventional parameterization is
reliable. The usual equations are based on approximations that
include the condition that distance is large compared to a molecular diameter (see 9 , p. 309), a condition certainly not met in
monolayer adsorption.

A special series of studies was made with ice as the solid.
Low temperature ice behaves as a non-polar adsorbent, belying its
high surface tension, but as expected if hydrogen bonding is not
important in interactions with the adsorbates used. At temperatures nearer the melting point, the behavior of ice as an adsorbent changes radically, however. Ethane, propane, and CO_2 adsorb
indefinitely on ice powder at temperatures around -80°C. (10)
The explanation is that dissolving with reorganization of the ice
structure to form the ethane, propane, clatherates occurred.

In the case of n-hexane, the molecule is too big to be
accommodated in the ice-clatherate cage, and, of course, no
clatherate formation was observed. At temperatures below -35°C,
the adsorption is "normal"; the heat adsorption is essentially
independent of surface coverage and close to the value for the
heat of condensation of hexane to the liquid state. (11) Above
-35°C, however, the first portions adsorbed give a relatively
high heat of adsorption, which decreases with increasing surface
coverage to the bulk condensation value. From this, as well as
the correspondingly contrasting behavior of the adsorption entro-
pies above and below -35°C, we concluded that n-hexane forms surface clatherate-like structures. The molecule may sit in created
surface pockets, somewhat as illustrated in Figure 3. Very similar behavior was observed by Ottewill and co-workers (12) in the
case of various organic vapors, including alkanes, adsorbing on
liquid water surface.

Thus both for ice above -35°C (whose surface region may be
liquid-like at this temperature), and for liquid water, the water
substance-hydrocarbon interface is highly restructured. The

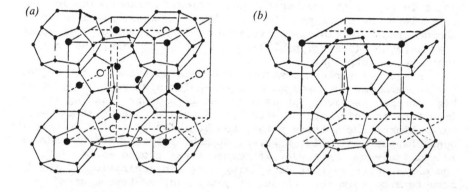

Figure 3. *(a) Structure of a gas hydrate or clatherate; (b) possible surface structure*

importance of dispersion interactions, which are non-directional, is least suspect; polar interactions such as between water and ethane bond dipoles, or of the dipole-polarizability type need to be examined more closely. Further, the conventional assumption of the geometric mean approximation can hardly be applicable at the water-hydrocarbon interface, and one should not neglect structural entropy contributions to the interfacial free energy.

The importance of surface restructuring was inferred as well from the contact angle behavior of various liquids on ice. As noted earlier in connection with Figure 1, one data group could be interpreted as giving a Zisman γ_c value for ice of about 28 erg cm^{-2}. Other liquids, however, wet ice even though their surface tension is well above this, an example being CH_3NO_2 whose surface tension is 40 erg cm^{-2}. The explanation advanced was that in such cases, surface restructuring is important. In analogy to the concept of hard and soft acids in inorganic chemistry (13) the designation of hard and soft interfacial systems was suggested. A hard or A type system is one that is not restructured in the presence of an adsorbed film or on contact with the bulk liquid. For it, equations such as Equations 1 and 2 should apply approximately. A soft or B type system shows significant surface restructuring. Additivity and geometric mean rules are likely invalid; the entropy of interface formation will be important. Schematic illustrations of the two cases are shown in Figure 4.

Adsorption and Contact Angle Behavior on Flat Polymer Surfaces: Potential - Distortion Model

Adsorption and contact angle (or, in general wetting) are in priciple related by the surface thermodynamics of a two component system. In particular, a complete statement of the interaction energy and entropy between a solid and a fluid (vapor or liquid) phase should contain or imply both the adsorption isotherm and the contact angle (or, if wetting occurs, the spreading coefficient, $S_{L/S} = \gamma_{SV} - \gamma_L - \gamma_{SL}$).

Such detailed description of interaction potentials is beyond present capabilities, in the case of systems of common importance, and a semi-empirical formulation that we have used is

$$RT\ln(P^\circ/P) = \varepsilon_o e^{-ax} - \beta e^{-\alpha x} + g/x^3 \qquad (5)$$

where x is the thickness of the adsorbed (or of the liquid) film, and P° is the saturation pressure, that is, the vapor pressure of the bulk liquid substance. The model is that of the Polanyi potential treatment. The attractive potential is exponential for small distances, a form supported by various experimental indications, (14) but with an inverse cube term taking over at large x values, corresponding to the dispersion (non retarded) interaction between a molecule and an infinite slab. Structural perturbation is allowed for by the middle term, the coefficient of which will

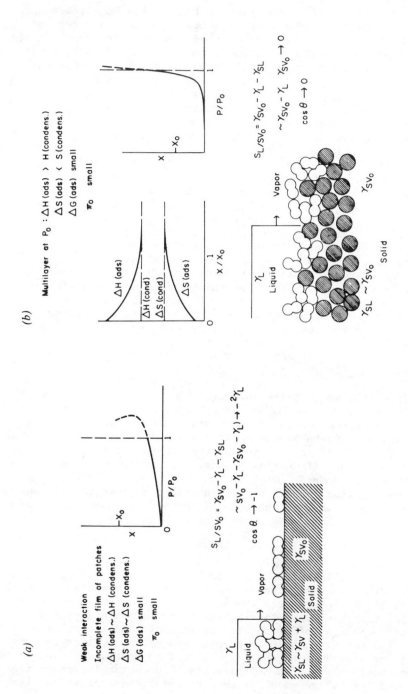

Figure 4. (a) Type A interface; (b) Type B interface

in general be temperature dependent, thus providing an entropy
contribution. It is expedient (although not necessary) to neglect
the inverse cube term in the case of adsorption,

$$Rt\ln(P^\circ/p) = \varepsilon_o e^{-ax} - \beta e^{-\alpha x} \tag{6}$$

and the equation leads to adsorption isotherms of the form shown
in Figure 5. For certain choices of the parameters, the isotherm
cuts the P° line, going through a maximum and minimum around this
line. The net area enclosed is, by the Gibbs adsorption equation,
just the spreading coefficient, $S_{L/S}$, which for a non-wetting
system, also gives the contact angle:

$$S_{L/S} = \gamma_L(\cos\theta - 1) \tag{7}$$

Use of Equation 6 requires adsorption data close to P°.
Because of this, the usual adsorption results obtained for powdered solids are not usable since in the important region near P°,
interparticle condensation dominates. (15) Accurate correction to
obtain the true absorption isotherm is virtually impossible. The
difficulty may be resolved if the adsorption is measured on a flat
surface, the adsorbed film thickness being obtained ellipsometri-
cally. (16) This approach has the further advantage that contact
angle measurements can now be made on the identical surface for
which the adsorption isotherm is determined. This is essentially
not possible in the case of a powdered adsorbent; it has generally
been necessary to assume that the surface properties of a bulk
sample of the adsorbent are the same as those of the powdered
form--a dangerous assumption.

We have reported on ellipsometric studies for water and
various organic vapors on a variety of surfaces. (17) The results
for water on optically smooth surfaces of PTFE and PE are of
special interest here. The isotherms are shown in Figure 6. Note
that the ellipsometric film thicknesses correspond to a layer of
two to five molecules thick as P° is approached. The film pressures given in Table 1 are calculated from the experimental isotherm, which covers only the region from $P/P^\circ = 0.8$ upwards; they
are thus minimum values since any low pressure adsorption which
might be present due to polar impurities, would add to give yet
larger values. Two interesting points emerge. The first is, that
contrary to usual assumption, non-negligible film pressures may be
present in a high contact angle system. Thus in Equation 3 the
left hand term is 2.5 and -10 erg cm^2 for PE and PTFE, respectively, as compared to the π° values of 14 and 9 erg cm^{-2}. Any estimate of $(\gamma_S - \gamma_{SL})$, such as for use in obtaining the $\gamma_P^{'}$ and $\gamma_W^{'}$
parameters in Equation 2 will thus be seriously in error if π° is
neglected. The second point is that even though PTFE and PE are
low energy surfaces, water does adsorb and does lower the interfacial free energy considerably. Thus, if we estimate the surface
tension, γ_S, of polyethylene to be about 30 erg cm^{-2}, the surface

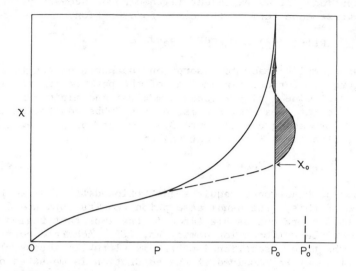

Figure 5.   Potential–distortion model: solid line, attractive potential only; dashed lines, with distortion term. Shaded area gives the spreading pressure.

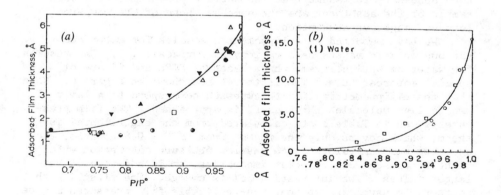

Journal of Colloid and Interface Science

Figure 6.   Adsorption of water on PE and PTFE. (a) PE: (◯) 21.5°C; (△, ▽) 16.5°C (two runs); (⊗) 9.7°C; (◑) −9°C; (◒) −24°C; shaded symbols denote desorption points (32). (b) PTFE, 19.8°C (17).

of the film covered surface, γ_{SV} drops to 21 erg cm^{-2} due to water
adsorption; the figure would be even lower were the full rather
than the minimum π° value used. This is not what would be expected were the adsorption forces dispersion only, and we have an
indication of a type B surface, with polar interactions and surface restructuring.

Table I gives the fitting parameters for Equation 6, the
values being chosen with the constraint that they also give the
experimentally observed contact angle. The interaction and distortion energies, ε_0 and β, correspond to only a few hundred cal
mole^{-1} for the process (bulk water) → (adsorbed film), and it is
important to realize that relatively minor energy effects are
determinative of the adsorption and contact angle behavior. The
energy relaxation distance, $1/\text{Å}$, is much larger for water adsorbed
on PTFE than on PE, while the distortion relaxation distance, $1/\text{Å}$,
is smaller. The picture is that water on PTFE is more weakly held
than on PE, but more distorted in structure relative to bulk
water.

Data for copper coated with a stearic acid monolayer are included in the Table, in illustration of another type of hydrocarbon surface--one now rich in CH_3 groups. As might be expected,
the adsorption and contact angle behavior resembles more that for
PE than for PTFE.

As another approach, the adsorption data fit a characteristic
isotherm, $x/_{x'} = f(P^\circ/P^{\circ\prime})$, where $P^{\circ\prime}$ and x' are fitting parameters. In effect, two adsorption systems are in corresponding
states if they are at the same $P^{\circ\prime}/P$. Not only do the data of
Table I form segments of a common characteristic isotherm, but so
also do the results for a number of other systems, as indicated in
Figure 7. The solid line in the figure is given by

$$\ln(P^{\circ\prime}/P) = x'/x \qquad\qquad (8)$$

$P^{\circ\prime}$ has the aspect of being an effective vapor pressure, it is
always higher than P°, and a physical interpretation is that $P^{\circ\prime}$
is the hypothetical vapor pressure of a bulk liquid adsorbate
whose structure is the same as that of the adsorbed film. On this
basis, RT $\ln(P^{\circ\prime}/P^\circ)$ is the local excess free energy of water in
the adsorbed film resulting from structural perturbations relative
to bulk water. The x' parameter functions as a characteristic
relaxation distance.

Structural perturbations are to be distinguished from what
are essentially phase changes. The perturbations represent relatively minor deviations from bulk liquid, and adsorption isotherms
are usually nearly invariant if $\underline{x}$ is plotted against P/P°. That
is, the heat of adsorption from the vapor phase is very close to
that of condensation. By contrast, and as an example, in the case
of water on PE, while this near invariance in isotherm shape holds
around 20°C, the isotherms at -9°C and -24°C are very flat, with
an x_0 of only about 1.5Å. It would appear that the adsorbed film

is now more ice-like than liquid water-like.

Table I includes the results of ellipsometric adsorption studies of water on a similar (and similarly prepared) PE to that of Ref. 14. (18) In addition, the adsorption was determined for a surface partially oxidized by treatment with an acid dichromate solution. As might be expected, the effect of the introduction of polar sites reduces θ, increases π°, and increases both ε_0 and β. The relaxation distance for ε_0 is increased, while that for the distortion term is decreased. Systematic studies of this type should lead to a much better understanding of the nature of adsorbed films near P°.

Table I. Adsorption and Contact Angle Behavior of Water on
Polymer Surfaces

		Surface			
	PTFE[a]	PE[a]	PE[b]	PE[c]	SA[d]
$\overset{\circ}{x}{}^\circ$	16	6	4.5	26	8
θ	98	88	91	63	80
π°	9	14	8	53	50
ε_o/RT	0.20	0.64	0.33	0.52	2.1
$1/\overset{\circ}{a}$	22	2.2	2.0	13	2.1
β/RT	0.11	0.04	0.038	0.11	0.035
$1/\overset{\circ}{\alpha}$	53	135	147	53	.32
$P^{\circ\prime}/P^\circ$	1.020	1.12			1.26
$\overset{\circ}{x}{}'$	0.37	0.71			1.79

(a) 20°C, see Reference 17. (b) Reference 18. (c) Partially oxidized PE, Reference 18. (d) Stearic acid coated copper.

Surface Topology

The above treatments, as well as those generally proposed for adsorption by non-porous adsorbents, explicitly or implicitly assume the adsorbent surface to be flat and smooth. Surface curvature is neglected as a potential contributor to the adsorption behavior. This type of assumption seemed especially valid in the case of the PTFE and PE surfaces of Table I. The slips were prepared by flow smoothing between optically flat glass plates under pressure and at 177°C in the case of the PE and 340°C in the

case of PTFE.   Enough flow occurred for the thickness of the slip
to be reduced by about half, and the resulting surfaces were
mirror smooth to the eye.   They acted as good reflecting surfaces
in the ellipsometric measurements; the adsorbed film thicknesses
reported are calculated on the basis of flat surfaces.

That appearances can be deceptive is illustrated by the scanning electron microscope (SEM) photograph shown in Figure 8.
Ripples and dimples abound.   Polymers studied in the literature
have not usually been subjected to a flow smoothing treatment,
and their surfaces were likely even rougher than that shown in
Figure 8.

An important possibility to consider is that, near P°, Kelvin
condensation occurs in ripple and dimple features, so that the
surface actually consists of multitudinous patches or "lakes" of
liquid adsorbate.   The model, for a single dimple, is shown in
Figure 9.   The dimple is taken to be a figure of revolution, of
profile shown, and at a given P/P°, liquid adsorbate is present to
a depth such that the curvature of the meniscus is defined by
$RT \ln(P^\circ/P) = 2\gamma V/R$, where V is the molar volume of the liquid and
R is the radius of curvature; the meniscus meets the dimple wall
at the macroscopic contact angle.   We find that isotherm shapes
cannot be reproduced by any single dimple profile, but could be if
a distribution of shapes were used.   At this point, however, the
number of fitting parameters is large, and it is difficult to
estimate the actual importance of the contribution from dimple
condensation.

An intermediate situation seems likely.   Most of the adsorption is not curvature influenced, but enough lakes may be present
to produce the contact angle hysteresis that is common for polymer
systems.   That is, advancing angle is mainly determined by ordinary film covered surface, while receding angle is smaller because
of the presence of lakes.

Ice-Polymer Adhesion

The adhesion of ice to polymer surfaces is a topic of some
practical importance as well as of theoretical interest, and a
number of studies have been made. ([19], [20], [21])   There is evidence
that the surface of ice near its melting point has a liquid-like
layer. ([22], [23], [24])   Our results on hexane adsorption on ice suggests that a transition from a rigid to a re-structurable surface
occurs at -35°C.   It might be supposed, therefore, that ice-polymer adhesion near 0°C is that of a liquid-like layer.   However,
the structure of the adsorbed film of water on PE at -9°C to -24°C
appears to have undergone a phase change relative to the state
20°C, as mentioned earlier above.   This now suggests that an
adsorbed water film on a polymer surface below 0°C is more solid
than liquid-like.   It is thus not at all certain what the boundary
layer region is in fact like at the polymer-ice interface.

Shear adhesion studies in this Laboratory have been made on

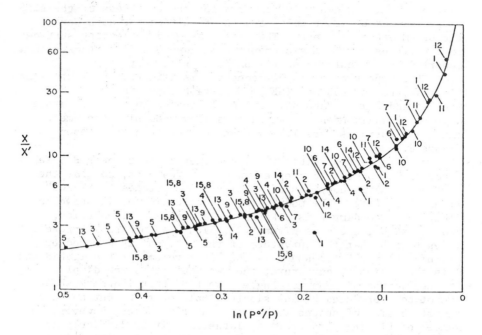

Figure 7. Characteristic isotherm plot (17). Data are for indicated vapor/solid system at 20°C: 1, $H_2O/PTFE$; 2, H_2O/PE; 3, $H_2O/stearic$-acid-coated copper; 4, $H_2O/pyrolytic$ carbon–silicon alloy; 5, $C_6H_5Br/PTFE$; 6, $CH_3NO_2/PTFE$; 7, $H_2O/pyrolytic$ carbon; 8, $C_6H_6/PTFE$; 9, n-$C_5H_{11}OH/PTFE$; 10, n-$C_4H_9OH/PTFE$; 11, n-$C_3H_7OH/PTFE$; 12, $C_3H_5OH/PTFE$; 13, $CCl_4/PTFE$; 14, n-$C_8H_{18}/PTFE$; 15, n-$C_6H_{14}/PTFE$.

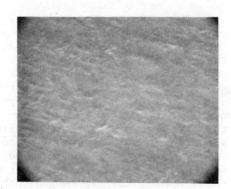

Figure 8. SEM photograph of flow-smoothed PTFE (×5000). The appearance of flow-smoothed PE was similar.

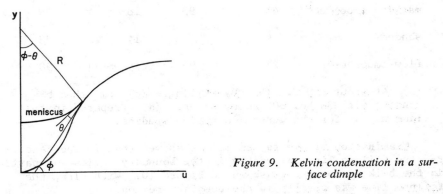

Figure 9.   Kelvin condensation in a surface dimple

single crystal ice, formed as a cylinder on a PTFE of PE surface.
The force applied parallel to the interface necessary to break the
adhesion was then determined at -15°C. (25) The results for
several different surface preparations are summarized in Table II.
They are similar to those reported in the literature. For untreated PE, our value is 65 lb in^{-2} as compared to literature
values of 37 lb in^{-2} at -12°C (26) and 117 lb in^{-2} at -6°C and
107 lb in^{-2} at -20°C. (27) We find 35 lb in^{-2} with untreated
PTFE, as compared to reports of 45 lb in^{-2} at -12°C, (26) 18 lb
in^{-2} at -6°C and 1 lb in^{-2} at -20°C, (27) and 21.5 lb in^{-2}. (28)

Table II.   Shear Adhesion of Ice to PE and PTFE[a]

Nature of Surface	PE		PTFE	
	lb in^{-2}	θ	lb in^{-2}	θ
untreated	65	88	36	112
machine smoothed[b]	43	93	18	109
sanded[c]			14	98
flow-smoothed[d]	27	95	very small	98

(a) At about -10°C. (b) By milling. (c) Smoothed by
sanding with 3M No. 600 emory paper. (d) Prepared in the
same way as for the vapor adsorption studies.

Examination of the failed joints showed that in the case of
PE, the failure line lay partly in the boundary region and partly
in the bulk ice phase, as shown in Figure 10. With PTFE, the
failure line was usually in the boundary region.
Jellenik (19) has considered the case of tensile adhesion,
for which the tensile stress due to a liquid-like layer between
the two adhering phases separated by distance d, is just $\Delta P = 2\gamma/d$. This picture is that of a Laplace tension resulting from a
meniscus of radius of curvature d/2. Alternatively, for a force
applied parallel to the interface the viscous resistance should be

$$\sigma = \eta v/d \tag{9}$$

where σ is the shear stress, η, the effective viscosity in the
boundary region, and v the relative velocity of the two phases.
Measurements were made of the creep at the ice-PE interface.
(25) At a shear stress close to the breaking point, or 7 kg force
per cm^2, the creep rate was about 10^{-4} cm min^{-1} at $\sim$ -10°C. If
the boundary region is taken to be of about the thickness of the
vapor adsorbed film, $x_0 \simeq 10A$, Equation 9 yields an effective
boundary region viscosity of 6×10^3 poise. This value, while

believable, has no independent confirmation.

From a thermodynamic point of view, the energy to separte two phases to give the separate equilibrium surface–vapor interfaces is just

$$W_{S_1 S_2 V} = \gamma_{S_1 V} + \gamma_{S_2 V} - \gamma_{S_1 S_2} \tag{10}$$

In the present case, S_1 is polymer and S_2, ice. If the interfacial properties of ice are taken to be the same as those of liquid water, Equation 10 can be written

$$W_{SLV} = \gamma_L (1 + \cos\theta) \tag{11}$$

where θ is the contact angle that a liquid of the same surface properties as ice would make against the polymer surface, presumably a value around 90°. W_{SLV} should thus be about γ_L, that is, the ice surface tension. To obtain the adhesive force, it is necessary to estimate the distance over which the reversible work develops, on separation of the two phases. It seems reasonable again to take this distance to be about that of the film thickness of adsorbed water, x_0. On this basis, the theoretical estimate of the adhesion of a ice–polymer interface becomes about 10^9 dyne cm^{-2} or about 1.4×10^4 lb in^{-2} (1000 kg cm^{-2}). A smilar estimate results from use of the Laplace pressure if d is similarly taken to be about the thickness of the vapor adsorbed film, x_0.

The above estimates are about two hundred times the observed breaking stress. This is a fairly common type of finding, and the usual conclusion is that, for various reasons, such as local stress concentrations, the practical adhesion is far smaller than the ideal. A more specific picture is that shown in Figure 11. Remembering the SEM photographs, if the actual ice–polymer boundary is wavy in contour, the local force, f, to break the interface is reduced by the slope of the interface relative to the general plane. If this slope is about 0.005 breaking forces of the order found would be predicted.

It is qualitatively consistent with the above picture that the breaking stress increased with increasing surface roughness (Table II). That is, increased roughness should increase the typical slope of the microscope contour and the work of interfacial separation now develops over a smaller displacement parallel to the general interfacial plane. This is illustrated in Figure 11b. On this interpretation, our measurements although nominally of shear adhesion are actually of the work of separating the phases as modified by the leverage factor on f. Also consistent with this picture is that x_0 for water on PTFE is about three times that for PE. The force to break a water–PTFE (and presumably then also an ice–PTFE) interface would therefore be expected to be correspondingly smaller than that for the water–PE (or ice–PE) interface.

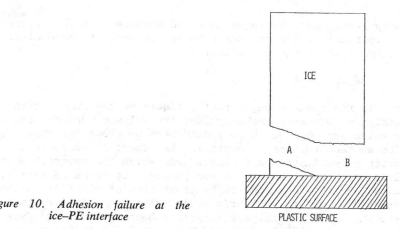

Figure 10. Adhesion failure at the
          ice–PE interface

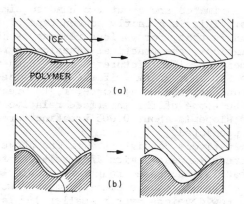

Figure 11.   Surface separation caused by creep displacement of rippled surfaces

Summary and Conclusions

This paper has emphasized two aspects. The first is that of developing reservations with respect to certain conventional assumptions. Thus, we believe the assumption that the water-hydrocarbon interaction is through dispersion forces only to be seriously in error. The studies of adsorption of hydrocarbons on ice and on liquid water strongly suggest that polar interactions of the bond dipole-bond dipole or the bond dipole-polarizability types play an important role. The adsorption data for water on PTFE and on PE cannot be accommodated by a dispersion only formulation of adsorption potential; in fact, an exponentially relaxing attractive term is indicated. Such a function could, of course, be no more than an empirical representation of how the integrated dispersion attraction behaves for distances of the order of a molecular diameter. The exponential form, however, is the theoretically predicted one for propagated polarizations, (9) and its usefulness in representing the adsorption potential is at least suggestive that polarization rather than dispersion is the dominant source of attraction.

A second common assumption challenged here is that structural perturbations in an interfacial region can be neglected (as a figure such as Figure 2 illustrates). The adsorption of hydrocarbons on ice and on liquid water are difficult indeed to explain unless structural perturbation occurs. While Figure 3 is not presented as in any way accurate in detail, it serves to underscore the point that structural changes in the interfacial region of the ice-hexane system are major. This conclusion is supported by the need for the type A and B classification of contact angle systems; the illustrations of Figure 4 again indicate that serious structural effects are envisaged. Finally, of course, finite contact angles cannot exist unless the adsorbed film in equilibrium with P° is in some way different from the bulk liquid adsorbate . This difference appears in the distortion term of Equation 5 or, alternatively, in the $P^{\circ\prime}$ parameter of Equation 8. Regardless of analytical form, the qualitative effect cannot be denied. That is, were there no structural perturbation, the thickness of an adsorbed film should increase without limit at P approaches P° because of the attractive potential, so that bulk liquid could never rest on a surface with a finite contact angle.

The second aspect of this paper is one of presenting a structural picture of the polymer-water interface, based on the potential–distortion adsorption model. Adjacent polymer chains interact by non-bonded, that is, van der Waals forces, so that, locally, the surface should show resemblances to that of a molecular solid. A type B situation is postulated, with local restructuring to from pockets into which water molecules have inserted themselves, as depicted in Figure 12. As the water activity (vapor pressure, if vapor phase is present) increases, a layer several molecules thick builds up. The layer is sufficient-

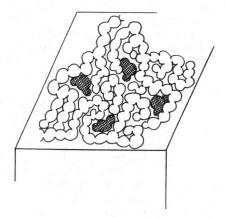

*Figure 12. Insertion of water into a
         polymer surface*

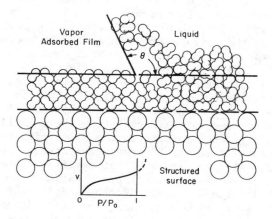

Figure 13.   Depiction of the interfacial film–bulk liquid transition

ly different in structure from that of bulk water that the latter
can co-exist with the equilibrium film at P°; finite contact angles are possible, as observed.

Contact angle hysteresis may have an alternative or at least
supplementary cause to that of surface lakes. Hysteresis has
been observed for liquids against an agar aquagel—a gel
whose composition was perhaps 99% water (see 9). While the gel
was relatively rigid macroscopically, it was essentially liquid
in the microscopic pockets of the gel mesh. The question at this
point is why should liquid-liquid interfaces, that is lenses,
show no angle hysteresis, yet the locally liquid-liquid system of
aqueous gel-organic liquid show hysteresis? A suggested explanation (9) is that lack of long range mobility, as in the case of
the gel surface, prevents the frictionless advance of a contact
angle front so that periodic barriers must be overcome. There
are similarities between this case and that of a polymer-water
interface, namely long range rigidity but relative mobility at
the molecular level. In effect contact angle hysteresis is now
assigned as a topological and rheological phenomenon rather than
one of surface heterogeneity.

An early formulation of contact angle or meniscus friction
adds or subtracts a frictional term to the driving force for contact angle, $\gamma_L \cos \theta$, for advancing and receding fronts, respectively. (29) On this basis, the correct contact angle is given
by

$$\cos \theta = \frac{1}{2} (\cos \theta_{adv.} + \cos \theta_{rec.}) \tag{12}$$

Use of such average θ values would significantly change the results of many literature qualifications of models such as that of
Equation 4.

Not so far answered experimentally is the nature of the transition between the interfacial film region and bulk water. That
is, if bulk water is resting on a polymer surface, there will be
a thin region of structurally perturbed water substance and a
transition from this region to bulk water structure. The situation is illustrated schematically in Figure 13. It seems reasonable to assume that the thickness of this transition region is
about that of the adsorbed film at P°, or about $\underline{x}_O$. The suggestion at this point is that adhesion between polymer and bulk
water is to a first approximation given by the interaction between
the film layer and bulk water. This is very close to the weak
boundary layer concept of Bikerman. (30) The ideal work of
adhesion for the ice-polymer interface is then related to the
free energy of the film layer-ice interface, which is probably
considerably smaller than that of the ice-vapor or liquid-vapor
interface. A value of 10 erg cm^{-2} would, for example, reduce the
discrepancy between theoretical and observed adhesion by about
ten-fold over that estimated earlier above. In terms of Figure
11, the required slope of surface ripples would now be about 0.05.

The potential-distortion model and its related emphasis on structural perturbation in the interfacial region suggest certain lines of further study. Much more information is needed about adsorption near $P°$ and on a variety of types of polymers having a variety of surface preparations. The emphasis would be on the effects of changing molecular geometry and polarity of functional groups. Temperature dependence studies both on adsorption and on contact angle are needed to obtain entropy contributions. (The potential-distortion model predicts a critical temperature of zero contact angle--a prediction for which there is insufficient existing data for proper testing.(31)) The behavior of molecularly smooth surfaces needs more attention; spectacularly different adsorption, contact angle, and adhesion properties may be found.

Abstract

Various aspects of the water-polymer and ice-polymer interface are discussed, with emphasis on polyethlene (PE), and polytetrafluoroethylene (PTFE). A rather different picture than the usual one is presented. Bond dipole type interactions are important. There is major surface restructuring of the polymer, and the adsorbed water film is significantly perturbed structurally from that of the bulk liquid. Wetting and adhesion behavior are strongly determined by the degree of these structural changes, and by the thickness and nature of the transition between the interfacial region and bulk liquid. In general, the nature of the polymer-water interface must be understood in terms of structure rather than in terms of non-specific disperison interactions.

Acknowledgement

The investigations reported here were supported by a grant from the U. S. National Science Foundation

Literature Cited

1.   Zisman, W. A.   Adv. in Chem., 1964, 43, 1
2.   Adamson, A. W.; Shirley, F. P.; Kunichika, K. T. J. Coll.
     Interface Sci., 1970, 34, 461.
3.   Ketcham, W. M.; Hobbs, P. V. Phil. Mag., 1969, 1161.
4.   Girifalco, L. A.; Good, R. J. J Phys. Chem., 1957, 61 904.
5.   Fowkes, F. M. J. Phys. Chem., 1960, 67, 2538; Ind. Eng. Chem.,
     1964, 56, 40.
6.   Panzer, J. J. Colloid Interface Sci., 1973, 44, 142.
7.   Kaelble, D. H., "Physical Chemistry of Adhesion", Wiley-
     Interscience, 1971.
8.   Dormant, L. M.; Adamson, A. W. J. Colloid Interface Sci.,
     1968, 28, 459.
9.   Adamson, A. W., "The Physical Chemistry of Surfaces", 3rd
     ed., Wiley-Interscience, 1976.

10. Adamson, A. W.; Jones, B. R. J. Colloid Interface Sci., 1971, 37, 831.
11. Orem, M. W.; Adamson, A. W. J. Colloid Interface Sci., 1969, 31, 278.
12. Jones, D.; Ottewill, R. J. Chem. Soc., 1955, 4076; Blank, M.; Ottewill, R. H. J. Phys. Chem., 1964, 68, 2206.
13. Huheey, J. E.. "Inorganic Chemistry", 2nd ed., Harper and Row, 1978.
14. Adamson, A. W. J Colloid Interface Sci., 1968, 27, 180, and 1973, 44, 273.
15. Wade, W. H.; Whalen, J. W. J. Phys. Chem., 1968, 72, 2898.
16. A gravimetric method is possible using a stack of metal foil. See Blake, T. D.; Wade, W. H. J. Phys. Chem., 1971, 75, 1887.
17. Hu, P.; Adamson, A. W. J. Colloid Interface Sci., 1977, 59, 605; Tse, J.; Adamson, A. W. J. Colloid Interface Sci., in press.
18. Hirata, M.; Ichikawa, M.; Iwai, S., Abstracts, ACS/CSJ Chemical Congress, Honolulu, April, 1979, and private communication.
19. Jellinek, H. H. G., U. S. Army, Ice and Permafrost Res. Estab., Res. Rept. 23, 1957; J. Coll. Sci., 1959, 14, 268.
20. Jellinek, H. H. G. Can. J. Phys., 1962, 40, 1294.
21. Jellinek, H. H. G. J. Appl. Phys., 1961, 32, 1793.
22. Weyl, W. A. J. Colloid Sci., 1951, 6, 389.
23. Jellinek, H. H. G. J. Colloid and Interface Sci., 1967, 25, 192.
24. Hosler, C. L.; Jensen, D. C.; Goldshlak, L. J. Meterol., 1957, 14, 415.
25. Tse, J. T., dissertation, University of Southern California 1978.
26. Landy, M.; Freiberger, A. J. Colloid and Interface Sci., 1967, 25, 231.
27. Ford, T. F.; Nichols, O. D., Naval Res. Lab. Rept. 5832, 1962.
28. Raraty, L. E.; Tabor, D. Proc. Roy. Soc., 1958, 245A, 184.
29. Adam, N. K., "The Physics and Chemistry of Surfaces", 2nd ed., Oxford, 1938.
30. Bikerman, J. J., "The Science of Adhesive Joints", 2nd ed., Academic Press, New York, 1968.
31. Adamson, A. W. J. Colloid Interface Sci., 1973, 44, 273.
32. Tadros, M. E.; Hu, P.; Adamson, A. W. J. Colloid Interface Sci., 1974, 49, 184.

RECEIVED March 17, 1980.

Thermodynamic and Related Studies of Water Interacting with Proteins

JOHN A. RUPLEY, P.-H. YANG, and GORDON TOLLIN

Department of Biochemistry, University of Arizona, Tucson, AZ 85721

It is recognized that important protein processes are controlled by the environment. In their study of an enzyme model reaction, for example, Bruice and Turner (1) have shown that change from water to a dioxane-water solvent alters the rate of carboxylate ion attack on substituted phenyl esters by four to six orders of magnitude. In spite of agreement on the importance of solvent, there is a lack of understanding of the basis of solvent control. Our interest in this problem came through attempting to relate the structure of active sites determined by x-ray diffraction analysis to the thermodynamics of binding of substrates and substrate analogs. These efforts were not successful. The difficulty of understanding the thermodynamics of protein reactions is exemplified through comparing protein folding with binding of a substrate analog (Table I). These reactions differ more than ten-fold in area of contact but by less than a factor of two in free energy and enthalpy. Although the way in which solvent participates may not be the origin of the unusual chemistry, it appears to be the least well understood aspect of protein reactions.

Investigations on protein-water interactions can be categorized according to whether protein solutions or protein films and powders were studied. Solid samples allow variation of water activity over a wide range. Because of this advantage, experiments on such samples are the focus of this paper. It is possible that the protein conformation may change with drying from the solution to the hydrated solid. Evidence given below indicates that the conformation in the dry state is not significantly different from that in dilute aqueous solution.

The literature and experiments discussed below describe the process of protein hydration, which is addition of water to dry protein to obtain the solution state. An understanding of this process is expected to increase understanding of water-protein interactions in solution. The protein hydration process, which can be described unambiguously by experiment, should be distinguished from the water of hydration concept, which is viewed

0-8412-0559-0/80/47-127-111$05.50/0
© 1980 American Chemical Society

Table I. Change in Surface Compared with Change in Thermodynamic Parameters[a]

	Change in Surface Area (Å^2)	ΔG° (J mol^{-1})	ΔH° (J mol^{-1})	ΔS° ($\text{J K}^{-1} \text{mol}^{-1}$)
Protein Folding Lysozyme	15,150	$-$ 33050	$-$ 94720	$-$ 205010
Substrate Binding Lysozyme + (GlcNac)$_3$	350	$-$ 29710	$-$ 57740	$-$ 92880

[a] Area calculations from Shrake and Rupley (17). Thermodynamic parameters for lysozyme unfolding from Tanford and Aune (33) and for substrate binding from Banerjee and Rupley (34).

differently according to the interests of the investigators and the experimental approach.

Thermodynamics of Protein-Water Interactions

Free Energy. The free energy of hydration is determined by sorption isotherms (2). These measurements show that the hydration process is stepwise. The isotherm for a protein typically has a "knee" at 0.05 h (g of water/g of protein), or 0.1 x (relative water vapor pressure) which represents binding of strongly interacting water to charged groups (3). Between 0.05 h and 0.3 h (0.1 and 0.9 x) there is a plateau in the isotherm. Above 0.9 x, which is close to the limit of sorption measurements, the amount of water bound increases sharply with increased pressure. Some models used in interpreting sorption data associate the plateau region with multilayer water. This cannot be correct, because even 0.4 g of water/g of protein is barely enough to cover the surface. The plateau region corresponds (see below) to binding of water at polar sites, such as carbonyl groups, and the rise at high relative pressure represents the start of the last stage in completion of monolayer coverage.

Enthalpy. The enthalpy of hydration has been determined from the temperature dependence of the sorption isotherm. The magnitude of the heat of sorption is about 80 kJ/mol of water at low coverage (the "knee" region at 0.05 h) and decreases to the heat of vaporization of water (44 kJ/mol) by 0.2 h. Because hysteresis is observed generally in sorption isotherm measurements, van't Hoff heats of sorption, calculated assuming thermodynamic equilibrium, may be incorrect. A calorimetric study for collagen(4) confirms the van't Hoff values. A Monte Carlo simulation of the lysozyme-water system (5) assigns a portion of the water at the protein surface an energy of interaction of 80 kJ/mol or greater. Water at the protein surface differs more in enthalpy than in free energy from bulk water (Table II); the magnitudes of both the heat and entropy of sorption decrease strongly with increased coverage of the surface, i.e. with decreased strength of interaction.

The dependence of the heat of sorption on the extent of coverage has been observed to be irregular, with an extremum in the knee region of the isotherm (6, 7). A calorimetric study (8) has demonstrated a similar irregularity in the hydration of polysaccharides. The extremum in the heat of sorption for lysozyme (6) corresponds with one in the heat capacity (see below) that reflects proton redistribution.

Volume. Volume has been measured as a function of the extent of hydration for ovalbumin (9) and for β-lactoglobulin in the crystal (10). The linear dependence of the volume on system composition shows that the partial specific volume of the solvent in

the solid sample is identical to that of bulk solvent at hydration levels above 0.2 h.

Heat Capacity. Measurements of the heat capacity of protein
systems are particularly interesting for several reasons: 1) they
reflect solvation of nonpolar elements in addition to other parts
of the protein surface and thus can be viewed as the most complete thermodynamic probe for water-protein interactions. 2) Heat
capacity can be measured conveniently for both solution and solid
samples; thus the two categories of protein hydration studies,
those on solutions and those on solid samples, can be correlated.
3) There is a substantial literature on the heat capacities of
small molecules and on additivity relationships, which appear to
be more accurate for heat capacity than for other thermodynamic
functions.

Heat capacity measurements for proteins are made using
differential scanning calorimeters, which give the heat capacity
as a function of temperature, and less frequently, by drop calorimetry, which gives the heat capacity at a fixed mean temperature.
Scanning calorimetric measurements of lightly hydrated protein
samples over the temperature range centered on 0 °C show that a
transition heat and change in heat capacity associated with
melting of water is observed only above a threshold level of
hydration. For example, measurements for collagen (11) show that
there is no transition for samples of 0.35 h; thus at least this
amount of water interacts so strongly with the protein that it
cannot freeze. Estimates of the water of hydration, here defined
as the amount of nonfreezing water, can be made by analysis of
data for samples of sufficiently high water content to show a
melting transition.

Scanning calorimetric measurements of the denaturation process as a function of extent of hydration have been reported for
several proteins (β-lactoglobulin: (12); lysozyme: (13)). The
melting temperature rises and other parameters of the denaturation transition change as the hydration level is decreased below
0.75 h. The changes are large below 0.3 h.

Heat capacities determined for a fixed mean temperature over
the full range of system composition, dry protein to dilute solution, have been reported for ovalbumin (9) and lysozyme (14).
Suurkuusk (15) has measured heat capacities of several proteins
at extremes of the composition range. The fixed-temperature heat
capacities obtained using a drop calorimeter are particularly
accurate. Measurements of this kind made at 25 °C, a temperature
between the freezing point and the onset of thermal denaturation,
define changes in the thermal properties of the system that were
not detected in scanning calorimetric measurements. Also, interpretation of the results is not complicated by the changes in
state, associated with fusion of the solvent or denaturation of
the protein, that are part of the scanning measurements.

Figure 1 shows the heat capacity of the lysozyme-water

Table II.   Thermodynamic Parameters for Water Interacting
with Lysozyme[a]

Hydration Range (g of water/g of protein)		$\overline{\Delta G}_1$ (J mol^{-1})	$\overline{\Delta H}_1$ (J mol^{-1})	$\overline{cp}_1$ (JK^{-1}g^{-1})
I	0.38 – (∞)	(0)	(0)	4.18
II	0.27 – 0.38 (0.325)	–	–	3.35
III	0.07 – 0.27 (0.17)	– 770	– 2100	5.8
IV	0 – 0.07 (0.035)	–6330	–70430	2.3

[a] $\overline{\Delta G}_1$ and $\overline{\Delta H}_1$ estimated from data of Hnojewyj and Reyerson (6); $\overline{cp}_1$ from Yang and Rupley (14); the hydration range is given for $\overline{cp}_1$; the hydration (given in parentheses) is for $\overline{\Delta G}_1$ and $\overline{\Delta H}_1$.

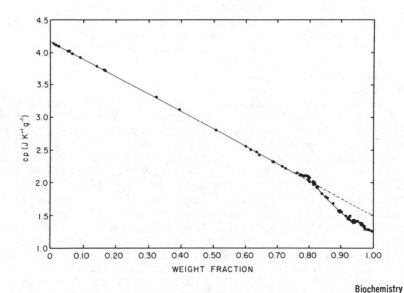

Biochemistry

Figure 1.   Specific heat of the lysozyme–water system from 0 to 1.0 weight fraction protein (14). Least-squares analysis of the linear portion of the heat capacity function from 0 to 0.73 weight fraction water gives $cp_2{}° = 1.483 \pm 0.009$ *J K^{-1}g^{-1}. The value of* $cp_2{}° = 1.26 \pm 0.01$ *J K^{-1}g^{-1}.*

system as a function of composition. The principal conclusions
are the following: 1) the linear response of the heat capacity
to change in composition from 0 to 0.73 weight fraction protein
means that 0.38 g of water/g of protein (300 water molecules/
protein molecule) is sufficient to complete the hydration process
as detected by changes in thermal properties. Addition of water
above this hydration level only increases the amount of bulk
water in the system.

2) The heat capacity of lysozyme in dilute solution is
greater than the heat capacity of the dry protein. The latter
can be calculated, assuming additivity, from heat capacities for
oligomers of glycine and for the several amino acids in the
crystalline state, as suggested by Hutchens et al. (16). The
heat capacity of the protein in solution can be calculated similarly from the dilute solution heat capacities of oligomers of
glycine and the amino acids, using in addition estimates based on
crystal structure studies of the extent to which backbone and side
chain elements are exposed to solvent in the native protein (17).
Results of a calculation of this sort for the dry protein and the
dilute solution state are given in Table III. Agreement with experiment is good. Also given in Table III are the results of a
calculation based on a different set of additive parameters, for
atoms or groups, given by Benson and Buss (18) for the solid
state and by Guthrie (19) for the solution. Again agreement with
experiment is good. The point to be made is that the heat capacity of the protein in the solid and solution states is consistent
with the heat capacities of small molecules. Specifically, there
is no evidence for there being a special contribution to the heat
capacity from vibrational modes that may be peculiar to the macromolecule. A contribution of this kind has been suggested by
Sturtevant (20).

3) The heat capacity varies nonlinearly with composition
below 0.38 $\underline{h}$. The data of Figure 1 for the water-poor region is
replotted in Figure 2, with coordinate transformations to apparent
specific heat capacity of the protein, ϕcp_2, versus the hydration
level expressed in units g of water/g of protein. The function
ϕcp_2 is a measure of the excess specific heat, normalized to unit
quantity of protein (14). The data of Figure 2 show that the
hydration process is stepwise. Four regions, designated I-IV
(Figure 2), can be distinguished in the specific heat profile.
Region I, comprising hydration levels above 0.38 $\underline{h}$, represents
addition of bulk water to the system. Thus within Region I ϕcp_2
(the normalized nonideality of the system) is constant and equal
to the dilute solution value. The rise and fall of ϕcp_2 within
Region IV and at the juncture of Regions III and II can be understood as reflecting heats of reaction. With regard to the reaction heat in Region IV, infrared measurements (21; see below)
have shown that interaction with water at the lowest level of
hydration results in proton transfer from carboxylic acid to
basic protein groups. The reaction heat at 0.27 $\underline{h}$ is understood

Table III. Calculations of Cp_2^o, $\overline{Cp_2^o}$, and ΔCp for Solvation
of Lysozyme[a]

	Calculation Method		
	Side Chain Contribution	Group Contribution	Experimental[b]
Cp^o (solid)	17533	17314	17961
ΔCp (solution – solid)	2572	3208	3146
$\overline{Cp_2^o}$ (solution)	20105	20522	21107

[a] Units $JK^{-1} mol^{-1}$; see text for description of calculations.

[b] Yang and Rupley (14).

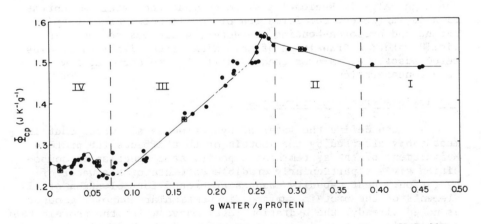

Figure 2. Apparent specific heat capacity of lysozyme from 0 to 0.45 g of water per gram of protein (14)

Curve is calculated with four regions of linear response to ligand composition (breaks at 0.07, 0.27, and 0.38 h) and two transitions (or reactions) centered at 0.05 and 0.26 g of water per gram of protein. Heat capacity measurements were made with lyophilized powders of lysozyme, appropriately hydrated, except for the four measurements indicated by the square symbols, for which the sample was a film formed by slowly drying a concentrated solution of lysozyme.

to represent condensation of water over the regions of the surface that interact most weakly, presumably the nonpolar elements.
A condensation process of this kind is predicted (22) for heterogeneous surfaces binding an adsorbate such as water that can
interact with other adsorbate molecules. Infrared measurements
(21; see below) associate Region III (Figure 2) with binding of
water to carbonyl oxygen and presumably other polar sites and
show that this process is essentially complete at the 0.27 h
transition. These observations support the following two conclusions: the 0.27 h transition and the hydration events of
Region II reflect principally changes involving nonpolar portions of the surface; the surface of the protein is completely
covered at 0.38 h.

4) Table II gives values of $\overline{cp}_l$ in each of the several
hydration regions described by Figure 2. $\overline{cp}_l$ for Region IV
is smaller than the heat capacity for liquid water and is close
to the value for ice. $\overline{cp}_l$ for Region III is greater than the
heat capacity for water. The heat capacity of liquid water is
about twice that of ice or water vapor and reflects a "configurational" contribution from the enthalpy change associated with
the temperature-dependent rearrangement of hydrogen bonding between water molecules. The low value of $\overline{cp}_l$ in Region IV thus
suggests that the water bound is dispersed about the surface, and
the high value in Region III suggests that the water concentration is sufficient for formation of mobile networks of varying
extent and hydrogen-bonding character, analogous to those in
liquid water; a transition of this kind, from dispersed to clustered adsorbate, is predicted by Hill (22) to occur at low adsorbate concentration.

Conclusions from Thermodynamics

One can define the water of hydration as all water that is
measurably affected by the protein or that affects the protein.
Measurement of the system heat capacity as a function of composition gives a particularly credible estimate of the water of
hydration. Heat capacity changes are found for solvation of all
elements of the protein surface, in particular nonpolar groups.
As noted already, the hydration level above which the protein heat
capacity is constant defines completion of the hydration process.
The value estimated for lysozyme is 0.38 g of water/g of protein,
equivalent to 300 molecules of water/molecule of lysozyme. With
regard to other thermodynamic measurements, the sorption isotherm
is not able to define completion of the hydration process, and
there can be difficulty in interpreting scanning calorimetric experiments in terms of completion of hydration, because different
states of the system are being compared (frozen and solution, or
native and denatured) and during a scanning calorimetric measurement the system is not at equilibrium, allowing reaction rates to
influence the response.

Three hundred molecules of water is a small amount for cover-
ing the surface of lysozyme, which is about 6000 $\overset{\circ}{A}{}^2$ in extent
(23, 17). The average area of one side of a cube equivalent in
volume to a molecule of liquid water is 10 $\overset{\circ}{A}{}^2$. Richards (10) has
noted the discrepancy between the approximately 600 molecules of
water predicted on this basis and the 300 or fewer molecules
determined by experimental probes of hydration. There are, how-
ever, water arrangements with a 20 $\overset{\circ}{A}{}^2$ area per water molecule,
for example, the planes of water molecules perpendicular to the
c-axis of ice I. The point to be made is that the hydration
estimated from the heat capacity data indicates that the water
about the protein must be locally ordered through interaction
with protein atoms, in order to obtain an area/water molecule
greater than in bulk water. In this regard, the average area per
polar or charged atom on the protein surface is approximately
20 $\overset{\circ}{A}{}^2$. Furthermore, multilayer water apparently is undetected by
heat capacity measurements; thus the protein with a monolayer
shell of water must mesh well with the bulk solvent.

Because the various thermodynamic parameters are measures of
the same molecular events, they should show parallel changes with
hydration. Steps in the hydration process are defined by the
sorption isotherm and the heat capacity-composition profile, and
the two measurements are in agreement. The knee in the sorption
isotherm corresponds to Region IV (Figure 2), the plateau region
of the isotherm corresponds to Region III, and the sharp rise in
coverage at high pressure corresponds to the transition observed
in the heat capacity at the juncture of Regions II and III.
Similarly, changes in the volume and enthalpy of the adsorbed
water correspond: the bulk water value is reached within Region
III.

Because the hydration process is stepwise, it is reasonable
to describe properties of the bound water in each region. This is
done in the concluding section below.

Comparison of Thermodynamic With Other Measurements of Hydration

The following paragraphs compare thermodynamic studies of
hydration with other static measurements: infrared, NMR, and x-
ray diffraction. Dynamic measurements are considered in the next
section.

Careri et al (21) have carried out a careful infrared spectroscopic examination of the hydration of lysozyme. Figure 3
compares the spectroscopic results with heat capacity measurements. The principal conclusions are the following: 1) the
first two steps in the hydration process, Regions IV and III, are
seen in the infrared measurements. The discontinuity at 0.07 h
observed in the dependence on hydration of the carboxylate,
amide, and water bands (Figure 3), corresponds to the juncture of
Regions IV and III. 2) The increase in carboxylate intensity
within Region IV means that proton redistribution follows

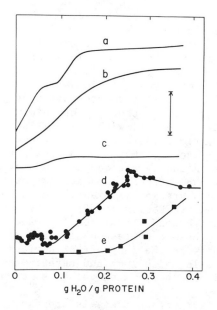

g H₂O / g PROTEIN

Figure 3.   Properties of the lysozyme–water system as a function of water content The correspondence between the vertical bar in the figure and the units of measurement is given separately for each curve.

(a) Absorbance at the carboxylate band maximum (1580 cm⁻¹), measured at 38°C; units, 0.35 Å. (b) Change in the amide I′ band, measured at 38°C as the difference between the negative (1690 cm⁻¹) and positive (1645 cm⁻¹) extremes about isobestic point of the differential spectrum; units, 0.17 Å. There is a small contribution from carboxylate ionization in the lower range of hydration. (c) Frequency of the highest intensity maximum of the OD stretching band of adsorbed D_2O, measured at 38°C. Total shift is from 2550 to 2580 cm⁻¹. (d) Apparent specific heat capacity of lysozyme, measured for 25°C; units, 0.2 J K⁻¹g⁻¹. (e) Enzymatic activity, measured at 25°C; units, 5. × 10⁻⁶ sec⁻¹. After Ref. 32.

adsorption of water, and as noted above this process serves to explain the reaction heat seen at 0.04 h. 3) Changes in the carbonyl and other infrared bands are essentially complete at 0.2 h. The significance of this observation was discussed in the preceding section. 4) Changes in the infrared bands above 0.07 h are continuous, which suggests that there are no major conformation changes associated with hydration above 0.07 h. Also, the effect of hydration upon the amide band can be understood as interaction of carbonyl oxygens with water without changes in peptide hydrogen bonding.

Kuntz and coworkers (2) have determined, using NMR spectroscopy, the amount of nonfreezing water associated with various macromolecules. Hilton et al (24), in a similar study on lysozyme, found the amount of nonfreezing water to be 0.35 g of water/g of protein, close to the hydration estimated from heat capacity measurements. There is generally good agreement between the amount of nonfreezing water detected by NMR and scanning calorimetric measurements (2). Kuntz (2) has concluded from NMR results obtained with model compounds that the nonfreezing water is predominantly associated with charged groups.

X-ray diffraction analysis has located water in crystals of various proteins (25). The analysis of rubredoxin (26) has given a particularly clear picture, in which 127 water molecules were identified. Because diffraction analysis gives a time-average picture, the occupancies of the water sites range from full to 0.3, and some sites are mutually exclusive. There are several extensive networks comprising hydrogen-bonded water and protein atoms. Most of the water in the model is within hydrogen-bonding distance of protein atoms. However, ¼ of the water is at 4 Å or greater distance, with a peak in the distribution function at 4 - 4.5 Å, the second nearest neighbor distance in water or ice. Water at 4 Å or greater distance is not in contact with protein atoms and represents multilayer water. We believe that multilayer water is not detected in thermodynamic measurements. Apparently water can be localized through hydrogen-bonding to other water at the protein surface and yet display thermodynamic properties indistinguishable from the bulk solvent (see concluding section below).

Dynamic Measurements

A description of the motions of water and other molecules at the protein surface is needed for an understanding of enzyme catalysis and other protein properties.

Hydrodynamics. Estimates of protein hydration water based on frictional properties are generally greater than estimates from thermodynamics (2). Squire and Himmel (27) found that the hydration of 21 proteins has a mean value 0.53 g of water/g of protein.

Relaxation Measurements. NMR measurements have been used to examine water-protein interactions in solution and in hydrated powders (2). Hilton et al (24) have shown that the motional properties of water in partially hydrated powders of lysozyme are best characterized as those of a viscous liquid. Dielectric relaxation spectra of water in lysozyme powders (28) distinguish two classes of water, one with a relaxation time near 10^{-9} s, the other, found at levels of hydration greater than 0.3 h, with relaxation time about 2×10^{-11} s, close to the relaxation time of bulk water. The break in the dielectric properties is at the point of completion of the hydration of the amide groups, seen in infrared measurements, and of the transition leading to completion of the monolayer, seen in the heat capacity measurements.

Electron Spin Resonance. ESR spectra (Figure 4) were measured for partially hydrated powders of lysozyme containing a nitroxide spin probe, TEMPONE. Figure 5 shows the change with hydration level of a parameter which characterizes the relative amplitudes of various spectral lines.

Estimates of the TEMPONE correlation time were made using spectrum simulations based on the following model: the bound TEMPONE is assumed to be distributed between two types of sites; TEMPONE bound at one of these is unaffected by hydration and has nearly rigid motional properties; the motion of TEMPONE at the second class of sites, designated variable, is affected by hydration above 0.2 - 0.25 h, and the fraction of the TEMPONE bound to these sites increases from 0.2 at 0.07 h to 0.5 at 0.3 h and higher hydration. In the absence of water, the correlation time for the TEMPONE at 25 °C is 6×10^{-8} s. At hydration levels below 0.2 h TEMPONE bound at variable sites has a correlation time of 3×10^{-9} s. Between hydration levels 0.2 and 0.4 h, the correlation time decreases by an order of magnitude; there is an additional smaller decrease in going to higher levels of hydration (Figure 6). The motional and spectral properties of TEMPONE at the highest levels of hydration studied were not identical with the properties found in dilute solution. The correlation time for the TEMPONE motion is 1 to 2 orders of magnitude greater than that for water at a corresponding level of hydration; it is much less than the characteristic times for enzyme processes.

The ESR properties show changes with hydration in agreement with the heat capacity and infra-red measurements: there are breaks at 0.05 to 0.1 h (Figure 5) and 0.2 - 0.3 h (Figure 6). The major change in motional properties is associated with completion of amide hydration and the start of the condensation of water over the weakly interacting portions of the surface. However, the motional properties continue to change at hydration levels above that for completion of the monolayer seen in the thermodynamic measurements.

Enzyme Properties. Enzymatic activity was measured for

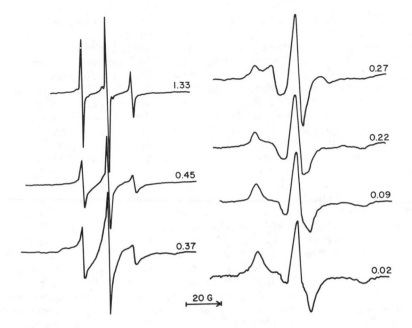

Figure 4. ESR spectra of Tempone noncovalently bound to lysozyme (mole fraction of Tempone, 0.018) at hydration levels of 0.02–1.33 g of water per gram of protein. All measurements were at 24°C.

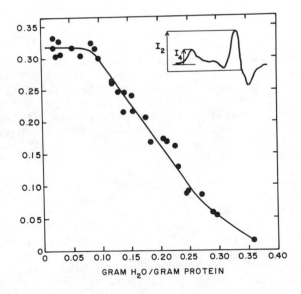

Figure 5. Ratio of heights of hyperfine lines (I_4/I_2) from 0.02 to 0.36 h for Tempone noncovalently bound to lysozyme (mole fraction 0.018)

partially hydrated samples of lysozyme. Under conditions of low
activity (high pH and 0 °C), a solution of lysozyme was mixed
with a solution containing an equimolar amount of substrate, the
hexasaccharide of N-acetylglucosamine; the mixture was lyophilized
to give a stable powder of the enzyme-substrate complex, which
could be hydrated as appropriate by isopiestic equilibration or
addition of water. The pH level was adjusted before lyophilization. It was necessary to vary pH in order to obtain reaction
times convenient for study of the lower levels of hydration. The
rate of enzymatic reaction was low (half-time of several days)
and less than the rate of hydration, which was complete within
one day. The rate of enzymatic reaction was determined by product analysis using charcoal columns.

Figure 7 shows the activity as a function of level of hydration for pH 8, 9, and 10. The decrease in rate of reaction with
increase in pH, observed for the powders, is in accord with the
behavior in dilute solution and presumably reflects the ionization of the active site residue Glu-35.

Enzymatic activity develops in two steps, as shown in Figure
8. The beginning of the first step, at about 0.18 weight fraction
water, equivalent to 0.22 $\underline{h}$, corresponds to the last stage in the
hydration process seen by the heat capacity, i.e. the completion
of the monolayer by condensation of water over the weakest interacting surface groups. The enzymatic activity at the end of the
first step is about 10% of the dilute solution value. The development of activity in the first step shows a 15th-order dependence on water activity. Thus the rate of reaction is not controlled by water being a reactant in the hydrolysis process, in
which case the order would be unity. The activity rises to the
dilute solution value with increase in hydration above 0.5 $\underline{h}$.

It has been shown that the structure of the enzyme-substrate
complex undergoes rearrangement in the rate-determining step of
the lysozyme reaction (29). This rearrangement may require the
mobility associated with completion of the water monolayer. It
is also possible that a network of water molecules participates
in the catalytic process. We favor the former alternative. Regardless of the explanation, it is important that not much water
is needed for enzymatic activity and that only the strongly
interacting sites must be filled before activity is observable.

Comments. Dynamic properties show changes at hydration
levels above 0.4 $\underline{h}$, the point of completion of the changes in
static properties. Because the later reflect a monolayer of
water about the protein, the additional water seen in the dynamic
measurements is "multi-layer" water. Furthermore, hydration affects the several rate properties differently. The more complex
hydration dependence of dynamic compared with static properties
is to be expected. Static properties, at least the thermodynamic,
have a single molecular basis. In contrast, the various transition states governing the rate processes are necessarily different.

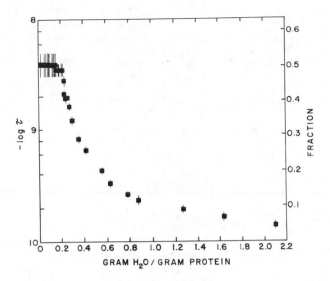

Figure 6. Values of τ for the variable Tempone environment as a function of hydration level. Error bar shows the range of values that gives acceptable simulated spectra. Fraction of Tempone in the variable environment is 0.5 ±0.2 at high hydration.

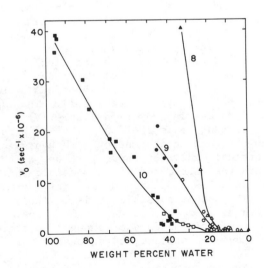

Figure 7. Enzymatic activity of lysozyme as a function of water content, at pH 8, 9, and 10. Open symbols: measurements on powders hydrated by isopestic equilibration. Closed symbols: solvent added to powder.

In fact, it is remarkable that the agreement between the dynamic and static measurements is so close in defining the sequence of events in the hydration process: as noted above, NMR, dielectric relaxation, ESR, and enzymatic measurements each define one or more of the steps in the hydration process seen in static measurements.

Absence of Conformation Changes with Hydration

Most of the studies described above were on partially hydrated protein powders or films. It is necessary to relate the conformation of the protein in this state to that in solution. There is a substantial body of evidence supporting the conclusion that the conformation of the protein does not change measurably between about 0.2 $\underline{h}$ and the dilute solution: 1) enzymatic activity is observable at 0.2 $\underline{h}$. 2) The infrared spectrum changes continuously with hydration above 0.1 $\underline{h}$ and is consistent with surface group hydration without change in protein conformation. 3) The temperature of the thermal denaturation transition increases with decreasing hydration below 0.7 $\underline{h}$. 4) The partial specific volumes of several proteins are the same in dilute solution and in the solid at hydration levels above 0.2 $\underline{h}$. 5) The circular dichroism spectrum of lysozyme in a film is closely similar to that in solution (30).

In order to compare the conformation in powders at very low hydration (0.02 $\underline{h}$) with that at higher levels, above 0.2 $\underline{h}$, ESR measurements were made on samples of lysozyme covalently labelled with succinimidyl-2,2,5,5-tetramethyl-3-pyrrolin-1-oxyl-3-carboxylate. The material contained two spin labels per protein molecule, sufficient for spin-spin interaction to be observed. Spectra were measured at -160 °C as a function of hydration. At this temperature, molecular motion makes no contribution to the spectrum, and an analysis of spectrum shape can be used to estimate the distance between spin centers (31). The line shapes were invariant to change in hydration (Table IV). The mean distance between spin centers is 26-28 Å. Thus to a resolution of 1 Å there is no change in conformation with hydration. This conclusion holds strictly only for those portions of the protein that have reacted with the label, and it is possible that there are changes greater than 1 Å in other portions of the molecule. However, we believe that this possibility is unlikely, because of the wide distribution of amino groups about the lysozyme surface and because of the cooperative nature of changes in conformation. It is also possible but, we believe, unreasonable that conformations different at various hydration levels at 25 °C will be brought at -160 °C to the same conformation. Thus the invariance of the ESR line shapes between low and high hydration levels is strong support for the invariance of the conformation.

To summarize, various measurements indicate that the conformation of the protein in the powder or film state is the same as

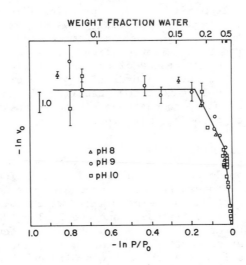

Figure 8. Logarithmic plot of the data of Figure 7, with arbitrary ordinate translations to bring curves for different pH into coincidence: (△) pH 8; (○) pH 9; (□) pH 10

Table IV. Average Distance Between Spin Centers at $-160\ {}^{\circ}\mathrm{C}$[a]

Hydration (g of water/g of protein)	$\dfrac{d_1}{d}$	Average Distance Between Spin Centers (Å)
0.02	0.54	26
0.21	0.54	26
0.22	0.53	28
0.30	0.53	28
0.41	0.53	28

[a] d_1 is the peak-to-peak height for the two outer extrema of the first derivative lines; d is the peak-to-peak height for the central field line; the average distance between spin centers corresponding to the ratio d_1/d was obtained from the curve given by Likhtenshtein (31).

in dilute solution. The ESR and enzymatic properties are sensitive to small changes in conformation (about 1 Å). Apparently,
removal of the solvent from about a protein, like incorporation
of a protein molecule into a crystal, does not affect the backbone conformation, although the arrangement of surface side chain
groups is expected to be altered slightly.

    In view of the importance of solvent for protein folding,
the absence of an effect of dehydration on conformation is surprising. It is possible that the stability of the native conformation at reduced water content has a kinetic basis, owing to
conformation changes being inhibited.

Picture of the Hydration Process

    The correspondences noted above between various measurements
of the hydration process suggest that the events in the hydration
of dry protein fit the following picture, which follows closely a
description by Careri, Gratton, Yang and Rupley (32) and which is
illustrated in Figure 9. A) The dry protein molecule has a conformation similar to that of the protein in solution, and in
films or powders it makes few contacts with neighboring molecules.
B) The first water added interacts predominantly with ionizable
groups and restores the normal pK order among groups perturbed
strongly through dehydration. This strongly bound water (Region
IV) is dispersed about the protein surface. It constitutes 25%
of the water of hydration, which is equal to the percentage of
the surface contributed by ionizable groups (17). C) At a concentration of near 0.1 g water/g protein, clusters develop, presumably centered on polar and charged protein surface atoms.
This change is evidenced by discontinuities in the infrared properties, in the heat capacity and the sorption isotherm, and in
ESR spectra. The high heat capacity found for the bound water
between 0.1 and 0.25 h (Region III) indicates that the hydrogen
bonding arrangements are variable, as they are in bulk water.
Time-domain measurements (NMR, dielectric relaxation, and ESR)
show that this water is restricted in motion relative to bulk
water. Presumably this type of structure for the water shell
(mobile, variable arrangements centered on polar and charged
surface sites) obtains also in the fully hydrated molecule.
D) Between 0.2 and 0.3 h, hydration of the hydrogen-bonding sites
is complete. E) Condensation of water over the weakest interacting portions of the surface, the nonpolar regions, leads to
monolayer coverage at 0.4 h. There must be special local arrangements of water at the protein surface, in order to obtain high
coverage per water molecule adsorbed. The condensation is a
major event in the hydration process; it is seen in the heat capacity, a static measurement, and is the point at which dynamic
properties (dielectric relaxation; spin probe correlation time;
enzymatic activity) show large changes. The mobility of the
protein-water system increases dramatically with completion of

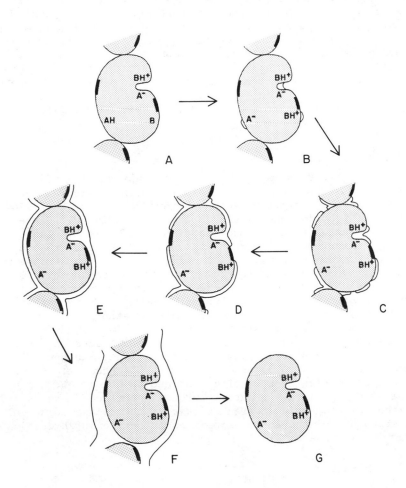

Figure 9.   Representation of the events of the hydration process. Letters correspond to discussion in the text.

the monolayer. F) The arrangement of the water within the monolayer is such that the protein at this hydration level meshes
well with bulk water. In this regard, there are no changes in
thermal properties with increase in hydration above 0.4 h. The
heat capacity is a sensitive probe of hydrogen bonding, and thus
would have detected changes produced by or upon the water shell.
Consequently, the special local arrangements of water suggested
in (E) above would have to be among those characteristic of liquid water. Thus it is not surprising that some multilayer water
should be located by x-ray diffraction analysis and yet have the
thermal properties of bulk water. G) Dynamic properties change
as the hydration is increased above 0.4 h. It is significant,
however, that these changes in motional properties appear to be
small compared to those occuring during the condensation process at 0.25-0.4 h.

Abstract

The interaction of water with proteins can be studied effectively using solid samples, because it is possible to control
water activity. Sorption isotherms have been reported for a wide
variety of proteins and related molecules, defining the free
energy and enthalpy as a function of hydration level. Heat
capacity measurements, which can be carried out over the full
range of system composition, serve to correlate the studies on
partially hydrated solid samples with studies on the dilute solution state. The thermodynamic results distinguish stages in the
hydration of dry protein and suggest a simple picture of the process. This picture accomodates results of other static measurements, such as infrared spectroscopic, and kinetic measurements,
such as ESR and enzymatic activity. Enzymatic activity is observed before completion of monolayer coverage, and the development of it appears to be associated with a condensation event.
Kinetic properties of the solid continue to change above a hydration level (ca. 0.4 g water/g protein) sufficient to establish
the thermal properties characteristic of the dilute solution.
The structure of the solid protein as a function of hydration
level is discussed.

Acknowledgements

We are grateful to Patricia Adams, for carrying out measurements of the enzymatic activity of lysozyme, and to Professor
Walter Kauzmann for stimulation and critical discussions. This
work was supported by NIH research grant GM-24760.

Literature Cited

1. Bruice, T.C.; Turner, A. J. Amer. Chem. Soc., 1970, 92, 3422-3428.
2. Kuntz, I.D.; Kauzmann, W. Adv. Protein Chem., 1974, 28, 239-345.
3. Leeder, J.D.; Watt, I.C. J. Colloid Interface Science, 1974, 48, 339-344.
4. Pineri, M.H.; Escoubes, M.; Roche, G. Biopolymers, 1978, 17, 2799-2815.
5. Hagler, A.T.; Moult, J. Nature, 1978, 272, 222-226.
6. Hnojewyj, W.S.; Reyerson, L.H. J. Phys. Chem., 1961, 65, 1694-1698.
7. Luscher, M.; Ruegg, M. Biochim. Biophys. Acta, 1978, 533, 428-439.
8. Bettelheim, F.A.; Block, A.; Kaufman, L.J. Biopolymers, 1970, 9, 1531-1538.
9. Bull, H.B.; Breese, K. Arch. Biochem. Biophys., 1968, 128, 497-502.
10. Richards, F.M. Ann. Rev. Biophys. Bioeng., 1977, 6, 151-176.
11. Mrevlishvili, G.M.; Privalov, P.L. "Water in Biological Systems," Kayushin, L.P., Ed., Plenum: New York, 1969, 63-66.
12. Ruegg, M.; Moor, U.; Blanc, B. Biochim. Biophys. Acta, 1975, 400, 334-342.
13. Fujita, Y.; Noda, Y. Bull. Chem. Soc. (Japan), 1978, 51, 1567-1568.
14. Yang, P.-H.; Rupley, J.A. Biochem., 1979, 12, 2654-2661.
15. Suurkuusk, J. Acta Chem. Scand., 1974, B28, 409-417.
16. Hutchens, J.O.; Cole, A.G.; Stout, J.W. J. Biol. Chem., 1969, 244, 26-32.
17. Shrake, A.; Rupley, J.A. J. Mol. Biol., 1973, 79, 351-371.
18. Benson, S.W.; Buss, J.H. J. Chem. Phys., 1958, 29, 546-572.
19. Guthrie, J.P. Can. J. Chem., 1977, 55, 3700-3706.
20. Sturtevant, J.M. Proc. Natl. Acad. Sci. USA, 1977, 74, 2236-2240.
21. Careri, G.; Giansanti, A.; Gratton, E. Biopolymers, 1979, 18, 1187.
22. Hill, T.L. J. Chem. Phys., 1949, 17, 762-771.
23. Lee, B.; Richards, F.M. J. Mol. Biol., 1971, 55, 379-400.
24. Hilton, B.D.; Hsi, E.; Bryant, R.G. J. Amer. Chem. Soc., 1977, 99, 8483-8490.
25. Finney, J.L. Phil. Trans. Roy. Soc. Lond., 1977, B278, 3-32.
26. Watenpaugh, K.D.; Margulis, T.N.; Sieker, L.C.; Jensen, L.H. J. Mol. Biol., 1978, 122, 175-190.
27. Squire, P.G.; Himmel, M.E. Arch. Biochem. Biophys., 1979, 196, 165-177.
28. Harvey, S.C.; Hoekstra, P. J. Phys. Chem., 1972, 76, 2987-2994.

29. Banerjee, S.K.; Holler, E.; Hess, G.P.; Rupley, J.A. J. Biol.
     Chem., 1975, 250, 4355-4367.
30. Chirgadze, Yu.N.; Ovsepyan, A.M. Mol. Biol., 1972, 6, 721-
     726.
31. Likhtenshtein, G.I. "Spin Labeling Methods in Molecular
     Biology," Wiley-Interscience: New York, 1974.
32. Careri, G.C.; Gratton, E.; Yang, P.-H.; Rupley, J.A.
     "Protein-Water Interactions. Correlation of Infrared
     Spectroscopic, Heat Capacity, Diamagnetic Susceptibility
     and Enzymatic Measurements on Lysozyme Powders," submitted
     for publication, 1979.
33. Tanford, C.; Aune, K.C. Biochem, 1970, 9, 206-211.
34. Banerjee, S.K.; Rupley, J.A. J. Biol. Chem., 1975, 250,
     8267-8274.

RECEIVED January 4, 1980.

PROTEINS:
THE MOBILE WATER PHASE

The Structure of Water in Polymers

C. A. J. HOEVE

Department of Chemistry, Texas A&M University, College Station, TX 77843

Water absorbed in polymers is important for mechanical, electrical and other physical properties. For proteins water is by far the most important solvent. It interacts so strongly with proteins that this water is commonly referred to as bound water. These concepts must, however, be critically examined in order that the molecular properties can be understood. Such questions as what the order is in these water clusters and how far they extend, need to be answered.

At first glance little reason seems to exist for water to be strongly bound to most of these polymers. The energy of interaction between water molecules in bulk is quite large. The absolute value of the hydrogen bond between water molecules is on the order of 5 Kcal/mol. If water molecules prefer binding to other groups the binding energy must considerably surpass this value. The energy differences of a hydrogen bond of water with itself and that with the hydroxyl and amide groups in the materials considered in this article are small, however. These small energy differences are not conducive to binding.

Some time ago (1,2) investigators pointed out that methane and other non-polar molecules are dissolved in water with negative enthalpy changes, contrary to the expected positive sign if hydrogen bonds between water molecules would be broken to accommodate the non-polar molecules. The explanation is that water molecules form cage-like structures to allow entrance of the guest molecule. The building blocks of these cages are water molecules that are mutually hydrogen bonded to a higher degree than in bulk water. Sometimes these structures are referred to as icelike, although it has become clear that they should not be considered to be static. As the temperature is raised these structures are gradually converted into less ordered bulk water. This gradual dissipation of structures is associated with a high heat capacity (3). The following consideration shows that only a small fraction of hydrogen bonds need to be broken in order to considerably increase the heat capacity. Comparing at 0°C the heat capacity of water to that of ice, we observe a

0-8412-0559-0/80/47-127-135$05.00/0
© 1980 American Chemical Society

twofold increase. Let us consider that no hydrogen bonds are broken with temperature in ice, and that the surplus 9 cal deg-1 mol^{-1} in heat capacity of liquid water over that of ice results from hydrogen bond breakage. If, furthermore, we adopt the value of 5 Kcal/mol for the energy of one H-bond and assume that at 0°C each water molecule forms two H-bonds, we estimate that only 0.1% of hydrogen bonds per degree need to be broken to account for the surplus heat capacity. This estimate is in excellent agreement with the value obtained by Luck (4) from the temperature dependence of the infrared absorption band intensities. It follows from these considerations that the heat capacity is a sensitive indicator for any structural changes occurring with temperature.

Although in this article we use the term hydrogen bond breakage, it should be mentioned that Pople (5) proposed that hydrogen bonds are not broken, but distorted. If the potential energy increase on hydrogen bond bending is relatively small, as believed (6), this description is to be preferred. For simplicity we shall use the former terminology. Of course, the large structural contributions to the heat capacity are expected on the basis of either mechanism.

If small amounts of water are absorbed in bulk polymers the cage-like water structures present in dilute solutions cannot be fully developed, since the required cavities of several tens of Ångstrom units do not exist in bulk polymers. The water structures must then deviate considerably from those found in dilute solution. We can expect, however, that the water molecules will tend to form vestiges of the cage-like structures, as far as space permits. The requirement for each water molecule to form four hydrogen bonds is especially difficult to satisfy if the polymer is glassy, or semicrystalline. In this case the polymer chains do not undergo appreciable conformational changes; compared to bulk water relatively few polar groups are present and these are generally situated inappropriately to satisfy the hydrogen bond requirements of water molecules. If the water molecules are sufficiently mobile, however, they can, at least, compensate for these deficiencies by additional hydrogen bond formation with each other. It follows from the foregoing considerations that water absorbed in rigid polymers should have a strong tendency to form mobile clusters. Only at extremely low concentrations are water molecules expected to be absorbed separately, without hydrogen bonding to each other.

NMR Methods. One of the best ways to investigate the mobility of water molecules in a fixed matrix is by NMR methods. In the following discussion the collagen matrix is emphasized, although we do not consider the state of water in collagen to be unique. Collagen fibers provide an excellent matrix for studying the water structure. Similar results have been obtained for

other hydrated matrices such as rayon (7) and layer silicates
(8). Moreover, collagen shares with other polymers the distinction that water absorbed in them is usually considered to
occur in different states, bound and unbound.

Collagen molecules consist of triple-stranded helices,
which for our purpose can be regarded as rigid rods of approximately 15 Å in diameter and 3000 Å in length. According to
x-ray diagrams the rods are packed parallel with the fiber axis
in an ordered lattice (9), the details of which are not important here. Furthermore, the lattice expands commensurate
with the water content without destroying the crystalline regularity. The interstices between the rods are available for
water absorption. Since the rod diameters are so small, the
diameters of the channellike interstices cannot be more than
several Ångstrom. Furthermore, as shown by the sharpness of the
x-ray spots (9), the water-filled channels must virtually be of
uniform size throughout the sample for water contents less than
1g of water per g of collagen. Thus collagen fibers provide us
with narrow, straight, channels of known orientation. Broadline NMR measurements of water in these channels show two absorption lines instead of one for bulk water (10-17). The
splitting of these lines depends on the orientation of the fiber
axis with respect to the field. This is direct evidence that
the water molecules are not isotropically rotating. They are,
however, rotating surprisingly rapidly considering the narrow-
ness of the channels between the rigid collagen rods. Most
investigators (13,14,15,16,17) interpret the results in terms
of two kinds of water, a fraction of water molecules specifically bound to collagen, and another fraction of essentially
free water molecules that rotate isotropically and have thermodynamic properties close to those of bulk water. In order to
explain the NMR spectrum the further assumption is made that
these two kinds of water molecules are rapidly interchanging,
on the order of 10^7 times per sec, so as to yield an NMR
spectrum averaged over both fractions. Disagreement exists,
however, regarding the amount and the orientation of the bound
fraction. Irrespective of how this fraction is bound, however,
it is difficult to see how isotropically rotating water molecules can exist with properties of those in bulk water. The
narrow collagen channels prevent a full development of tetrahedral clusters, so characteristic of bulk water. In that case
neither the thermodynamics nor the rate of rotation of water
molecules and the isotropic character can be expected to be
similar to that in bulk water. It is more likely that water
molecules form the maximum number of hydrogen bonds with collagen and with each other, that all water molecules diffuse in
a liquid-like fashion, but that none rotate isotropically.
Whether to call this water bound or not, is a matter of se-
mantics. In regard to its relatively large mobility and its

heat capacity its properties appear to be closer to bulk water
than to those in ice or in crystalline hydrates.

Sometimes (18) the water molecules that fail to freeze on
lowering the temperature are denoted as bound. This notion is
open to criticism, however. It is true that collagen shares
with other polymers the property that a considerable fraction
of water remains unfrozen on lowering the temperature. On the
basis of the number of grams of water per g of polymer the
values are 0.5, 0.3 and 0.3 for collagen, elastin, and methylcellulose, respectively. For different reasons the polymer
chains are essentially immobile. For collagen the crystalline,
rodlike, molecules are apparently in close contact with each
other and in elastin and methylcellulose the amorphous polymers
are in the glassy state. At temperatures below 0°C ice is the
stable phase in bulk. In the narrow, fixed, polymer interstices, however, space requirements are insufficient to form
three-dimensional ice crystals. Other options available to
the water molecules are to remain in the interstices in liquid
form, or to form ice outside the polymer as a separate phase.
The latter option involves breakage of hydrogen bonds between
water molecules and polymer groups. The fixed positions of the
latter groups prevent formation of hydrogen bonds between the
polymer groups which provides compensation in the case of
polymer solutions. Since the energy of hydrogen-bond rupture
is so high, a lower (free) energy is obtained if liquid-like
water remains in the interstices. If at higher water contents
sufficient polymer mobility exists, water diffuses out of the
interstices, and forms ice as a separate phase; while the
polymer shrinks and forms hydrogen bonds with itself dehydration
occurs without a net loss of hydrogen bonds. Dehl's results
(19) on collagen with high water contents show indeed that upon
ice formation the wideline NMR spectrum of the unfrozen fraction
of water becomes identical to that of samples with 0.5 g of
water per g of collagen, the maximum unfreezable amount. Recently (20), x-ray diagrams have confirmed that ice is formed
in large three-dimensional crystals that cannot be accommodated
in the narrow channels. We see from this general pattern that
the amount of unfreezable water is dependent on the mobility
of the polymer and the size of its interstitial spaces, rather
than on the strength of the bonds that can be formed with water.
If so, bound water is an unfortunate term if it is used to denote unfreezable water.

Dielectric Measurements. The dielectric properties of ice
can be understood on the basis of Pauling's model (21) in which
a central water molecule is tetrahedrally surrounded by four
others with which hydrogen bonds are formed. Depending on how
its two protons are directed towards the four neighbors, the
central molecule can occupy six positions. On applying an

electric field the neighboring dipoles do not reorient independently. In order to preserve all hydrogen bonds molecular rotations must be cooperative. This cooperative effect increases the dielectric constant approximately by the factor g (22). For ice g is close to 3 (23). For bulk water this factor cannot deviate much from this value, since, locally, the water and ice structures are similar. Indeed at 0°C the values of the dielectric constant of ice and liquid water are approximately equal. As we have seen, however, for lack of space tetrahedral structures cannot form if small amounts of water are absorbed in polymers. In that case the g-factor must considerably deviate from that in water and ice. If linear water clusters occur, as seems likely, especially in collagen, large g-factors and thus large dielectric constants are expected (24). Even when the water molecules are not bound, their dielectric behavior must greatly deviate from that of the bulk water and ice phases. Potentially, valuable information is available about these water structures by studying their dielectric properties as functions of temperature, frequency and concentration.

Unfortunately, the dielectric behavior of absorbed water is complicated by another effect. Besides the dipolar reorientation, just discussed, ionic effects are often present. Under the influence of an electric field positive and negative charges diffuse to the oppositely charged condenser plates, giving rise to space charges. These space charges also contribute to the measured values of ε' and ε''. Their contributions are known as the Maxwell-Wagner effects (25). In principle these effects can be suppressed by performing measurements at high frequency, low temperature, or by using deionized samples. As a result of these complications a quantitative interpretation of the results has not yet been given. Whatever the exact molecular interpretation, however, both dipolar reorientation and Maxwell-Wagner effects are dependent on the mobility of the water molecules. It is interesting to study this mobility at lower temperatures, where the water molecules become sluggish and the possibility of vitrification exists.

Although the nature of the molecular processes associated with the glass transition is still incompletely understood (26), it has been proposed that at lower temperature the free volume decreases. Consequently, the molecules do not diffuse individually, but their molecular motions become cooperative. Apparently, clusters of molecules become the kinetic units. The broad relaxation spectrum usually observed must then reflect the range of cluster sizes. Often, the activation energy reaches large values as the glass point is approached, indicative of a high degree of cooperativity of the motions. Conversely, if a wide spectrum of relaxation times is found with a strongly increasing activation energy as the temperature is lowered a glass transition is indicated.

For water in collagen, even at low percentages (7 per cent by weight), a wide distribution of relaxation times for the time-dependent dielectric properties ε' and ε'' was found and, furthermore, the activation energy was found to increase rapidly at lower temperatures (24). The data fitted the Williams, Landel and Ferry equations (26), widely used for the mechanical properties of polymers to describe molecular motions as the glass point is approached. On this basis a glass point of 174°K (24) was deduced for water in collagen. This value must be considered as somewhat tentative, in view of the extrapolation to lower temperatures and the assumed analogy between mechanical and dielectric properties; the mechanical and dielectric relaxation mechanisms do not always follow the same pattern. More important than the exact value, however, is the concept that water absorbed at such low concentrations in collagen forms a glass. This conclusion is a natural one if water is considered to be one mobile liquid-like phase. Necessarily, when freezing of water does not occur, a glass transition must then result at lower temperatures. Since bulk water displays a glass transition at 135°K (27) and water in collagen is less mobile, a glass point near 174°K appears reasonable. In contrast, a glass transition is incompatible with the concept of specifically bound water molecules. Surely, if at room temperature water molecules are individually bound at specific adsorption sites, they must be even more strongly bound at low temperatures. In that case, however, their mobility and thus their contribution to the dielectric properties should be small. If bound individually, they certainly cannot perform mutually cooperative motions, indicated by the glass transition of water in collagen samples; amazingly, this glass transition in water is observed in collagen samples containing only 7 per cent of water (24).

Recently, Grigera et al. (28) performed dielectric measurements by the cavity perturbation technique in the region of 8-23 GHz near room temperature. From their finding that the dielectric loss in the dehydrated collagen samples was found to be less than that in the water-containing samples, they concluded that hydration water with a relaxation time close to that of bulk water (approximately 10^{-10} sec) must be present. This water would correspond to the fraction of unbound water. They did not estimate the amount of this fraction, however. It is difficult to comprehend how a considerable number of water molecules in the narrow collagen channels, even when unbound, can be as mobile as those in bulk.

That collagen is not unique in inducing water to remain unfrozen at low temperatures is evident from results obtained for elastin. Elastin is an amorphous (29,30,31) protein occurring in ligaments and arteries, where it contains approximately 0.5 g of water per g of elastin. Approximately 0.3 g of water per g of elastin is unfreezable. At these low water contents elastin is in the glassy state (30). Dielectric measurements (32) on hydrated elastin, similar to those performed on collagen, indicated a relatively high mobility of the water molecules, decreasing rapidly at lower temperatures. An analysis like that for collagen indicated again a glass point at low temperatures (130°K).

Recent dielectric measurements (33) on hydrated methylcellulose at lower temperatures, down to 90°K, are noteworthy. Like elastin, methylcellulose is an amorphous glassy polymer at low water contents (less than 30 per cent on a weight basis). At temperatures below 150°K the values of ε' and ε'' for the hydrated (8 per cent of water) and dehydrated polymer are indistinguishable. Above 150°K, however, both values drastically increase for the hydrated sample. For a given frequency an extremely broad maximum is observed for ε'' near 200°K. From the frequency dependence of these curves, the activation energy can be estimated. Values obtained in the ranges of 230 and 180°K are approximately 10 and 100 Kcal/mol, respectively. The extremely high values obtained for the activation energy at the low temperature is evidence for a glass transition of water in this range.

We conclude that our dielectric results obtained for water absorbed in the rigid polymers collagen, elastin, and methylcellulose indicate that at room temperature the water molecules are rather mobile, that their mobility decreases at lower temperatures, until not separately, but collectively, they become immobilized in a glassy state in the range from 130 to 170°K.

Heat Capacity Measurements. As discussed in the introduction, the heat capacity of water is higher than that of ice, since hydrogen bond breakage occurs with temperature in water. This property can be used to investigate the mobility of water molecules. If water molecules are immobile, their heat capacity contribution might be expected to be close to that of ice. Indeed, by comparing the heat capacity of crystalline hydrates with that of corresponding dehydrated crystals, the heat capacity contribution of water was found to be within 10% of that of ice over a temperature range from 90 to 300°K (34). Although the nature of the bonding of water molecules in the different lattices is quite different, the combined contributions of the three librations and three translations of the water molecules are approximately equal. The value of the heat capacity of water as a function of water content in polymers is, therefore, a valuable indication of the amount of bound water. If at low water contents water would be specifically bound, like in the crystalline hydrates, its contribution to the heat capacity should be low, equal to that of ice. If at high water contents mainly bulk water would exist, the water contribution should be representative of bulk water, a value approximately twice that of ice. In the intermediate region the heat capacity might be expected to rise to a maximum if on increasing the temperature bound water with a low energy is converted to free water with a higher energy. Near room temperature the heat capacity was determined for hydrated samples of collagen (34,35), elastin (30), poly[2-(2-hydroxyethoxy)ethyl methacrylate] (36) and methylcellulose (37). The values of the heat capacity C_p of 1 g of polymer and y g of water as a function of y are schematically shown in Figure 1. In all cases the experimental data are represented by straight lines. The slopes of these lines are indicative of the partial molar heat capacity, $\bar{C}_p$. The values obtained for the heat capacity contribution are in the range of that of bulk water (18 cal deg^{-1} mol^{-1}). This shows unambiguously that water absorbed in these samples should be considered to be a single liquid-like phase in the thermodynamic sense. The values for the partial molar heat capacity at room temperature are much closer to that of bulk water than that of ice; this suggests that also the properties of absorbed water are close to that of bulk water. Given the experimental errors, at most, a few per cent of bound water can be present at room temperature. The similarity of the results obtained for quite different polymers suggests that the often proposed concept that absorbed water can be subdivided into bound and free water needs reexamination.

Equally interesting are the results at lower temperatures. For less than 50 per cent of water absorbed in collagen the partial molar heat capacity was found (38) to be independent of water content at any temperature in the range from 180 to 300°K. This value gradually decreased until near 180°K (the instrumental lowest limit) it converged to the value of ice. It follows from the foregoing considerations that the configurational contributions of water to the heat capacity have then disappeared. It is noted that this observation is in excellent agreement with the deductions from the dielectric properties that water absorbed in collagen is liquid-like with a decreasing mobility as the temperature is lowered until a glass results at 173°K.

Recently, we have measured the partial molar heat capacity $\bar{C}_p$ of hydrated samples of methylcellulose and elastin at temperatures ranging from 110°K to 330°K (37). For both, just as for collagen, at any temperature, the partial molar heat capacity of water appears to be independent of water content. Its values is equal to that of ice from 110°K to 150°K, where it starts to increase over that of ice and then increases virtually linearly with temperature to values close to that of liquid water at room temperature. Schematically, the results are given in

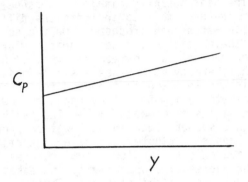

Figure 1. Heat capacity, C_p, of 1 gram of polymer and y grams of water at fixed temperature

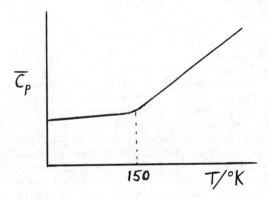

Figure 2. Partial molar heat capacity, $\overline{C}_p$, of water as function of temperature

Figure 2. Thus, the heat capacity results for collagen, elastin, and methylcellulose confirm the dielectric results that a glass point of absorbed water occurs near 150°K.

Conclusions. For a number of different immobile polymer matrices the results of dielectric and heat capacity measurements show that at room temperature water is absorbed as a mobile phase. The heat capacity data over a range of water contents and temperatures fail to indicate the different states of water that have often been proposed. On the contrary, the independence of the partial molar heat capacity for water as a function of water content, at any temperature, is direct evidence that in the thermodynamic sense a single phase of water is present. This single phase is supposed to consist of relatively large clusters of mutually hydrogen bonded water molecules (24). The diffusion of water molecules is then more cooperative, and thus slower, than in bulk water. Furthermore, depending on the range of cluster sizes, a wide distribution of relaxation times exists. At lower temperatures the rate decreases and the degree of cooperativity increases until near 150°K all motion ceases and the clusters vitrify. We note that the cooperativity of motions is in sharp contrast with the individual motions to be expected of specifically bound molecules. It is true that specifically bound molecules could perform coordinated motions, if the polymer chains to which they are attached become mobile. This mechanism is, however, unlikely in view of the same glass transition regions indicated for water absorbed in different polymer matrices.

The model proposed does not imply that at a given moment the environment of all water molecules is identical. Indeed, some water molecules may, in addition to being hydrogen bonded to each other, also form hydrogen bonds with the polymer, whereas others may be exclusively hydrogen bonded to each other. At any instant long, essentially linear, chains of H-bonded water molecules exist. Diffusion of such a cluster occurs if some water molecules break H-bonds with the polymer and, simultaneously, others in the same cluster form H-bonds with the polymer, without breakage of H-bonds between water molecules. In this manner the water molecules move cooperatively instead of individually. The cluster, instead of an individual water molecule, becomes the kinetic unit. As the entire cluster diffuses through the fixed matrix these different water molecules experience different environments.

In view of the similarity of the results obtained for a variety of polymers, no need exists to assume that water molecules are specifically bound in collagen. In this connection it is interesting that the locations of the water molecules could not be obtained from x-ray patterns (39), in qualitative agreement with our proposal of liquid-like water clusters. It should also be noted that the splitting of the NMR lines for absorbed water molecules is a direct measure only of the average rotational anisotropy. No requirement exists to introduce two phases. Our proposal of a single anisotropic phase is, of course, equally consistent with the NMR results.

Acknowledgment

Support by the National Science Foundation through Grant #PCM-78-08580 and by the Office of Naval Research through Grant #N00014-79-C-0267 is greatfully acknowledged.

Literature Cited

1. Frank, H. S.; Evans, M. W. J. Chem. Phys., 1945, 13, 507.
2. Kauzmann, W. Adv. Protein Chem., 1959, 14, 1.
3. Goddard, E. D.; Hoeve, C. A. J.; Benson, G. C. J. Am. Chem. Soc., 1957, 61, 593.
4. Luck, W. A. P. Progr. Colloid Polymer Sci., 1978, 65, 6.
5. Pople, J. A. Proc. Roy. Soc. London, 1951, A 205, 163.
6. Eisenberg, D.; Kauzmann, W. "The Structure and Properties of Water", Oxford University Press, Oxford, England, 1969, Chapter 4.
7. Dehl, R. E. J. Chem. Phys., 1968, 48, 831.
8. Ducros, P. Bull. Soc. Franc. Mineral. Crist., 1960, 83, 85.
9. Rougvie, M. A.; Bear, R. S. J. Amer. Leather Chem. Ass., 1953, 48, 735.
10. Berendsen, H. J. C. J. Chem. Phys., 1962, 36, 3297.
11. Berendsen, H. J. C.; Migchelsen, C. Ann. N. Y. Acad. Sci., 1965, 125, 365.
12. Dehl, R. E.; Hoeve, C. A. J. J. Chem. Phys., 1969, 50, 3245.
13. Chapman, G. E.; McLauchlan, K. A. Proc. Roy. Soc., 1969, Ser. B 173, 223.
14. Chapman, G. E.; Danyluk, S. S.; McLauchlan, K. A. Proc. Roy. Soc., 1971, B 178, 465.
15. Fung, B. M.; Trautmann, P. Biopolymers, 1971, 10, 391.
16. Fung, B. M.; Wei, S. C. Biopolymers, 1973, 12, 1053.
17. Migchelsen, C.; Berendsen, H. J. C. J. Chem. Phys., 1973, 59, 296.
18. Andronikashvili, E. L.; Mrevlishvili, G. M.; Japaridze, G. Sh.; Sokhadze, V. M.; Kvavadze, K. A. Biopolymers, 1976, 15, 1991.
19. Dehl, R. E. Science, 1970, 170, 738.
20. Nomura, S.; Hiltner, A.; Lando, J. B.; Baer, E. Biopolymers, 1977, 16, 231.
21. Pauling, L. J. Am. Chem. Soc., 1935, 57, 2680.
22. Kirkwood, J. G. J. Chem. Phys., 1939, 7, 911.
23. Gobush, W.; Hoeve, C. A. J. J. Chem. Phys., 1972, 57, 3416.

24. Hoeve, C. A. J.; Lue, P. C. Biopolymers, 1974, 13, 1661.
25. van Beek, L. K. in Progress in Dielectrics, J. B. Birks, Ed. Heywood, London, 1967, p. 71.
26. Ferry, J. D. in Viscoelastic Properties of Polymers, Wiley, New York, Chapter 11, 1970.
27. Sugisaki, M.; Suga, H.; Seki, S. Bull. Chem. Soc. Japan, 1968, 41, 2591.
28. Grigera, J. R.; Vericat, F.; Hallenga, K.; Berendsen, H. J. C. Biopolymers, 1979, 18, 35.
29. Hoeve, C. A. J.; Flory, P. J. Biopolymers, 1974, 13, 677.
30. Kakivaya, S. R.; Hoeve, C. A. J. Proc. Nat'l. Acad. Sci. U.S., 1975, 72, 3505.
31. Grut, W.; McCrum, N. G. Nature, 1974, 251, 165.
32. Kakivaya, S. R.; Hoeve, C. A. J. J. Macromol. Sci. Phys., 1977, B 13, 485.
33. To be published.
34. Hoeve, C. A. J.; Kakivaya, S. R. J. Phys. Chem., 1976, 80, 745.
35. Fung, B. M.; Cox, J. A. Biopolymers, 1979, 18, 489.
36. Pouchlý, J.; Biroš, J.; Beneš, S. Makromol. Chem., 1979, 180, 745.
37. To be published.
38. Hoeve, C. A. J.; Tata, A. S. J. Phys. Chem., 1978, 82, 1660.
39. Rich, A.; Crick, F. H. C. J. Mol. Biol., 1961, 3, 483.

RECEIVED January 3, 1980.

Water-Protein Interactions

Nuclear Magnetic Resonance Results on Hydrated Lysozyme

ROBERT G. BRYANT and WILLIAM M. SHIRLEY

Department of Chemistry, University of Minnesota, Minneapolis, MN 55455

Interest in water at protein surfaces and other surfaces arises from a desire to understand structural, functional, and dynamic factors as well as their interrelationships. Nuclear magnetic resonance (NMR) spectroscopy provides both structural and dynamic information. This presentation will focus on dynamical aspects of the water-protein interaction. In particular, the phenomenon of cross relaxation between the water and protein proton systems will be discussed and new evidence will be reported. Failure to recognize the importance of cross relaxation effects leads to incorrect conclusions about the dynamics of water at protein surfaces.

Background

The underlying strategy for extracting dynamical information from NMR relaxation data is based on the equations for either longitudinal (T_1^{-1}) or transverse (T_2^{-1}) relaxation rates. If relaxation is dominated by the magnetic dipole-dipole interaction between like-spin nuclei, then

$$1/T_1 = 3/2 \ \gamma^4 \ \hbar^2 \ I(I + 1)\{J(\omega) + J(2\omega)\}, \tag{1}$$

$$1/T_2 = 3/8 \ \gamma^4 \ \hbar^2 \ I(I + 1)\{J(0) + 10J(\omega) + J(2\omega)\} \tag{2}$$

where γ is the nuclear magnetogyric ratio, I the nuclear spin, and $J(\omega)$ the spectral density at the resonance frequency, ω. Spectral densities are obtained from the Fourier transform of the autocorrelation function describing reorientation of the internuclear vector, $\underline{r}$. The autocorrelation function is usually assumed to decay exponentially with a correlation time τ_c. With the additional assumption of isotropic rotational motion, the relaxation equations become (1)

0-8412-0559-0/80/47-127-147$05.00/0
© 1980 American Chemical Society

$$\frac{1}{T_1} = \frac{2}{5} \gamma^4 \hbar^2 \frac{I(I+1)}{r^6} \{ \frac{\tau_c}{1 + \omega^2 \tau_c^2} + \frac{4\tau_c}{1 + 4\omega^2 \tau_c^2} \} , \tag{3}$$

$$\frac{1}{T_2} = \frac{1}{5} \gamma^4 \hbar^2 \frac{I(I+1)}{r^6} \{ 3\tau_c + \frac{5\tau_c}{1 + \omega^2 \tau_c^2} + \frac{2\tau_c}{1 + 4\omega^2 \tau_c^2} \} . \tag{4}$$

In general, care must be taken to recognize both intra- and intermolecular contributions to relaxation; however, intermolecular relaxation is often neglected in discussions of liquids at surfaces. With this assumption there is direct access to characterization of liquid dynamics at a surface (2,3,4). A study of the relaxation rate at different temperatures or frequencies provides a measure of the correlation time for water at the surface of a protein if the interproton distance, r, in the water molecule is known. The temperature dependence for longitudinal and transverse relaxation times predicted by equations 3 and 4 is shown schematically as the solid lines in Figure 1. At the position of the minimum in the longitudinal relaxation time, $\omega\tau_c$ is about 0.616 and T_1/T_2 is about 1.6. Since ω is known, the correlation time is determined at the minimum in T_1.

The difficulty applying this simple approach in practice has been that the experimental results for a variety of systems do not fit the theory (2,3,5,6). Though there are differences in detail, the basic features of the observations are summarized in Figure 1 as the dashed lines. The problems are clear: the T_1 values usually obtained are much too long relative to the T_2 values and the T_1 minimum is much broader than is expected.

One way to eliminate this apparent discrepancy is to assume that there is a distribution of correlation times experienced by the liquid molecules in the vicinity of the surface (7,8,9). With increasing width of the distribution of correlation times, the T_1 calculated from an equation of the form 3 and T_2 calculated from an equation of the form 4 does approach a fit to the data. One difficulty is that the distribution is often so broad that a rigid lattice correction must be made for very slow moving solvent molecules (8). Extrapolation of protein results to solution situations appears to require a slow moving fraction of water at the surface, but irrotationally bound, slow moving water molecules in protein solutions are inconsistent with solution phase NMR results (10,11). While it appears to be of great value for some systems, this theoretical apparatus neglects an important feature of the water proton relaxation. Namely, that the longitudinal relaxation is generally not described by a single time constant.

Nonexponential NMR relaxation was reported several years ago for water protons in protein crystals (12). This report has been largely ignored apparently because of a less than adequate explanation that leaned heavily on a chemical exchange model. There are two rather different hypotheses that may account for such behavior. Nonexponential relaxation may result if there is a slow exchange of the observed species between two populations, water populations in this case, each characterized by substantially different NMR relaxation rates (13). Such a model is pulse width independent and also largely independent of the nucleus observed in the exchanging molecule. Comparison of the relaxation curves shown in Figure 2 indicate that the apparent shape is significantly pulse width dependent. Deuterium relaxation data also fail to show the rather striking nonexponential character indicated here (14). For these reasons a chemical exchange model must be dismissed as inadequate for the present situation and hence the difficulty of explaining the very long lifetimes implied by such a model is eliminated (12,15,16).

Nonexponential NMR relaxation may also be caused by cross relaxation involving the exchange of spin magnetization between different spin systems through mutual proton spin-flips. Cross relaxation between the water phase and the protein phase has been studied in hydrated proteins (17,18) and protein solutions (19). The exchange of magnetization between the rotating methyl protons and the other protons of a macromolecule also involves a cross relaxation. Nonexponential behavior of proton relaxation has been demonstrated for proteins dissolved in D_2O (20,21). However, there is a distinct difference between this cross relaxation involving only protein protons and the cross relaxation described by Edzes and Samulski for hydrated collagen (17). In the former case the sample is treated as a single system of coupled spins that behave according to a generalization of Solomon's treatment of a spin pair (22). In the hydrated protein case, however, the cross relaxation model assumes two separate thermodynamic systems, the water phase and the protein phase, each able to achieve a well defined temperature in a time short compared with the relaxation times measured.

The time dependence of the magnetization for the proton systems in a hydrated protein may be described heuristically by two coupled equations containing three relaxation rates: R_{1w}, the longitudinal relaxation rate for the water in the absence of the protein proton interaction; R_{1p}, the longitudinal relaxation rate for protein protons in the absence of a relaxation path provided by water protons; and R_t, a rate of magnetization transfer between the two spin systems. The equations then become

$$dM_w/dt = -(R_{1w} + R_t)M_w + R_t M_p \tag{5}$$

$$dM_p/dt = -(R_{1p} + R_t/F)M_p + R_t M_w/F \tag{6}$$

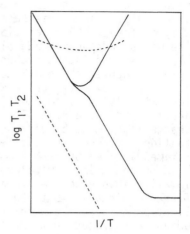

Figure 1. Schematic of the temperature dependence of the longitudinal and transverse relaxation times for water protons. Solid lines are predicted by Equations 3 and 4, while the dashed lines indicate the dependences often observed.

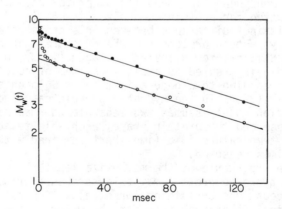

Figure 2. Water proton relaxation in hydrated lysozyme powder (0.17 g H_2O/g lysozyme) at 57.5 MHz at 253 K. Amplitudes were measured after the second pulse of a $180°$–τ–$90°$ sequence with the $180°$ pulse width either (●) 8.6 μsec or (○) 55 μsec.

where M_w and M_p are the water and protein normalized, reduced magnetizations,

$$M_i(\tau) = \{S(\infty) - S(\tau)\}/nS(\infty) \qquad (7)$$

where S is the free induction decay amplitude after the second pulse of a $180°$-τ-$90°$ experiment (n = 2) or a $90°$-τ-$90°$ experiment (n = 1). F is the ratio of protein protons to water protons. The solution of these equations is a sum of exponentials corresponding to the fast, R_{1f}, and slow, R_{1s}, components seen in Figure 2 (17,18). It is critical to note that while the appearance of the relaxation curve is significantly pulse width dependent, the limiting slopes for the fast and slow components are not. Due to the pulse width dependence and the rapid decay of the fast component, the double exponential nature of the curve is very easy to miss experimentally and there are reports that treat only the slow component. It is clear that any attempt to interpret the slow component as a simple relaxation time such as that described by equation 3 neglects a significant feature of the relaxation.

The complete solutions of the coupled equations are given by Edzes and Samulski (R_t becomes k_w and R_t/F becomes k_m in their notation) (17). The water and protein proton relaxation curves are completely described by R_{1p}, R_{1w}, R_t, F, $M_w(0)$, and $M_p(0)$. By observing both water and protein curves at different pulse lengths all the parameters may be obtained. However, R_{1p} is generally expected to be too small to be obtained accurately from the relaxation curves so that the simpler procedure of finding R_{1w}, R_t, $M_w(0)$, and $M_p(0)$ from a single water relaxation curve using estimates of R_{1p} and F seems to be reasonable.

The aim of the present investigation is to study water dynamics in hydrated proteins while testing the cross relaxation model. According to equation 5, the temperature dependence of the observed water relaxation components could arise from changes in any of the three fundamental rate constants R_{1w}, R_{1p}, and R_t. Thus, extraction of R_{1w} from the observed R_{1f} and R_{1s} in order to find its temperature dependence is necessary before a detailed interpretation in terms of water motion is attempted.

Experimental

Three times crystallized, dialyzed, and lyophilized hen egg white lysozyme powder from Sigma was used after a subsequent dialysis and lyophilization. Hydration of the dry powder was accomplished through the vapor phase by exposing the sample to a constant relative humidity for 5 days. In preparing the lysozyme-D_2O sample, the lysozyme was twice dissolved in D_2O and held at 40 deg C for 24 h both times before lyophilization. The D_2O was treated with Chelex-100 before use; both were obtained from Bio-Rad Laboratories. As in the hydration, D_2O was deposited on the dry protein through the vapor phase. The water and deuterium oxide contents were determined by Karl Fischer titrations.

The NMR measurements were made at 57.5 MHz on a pulsed NMR spectrometer that included a 12-inch Varian electromagnet and a Nicolet NMR-80 data system interfaced with a Biomation 805 wave form recorder. Unattenuated 90° pulse widths were 4 to 5 microsec using a 10 watt ENI power amplifier. At best, recovery time for the home built receiver was 10 microsec. Nitrogen gas boiled from a liquid nitrogen dewar was used to cool the sample. The temperature was measured to within 2 deg with a diode thermometer.

Spin-lattice relaxation times for the protein protons were measured using the 90°-τ-90° pulse sequence. Free induction decay amplitude was measured 15-20 microsec after the end of the second pulse by averaging 30 repetitions. To obtain the obvious double exponential relaxation behavior from the water protons, the first pulse of a 180°-τ-90° sequence was attenuated so that the 180° pulse width was about 55 microsec while the second pulse remained near 4 microsec. Experimental considerations led us to believe the errors for the protein T_1 values and the slow component of the water T_1 curve are about 5% although linear least squares fits indicate better precision.

A nonlinear least squares fit of the water relaxation data was used to find the four parameters R_{1w}, R_t, $M_w(0)$, and $M_p(0)$ taking R_{1p} from the lysozyme-D_2O data and setting F = 3.5. The standard deviations from R_{1w}^{-1} and R_t^{-1} are indicated in Figure 3 when they exceed 5%. According to these calculations, immediately following the 55 microsec pulse, the water magnetization, $M_w(0)$, was about 0.9 while the $M_p(0)$ was near 0.5. The only exception is the 207 K measurement which shows $M_w(0) = 0.51$ and $M_p(0) = 0.24$. Standard deviations for the $M_w(0)$ and $M_p(0)$ were about 1% except at 207 K where it was 5%.

Results and Discussion

The temperature dependence of the longitudinal relaxation of protein protons in a sample containing 0.07 g $D_2O/$ g lysozyme is shown in Figure 3. Despite the presence of a small deuterium dipole-dipole contribution and a possibly changed protein in response to the substitution of deuterium oxide for water, these values are thought to be good estimates of R_{1p} for the hydrated lysozyme. Previously published data for dry lysozyme shows a similar curve reaching a minimum near 180 K (23). Also shown in this figure is the temperature dependence of the proton T_1 for the protein in hydrated lysozyme (0.17 g H_2O/g lysozyme). The presence of water on the protein is shown to cause a substantial increase in the relaxation rate. Thus, in the hydrated protein,

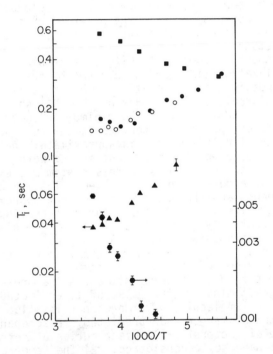

Figure 3.   Temperature dependence of various proton relaxation times in lysozyme powder samples at 57.5 MHz. (■) indicates T_1 for the lysozyme–D_2O sample; other symbols refer to the lysozyme–H_2O powder; (●) protein R_{1s}^{-1}; (○) water R_{1s}^{-1}; (▲) water R_{1w}^{-1}; (⬡) water R_t^{-1}.

the bulk of the protein relaxation occurs through the water phase
and not through the rapidly rotating methyl groups that dominate
relaxation in a dry protein system ($\underline{23}$). No attempt was made to
find the double exponential protein proton relaxation curve and
only the slow component, R_{1s}^{-1}, was measured.

Other data points in Figure 3 represent proton relaxation
times of water on the hydrated lysozyme as calculated from
relaxation curves with obvious double exponential character.
Since the values of R_{1s}^{-1} obtained from the water proton data agree
with those obtained from the protein data, the present experiments strongly support the dominance of cross relaxation. As
anticipated, R_t^{-1} appears to be on the order of or slightly longer
than T_2 and thus indicates efficient spin transfer between water
and protein protons.

Although the slow component, R_{1s}^{-1}, is a combination of more
fundamental relaxation times and should not be treated using
equation 3, a detailed interpretation of R_{1w} is still not a
straightforward problem. The values of R_{1w}^{-1} shown in Figure
3 suggest that the curve lies at the minimum or on the low
temperature side of the minimum for the temperature range studied.
The water content for this sample is approximately half the
value usually associated with the "nonfreezing water" on lysozyme
($\underline{24},\underline{25}$). Therefore, the motions in this system are expected to
be slower than in a more water rich system. The increase in
resonance frequency and the lower water content approximately
account for the shift in the minimum from that previously
reported for hydrated lysozyme ($\underline{18}$). If the R_{1w}^{-1} minimum lies
near 273 K as seems possible, equation 3 suggests that the
correlation time at the protein surface in this relatively dry
sample is on the order of nsec at this temperature.

While recognizing the uncertainty about the presence and
position of the R_{1w}^{-1} minimum, the value of T_1 estimated from
equation 3 for intramolecular water proton relaxation indicates
that R_{1w}^{-1} is somewhat larger than predicted. Discrepancies could
arise from several factors: 1) A distribution of correlation
times may still have to be considered. 2) The temperature
dependence of the water signal amplitude is difficult to monitor
when transverse relaxation rates become large at low temperatures
because the water signal is increasingly difficult to resolve
from the solid protein proton signal. Freezing out of water at
low temperature would distort the shape of the T_1 plot by raising
T_1 of the apparent minimum and shifting it to higher temperatures.
3) Equation 3 neglects effects of anisotropic motion on both
longitudinal and transverse relaxation rates ($\underline{2}$). Recent
experiments using deuterium NMR on samples similar to those
studied here show significant nuclear electric quadrupole
splittings that imply an anisotropic component in the water
molecule motion ($\underline{26}$). Such motional anisotropy will depress T_2
and elevate T_1.

Conclusion

The present study has provided a critical test of the importance of cross relaxation involving both water and protein protons in hydrated protein systems. In addition it has successfully demonstrated that the temperature dependence of both the water and protein relaxation is dominated by motions in the water phase and not by motions in the solid phase such as methyl group rotations. While a detailed analysis has not yet been attempted, it appears that a picture of water in the interfacial regions around a protein that is consistent with the NMR relaxation data is one characterized by fast if slightly anisotropic motion. Structural models for water-protein interactions must be consistent with this very fluid character of water in this interfacial region; however, it is also important to recognize that the time scale appropriate to the present experiments is still long when compared to the rotational correlation times or diffusion times usually associated with water in the pure liquid state.

Literature Cited

1.   Abragam, A., "Principles of Nuclear Magnetism", Clarendon Press, Oxford, 1961, Ch. 8.
2.   Pfeifer, H., NMR Basic Principles and Progress, 1972, 7, 53.
3.   Resing, H. A., Adv. Mol. Relax. Proc., 1972, 3, 199.
4.   Woessner, D. E., Mol. Phys., 1977, 34, 899.
5.   Packer, K. J., Philos. Trans. Roy. Soc., London, Ser. B, 1977, 278, 59.
6.   Resing, H. A.; Wade, C. G., Eds, "Magnetic Resonance In Colloid and Interface Science", ACS Symposium Series No. 34, American Chemical Society, Washington, D.C., 1976.
7.   Odajima, A., Prog. Theor. Phys., Suppl., 1959, 10, 142.
8.   Resing, H. A., J. Chem. Phys., 1965, 43, 669.
9.   Lynch, L. J.; Marsden, K. H.; George, E. P., J. Chem. Phys., 1969, 51, 5673.
10. Koenig, S. H.; Hallenga, K.; Shporer, M., Proc. Natl. Acad. Sci. USA, 1975, 72, 2667.
11. Bryant, R. G., Ann. Rev. Phys. Chem., 1978, 29, 167.
12. Jentoft, J. E.; Bryant, R. G., J. Amer. Chem. Soc., 1974, 96, 297.
13. Zimmerman, J. R.; Brittin, W. E., J. Phys. Chem., 1957, 61, 1328.
14. Hsi, E.; Bryant, R. G., Arch. Biochem. Biophys., 1977, 183, 588.
15. Hsi, E.; Jentoft, J. E.; Bryant, R. G., J. Phys. Chem., 1976, 80, 412.
16. Hsi, E.; Mason, R.; Bryant, R. G., J. Phys. Chem., 1976, 80, 2592.

17. Edzes, H. T.; Samulski, E. T., J. Mag. Reson., 1978, 31, 207.
18. Hilton, B. D.; Hsi, E.; Bryant, R. G., J. Amer. Chem. Soc., 1977, 99, 8483.
19. Koenig, S. H.; Bryant, R. G.; Hallenga, K.; Jacob, G. S., Biochemistry, 1978, 17, 4348.
20. Kalk, A.; Berendsen, H. J. C., J. Mag. Reson., 1976, 24, 343.
21. Sykes, B. D.; Hull, W. E.; Snyder, G. H., Biophys. J., 1978, 21, 137.
22. Solomon, I., Phys. Rev., 1955, 99, 559.
23. Andrew, E. R.; Green, T. J.; Hoch, M. J. R., J. Mag. Reson., 1978, 29, 331.
24. Kuntz, I. D.: Brassfield, T. S.; Law, G. D.; Purcell, G. V., Science, 1969, 163, 1329.
25. Hsi, E.; Bryant, R. G., J. Amer. Chem. Soc., 1975, 97, 3220.
26. Cygan, W.; Bryant, R. G., 1979, unpublished results.

RECEIVED January 4, 1980.

The Dynamics of Water–Protein Interactions

Results from Measurements of Nuclear Magnetic Relaxation Dispersion

SEYMOUR H. KOENIG

IBM Thomas J. Watson Research Center, Yorktown Heights, NY 10598

There can be little question that the concept of "protein hydration" is a meaningful and even useful one, particularly if one does not attempt to define it too precisely. Water in all its states interacts with protein to the extent of forming an identifiable layer on the surface of the protein molecules. Thus, water vapor condenses onto dried protein powders with the first layer bound quite strongly. The quantitative data on adsorption isotherms shows this clearly: a monolayer forms with a binding enthalpy of about -15 kcal/mole; subsequent layers bind progressively more weakly (cf. (1) for a more extensive discussion and a comprehensive bibliography). Analogously, protein molecules in ice are surrounded by a monolayer or so of water that remains liquid (with restricted rotational mobility) at temperatures well below freezing (2). This indicates that the interaction of the water molecules with the protein is large, indeed large enough to overcome the entropy of ordering that determines the heat of fusion.

Thus protein immersed in gaseous and in solid water interacts strongly with roughly a monolayer of water molecules; loosely speaking, the protein is "hydrated" under these conditions. Moreover, the structure of the water of hydration is different from that of either the gaseous or the solid water phase. It is roughly of liquid density, and more or less in the liquid state as judged from (or perhaps, defined by) the motional freedom of the water molecules (cf. 1). The question then arises as to what "protein hydration" might mean for proteins in solution: Is there a layer of water surrounding solute protein that is "liquid-like", but associated with the protein in an identifiable fashion? And if so, what dynamics describe it; how rapidly does the hydration water exchange with solvent; and are the properties of the solvent altered by the phenomenon of hydration in a measurable way?

Again, so long as one speaks qualitatively, it is well established that solute protein is hydrated in water solution (cf. 1). Results from measurements of the hydrodynamic properties of protein solutions and the hydrodynamic properties of proteins molecules in solution, from the early work on viscosity, flow birefringence, and dielectric relaxation (cf. 1) to the more modern work on translational and rotational diffusion measured by inelastic

0-8412-0559-0/80/47-127-157$05.00/0
© 1980 American Chemical Society

laser-light scattering (3), all indicate that the inertial properties of protein molecules in solution are those of protein molecules with an "attached" layer or two of water. Moreover, measurements of buoyant density show that the density of this water of hydration is essentially that of liquid water (cf. 1).

Similarly, with improvements in the techniques of X-ray crystallography, including the improvements in the mathematical and computational procedures for refining the data subject to known geometric constraints imposed by our knowledge of chemical bonding (4), water molecules are beginning to show up in the analysis of the data for protein crystals. In lysozyme, for example, essentially a monolayer, corresponding to some 150 water molecules, has been identified (5). It must be realized that protein crystals are about one-half water or "mother liquor", and that for the present considerations may be regarded as concentrated protein solutions.

Thus the hydrodynamic and X-ray data show that proteins are "hydrated" in solution. However, X-rays travel with the speed of light and take but 10^{-17} s or so to traverse the few unit cells needed to produce a coherent X-ray diagram. Thus these data can in principle give little information regarding the (much slower) dynamics of the waters of hydration. They only show that the hydration layer is ordered with respect to the protein structure; the lifetime of a water molecule in this ordered layer is not something that can be determined straightforwardly (if at all) from X-ray data. However, the lifetime of a water molecule in the hydration layer, or at least an indication of its order of magnitude, - is it a second, a microsecond, a nanosecond, or less? - is what we are after as a way of arriving at a first order description of the dynamics of water-protein interactions.

A naive interpretation of the hydrodynamic data might suggest that, since the hydration layers contribute to the inertia of the protein molecule (in solution), the lifetime of a water molecule in these layers must be long compared to some time characteristic of the motion of the protein molecules. For example, since the orientational relaxation times of large protein molecules, as measured, say, from the frequency dependence of the dielectric constant or computed from Stokes' Law, can be as long as 10^{-5} s, then the lifetime might be expected to be significantly longer. However, this view is incorrect. To explain the hydrodynamic data, it is sufficient that a water molecule be in a well-defined position when hydrating the protein. If this condition is satisfied (and to satisfy it a water molecule must abruptly change both its linear and angular momentum as it enters or leaves the solvent), then the water lifetime can be short compared to any time that characterizes the motion of the protein and still contribute to the inertia of the protein molecules. That this requirement is met is indicated by the X-ray data, which show that many of the water molecules in the hydration layer are in well-defined positions.

From the above discussion then, it is clear that X-ray diffraction data together with the large body of hydrodynamic results indicate that protein molecules in water solution are surrounded by a monolayer or so of water molecules of hydration located in well-defined positions on the surface of the protein molecules. Indeed, the X-ray data show these waters (really, the oxygen atoms of the water molecules) to be positioned to form one or two hydrogen bonds with the surface residues (5). However, the point we emphasize here is that these data give little information on the dynamics of the interaction.

It is our view that the lifetime of a water molecule in a hydration layer is of order 10^{-9} s, a time about 100-fold longer than the time required for a solvent water molecule to break and reform the few hydrogen bonds that restrain its motion in the pure solvent, but nonetheless short enough to be considered characteristic of liquid motions. The explication of this view, and other aspects of the dynamics of both water-protein and protein-water-protein interactions in protein solutions, is the subject of this paper. In what follows we present data and deductions resulting from a powerful experimental technique that, though no longer novel, is pursued in very few laboratories other than our own. We measure the magnetic field dependence of the magnetic spin-lattice relaxation rate of solvent water nuclei in protein solutions, a technique to which we have given the acronym NMRD (for nuclear magnetic relaxation dispersion). NMRD experiments have shown that the rapid rotational Brownian motion of solvent water molecules has superposed on it, by an interaction mechanism not yet understood, a very small component that mimics the much slower rotational motion of the protein molecules (6, 7). Moreover, the NMRD experiments measure the averaged properties of all solvent molecules so that the lifetime of a water molecule in the hydration layer enters as a natural parameter in many models that explain the data. In addition, since NMRD data indicate the orientational Brownian motion of the protein molecules rather directly and rapidly, it becomes possible to study the microscopic viscosity of the solvent in the neighborhood of a protein molecule under a variety of conditions of buffer, pH, etc., not always amenable to other methods.

A Summary of NMRD Phenomena in Diamagnetic Systems

^{1}H NMRD. The figures in this section present the phenomenology of NMRD for solutions of diamagnetic proteins, without consideration of the underlying mechanisms that give rise to the phenomena.

Figure 1 shows NMRD data for solvent protons in a solution of alkaline phosphatase, a "typical" globular protein. Relaxation rate is shown as a function of magnetic field, the latter expressed in units of the proton Larmor frequency. A discussion of the instrumentation and the methodology used to obtain the data can be found elsewhere (6, 7).

There are two separate contributions to the solvent relaxation rate that are introduced by the presence of solute protein, both of which add to the field-independent relaxation rate of protons in pure solvent: the larger NMRD contribution, labelled A, disperses at low fields with an inflection generally between 0.1 and 10 MHz; the smaller contribution, labelled D, is known to

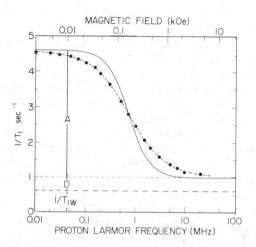

Figure 1.   Dispersion of $1/T_1$, the magnetic relaxation rate of solvent water protons, for a 65 mg/mL solution of alcohol–dehydrogenase from yeast, 160,000 Daltons, at 5.9°C.

Solid circles are the data points; dashed line through the points is from a least-squares comparison of the Cole–Cole expression (Equation 1), with the data.   Two of the parameters of the fit are indicated: A, the amplitude of the major dispensive contribution; and D, the residual high field contribution. $1/T_{1W}$ is the proton relaxation rate in protein-free solvent.   The solid line that intersects the dashed line is a Lorentzian curve (Equation 2), obtained using the same values for A, D, $1/T_{1W}$ and v_c as for solid line.   After Ref. 7.

disperse at fields in the range 100 to 300 MHz, beyond the range of our present automated instrumentation (7, 8).

The dashed line through the data points is a fit to the data, by analogy with procedures used for dielectric relaxation dispersion, using the Cole-Cole expression (9):

$$1/T_1 = (1/T_{1W}) + D + ARe[1 + i\nu/\nu_c)^{\beta/2}] \tag{1}$$

$$= (1/T_{1W}) + D + \frac{A(1 + (\nu/\nu_c)^{\beta/2} \cos(\pi\beta/4))}{1 + 2(\nu/\nu_c)^{\beta/2} \cos(\pi\beta/4) + (\nu/\nu_c)^{\beta}}$$

Here $1/T_{1W}$ is the relaxation rate of the water nuclei in the protein-free buffer, ν the Larmor precession frequency of the nuclei at field H_o, and D, A, ν_c, and β are parameters that are determined by fitting Equation 2 to the data. "Re" stands for "the real part of".

For $\beta = 2$, Equation 1 reduces to a constant plus a Lorentzian dispersive term:

$$1/T_1 = (1/T_{1W}) + D + A/(1 + (\nu/\nu_c)^2) \ . \tag{2}$$

The dispersive part of the Cole-Cole expression, like the Lorentzian, drops to half its maximum value of A at $\nu = \nu_c$, the precession frequency at which the curve inflects. However, for $\beta < 2$, the Cole-Cole expression has a slower variation with ν than the Lorentzian. A simple two-site model of water exchange between bulk solvent and protein (now believed to be inadequate (10)) predicts a Lorentzian contribution to the relaxation, whereas the experimental data are known to vary more slowly with ν (11, 12). The Cole-Cole expression should be regarded as a heuristic equation that, while it represents the relaxation dispersion data very well, has no *a priori* validity. We use it here as a way of cataloging the NMRD data and characterizing them in terms of four parameters.

NMRD data for solutions of several proteins of differing molecular weights are shown in Figure 2. There are several points to note: with increasing molecular weight, the limiting low field value of $1/T_1$ increases, and ν_c decreases; for fixed molecular weight, comparing the immunoglobulin and alcohol dehydrogenase data, the more anisotropic protein behaves as though it were the heavier, and it has a somewhat broader dispersion profile. The solid lines through the data points are least-squares fit of Equation 1 to the respective data.

The numerical magnitude of ν_c (Figure 1), its variation with protein molecular weight and anisotropy (Figure 2), its dependence on temperature (7), and the qualitative aspects of its dependence on concentration (6, 13), all suggest that ν_c is related to the rotational relaxation rate of the protein molecules. This empirical view is further substantiated in Figure 3 where experimentally-derived values of ν_c for proteins with a range of molecular weight spanning three decades are compared with computations of the rota-

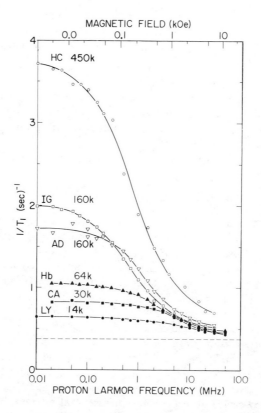

Figure 2.   *Dispersion of $1/T_1$ for 50 mg/mL solutions of proteins with a range of molecular weights, at 25°C.*

Abbreviations are: HC, hemocyanin from Helix pomatia; IG, nonspecific human γ-immunoglobulin; AD, alcohol dehydrogenase from yeast; Hb, human adult carbonmonoxy-hemoglobin; CA, human erythrocyte carbonic anhydrase B; and LY, hen egg-white lysozyme. The numbers next to the abbreviations give the molecular weights in units of thousands of Daltons. The relaxation rate of the protein-free solvent is indicated by the horizontal dashed line. After Ref. 7.

tional relaxation time τ_R using Stokes' Law for spherical particles, and assuming a 3.5 Å hydration layer (i.e., one monolayer) as part of the macromolecules. The relationship used is

$$\nu_c \simeq \sqrt{3}/(2\pi\tau_R) \; ; \; \tau_R = 4V\eta_o/kT \tag{3}$$

where V is the volume of a protein molecule. There are no adjustable parameters in this computation; the only subtleties are that η_o is taken as the viscosity of neat water, and (a technical point) the factor $\sqrt{3}$ (an excellent approximation to the exact value (6)) enters because magnetic relaxation involves the relaxation of a second order Legendre polynomial (a d-like term) whereas τ_R describes the orientational relaxation of a direction-vector (a p-like term).

There can be little question that, based solely on the phenomenology of NMRD data for solvent protons in protein solutions, the A term, Figure 1, is a contribution with a dispersion that is related to the rotational Brownian motion of the solute macromolecules. Further, as can be seen in Figure 3, two highly anisotropic proteins (indicated by open squares) behave as though they are "heavier" than their molecular weights suggest; and t-RNA, with a surface completely different from that of proteins, behaves as do the proteins. Moreover the proteins in Figure 3 have quite different isoelectric points; this together with the t-RNA result shown, as well as other data for a given protein over a wide range of pH values (6), indicates that the phenomena of NMRD depends little on the nature of the surface and the surface charge of the solute macromolecules (other than the influence the charge has on the Brownian motion of the protein molecules at high concentrations). Rather, it would appear to be due to a hydrodynamic interaction of the solute molecules with solvent, judging from the nonspecific nature of the proton NMRD data.

^{2}H and ^{17}O NMRD. From the most general considerations of the mechanisms of magnetic relaxation (14), one can conclude that the observed solvent proton relaxation must arise (in the main) either from magnetic dipolar interactions of a given proton with its neighbor on the same water molecule, or with protons of the (hydrated) solute protein (or both). In the first case, every water molecule (on average) would have to sense the rotational motion of the protein molecules, either via a (long range) hydrodynamic interaction or by spending a fraction of its time attached in some fashion to solute protein, and reorienting with it. If the interaction were strictly intramolecular, so that the NMRD data were due totally to an influence on the time-averaged kinetic history of the solvent molecules, then two things should be true: the NMRD spectra of ^{2}H in deuterated solvent (which is readily measured) and ^{17}O (which is very difficulty to measure) should, when normalized to the relaxation rate of the neat solvent, mimic the proton NMRD data; and the magnitude of A and D, Equation 1, should decrease for protons as they are diluted by deuterons, whereas there should be no dependence of the ^{2}H or ^{17}O NMRD on nuclear concentration (other than signal-to-noise ratio). This is because

Figure 3. Variation with molecular weight of the inflection frequency v_c (Equation 1) of the relaxation dispersion of solvent protons for solutions of macromolecules at 25°C.

Abbreviations are: LY, hen egg-white lysozyme CON A, demetallized concanavalin A; TP, demetallized porcine trypsin; tRNA, nonspecific yeast transfer ribonucleic acid; CA, human erythrocyte carbonic anhydrase B; Hb, human adult carbonmonoxyhemoglobin; AP, E. coli alkaline phosphatase; TF, demetallized human transferrin; IG, human nonspecific γ-immunoglobulin; AD, alcohol dehydrogenase from yeast; CP, human ceruloplasmin; HC 1/20, 1/10(L), 1/10(C), 1/2, 1/1, various states of association of Helix pomatia hemocyanin. Dashed line was calculated using Equation 3 with no adjustable parameters, using the viscosity of pure water to compute v_c. The proteins were assumed spherical, and a 3.5-Å hydration layer was included in computing the hydrodynamic radii. After Ref. 7.

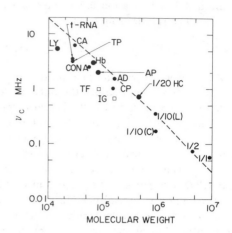

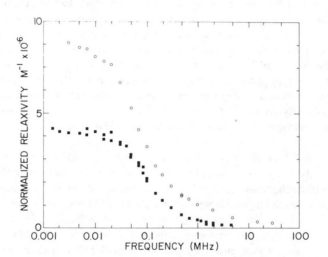

Figure 4. Protein contribution to the nuclear magnetic relaxation rates, T_1^{-1} of solvent (○) ^1H and (■) ^2H in aqueous solutions of hemocyanin (containing 11.5 and 30.6 mg protein per mL, respectively) at 25°C, as a function of the Larmor frequency of the respective nuclei.

Data are normalized both with respect to the deoxygenated neat solvent rate and to the protein concentration. The ^1H rates were measured to a 100% ^1H_2O solvent and the ^2H rates in a mixture of 80% ^2H_2O/20% ^1H_2O. Both mixtures contained 0.1M phosphate buffer at pH 7.0. Hemocyanin is known to associate into compact polymers of 9×10^6 Daltons in 100% H_2O solvent in the pH range 5–7 at ionic strength of 0.1. After Ref. 7.

both ^2H and ^{17}O relax by quadrupolar interactions with the gradient of the electric fields at these nuclei, rather than by interactions with other nuclei. On the other hand, any contribution to the NMRD of (solvent) protons that depended on their interactions with solute protein protons would not contribute measurably to ^2H or ^{17}O relaxation; the magnetic moments of these nuclei are too small for magnetic interactions with protons to be significant compared to the much larger intramolecular quadruple interactions responsible for their relaxation.

Proton and deuteron NMRD spectra for solutions of hemocyanin ($\underline{7}$) are compared in Figure 4, from which it can be judged, in terms of the foregoing, that only about half the proton relaxation is intramolecular; there is a comparable contribution that must be due to solvent-solute proton interactions. (Unpublished results for solutions of demetallized concanavalin A (54,000 Daltons) show similar behavior.) The conclusion regarding the importance of solvent-solute proton interactions is further confirmed by the variation of A, Equation 1, for protons as they are diluted with deuterons ($\underline{15}$); a particularly striking example is shown in Figure 5 where it is seen that (loosely speaking) as one interaction is progressively removed, the relaxation rate decreases slightly, and then increases substantially. Clearly the solvent protons must be interacting with solute protons. Reference to the original work ($\underline{15}$) will show that, by contrast, both ν_c for protons and the deuteron NMRD spectra remain unchanged in these experiments in partially deuterated solvents. In addition, it is shown there that the minima in A can be explained in terms of additivity of intermolecular and intramolecular rate processes, but not in terms of additivity of their separate contributions to A; the measured relaxivity is related to these interactions via a pair of coupled differential equations whose eigenvalues are the relaxation rates.

The major phenomenological aspects of NMRD of solvent nuclei in protein solutions have been presented above, from which we infer that from measurements of solvent nuclei it is possible to obtain information about the Brownian motion of solute protein. Though little has been concluded about the nature of the underlying interactions, it is clear that the time-averaged kinetic history of a typical solvent molecule must have a component that reflects the motion of the protein: either the solvent molecule spends some time on the protein (an exchange, short range model) or, via a (long range) hydrodynamic interaction, moves in a correlated fashion. In addition, it is found that there is an interaction between solvent and solute protons whose contribution to A is comparable in magnitude to the kinetic effects. This interaction can arise either by solvent exchange or by magnetic interactions as solvent diffuses near the protein surface.

Other Phenomena. Figure 6 shows proton NMRD data for two suspensions of red blood cells (erythrocytes), one from a normal adult, the other from a person with sickle-cell disease. The density of cells is sufficiently low so that the NMRD data arise from the protons of the water in which the cells

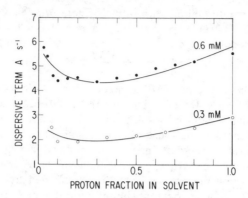

Figure 5. A, the magnitude of the dispersive term of Equation 1, for E. coli alkaline phosphatase at two concentrations: (●) 0.6mM; (○) 0.3mM, as a function of solvent proton fraction. Solid lines through the data points result from a least-squares comparison of the data with the model for cross-relaxation given in Ref. 15.

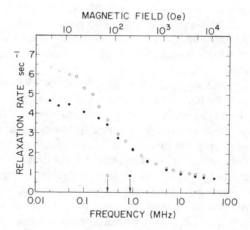

Figure 6. Proton relaxation dispersion data for representative suspensions of washed and oxygenated AA and SS erythrocytes at 35°C.

Cells are from fractions prepared by ultracentrifugation to restrict the range of mean corpuscular hemoglobin content (MCHC): (○) SS cells, MCHC = 5.8mM, pH 6.95; (●) AA cells, MCHC = 5.9mM, pH 7.07. The inflection frequency v_c of each dispersion curve is indicated by arrows along the horizontal axis. After Ref. 13.

are suspended; the external and not the intracellular water. Clearly, these two classes of water molecules must be in reasonably rapid exchange such that, to first order at least, the cell membrane may be regarded as a somewhat porous sac that contains the protein but allows rapid exchange of solvent. A natural question that arises is whether the hydrodynamics of the protein is influenced by containment in the cell, or whether a "porous sac" model is adequate to account for the NMRD spectra of suspensions of erythrocytes (12).

Figure 7 shows the variation of τ_R (related to ν_c in Equation 3) with hemoglobin concentration for a series of solutions of normal adult hemoglobin, and for three suspensions of normal erythrocytes with differing concentrations of hemoglobin (obtained by centrifugation), all liganded with either oxygen or carbon monoxide. It is seen that, for the present purposes, the erythrocytes may be regarded simply as containers of highly concentrated hemoglobin solutions. Figure 8 compares data for analogous samples of oxygenated sickle-hemoglobin solutions and sickle-cell suspensions with the data of Figure 7. It is seen that τ_R for sickle hemoglobin solutions is systematically greater than for normal hemoglobin: inorganic phosphate, and organic phosphate in lesser amounts, produce additional increases in τ_R. Confinement in cells increases τ_R still more. It had long been thought that the hydrodynamic properties of oxygenated sickle-hemoglobin (as well as most other properties) are identical to those of normal liganded hemoglobin. In particular, the (macroscopic) viscosity of oxygenated solutions of normal and sickle hemoglobin are identical at all concentrations (16), including conditions under which the τ_R values ("microscopic" viscosity), Figures 7, 8 differ.

Some Qualitative but Rigorous Deductions

We first consider the A contribution, Equation 1, and an explanation in terms of a two-site model: i.e., a model in which a water molecule exchanges between solution and sites (or class of sites) on or near a protein molecule such that at least one direction fixed in the water molecule is constrained to move rigidly with the protein molecule. In the simplest case, a water molecule attaches rigidly to the protein, moves with it for a while, and then leaves. In a somewhat more complex case, the attachment may be less rigid so that the water molecule is free to rotate about an axis fixed with respect to the protein. Additionally, a situation in which water molecules partially orient in the electric fields near the protein surface because of their electric dipole moments would also be a two-site model. Characteristic of a two site-model is that a time τ_M, or a distribution of such times, can be defined that measures the mean lifetime of a water molecule in the protein-associated state. Moreover, such a time is in principle a measurable quantity, and its value must satisfy two criteria: it must be at least comparable to if not longer than τ_R, otherwise the nuclei of the bound water molecules could not sense the rotational motion of the protein molecules; and it must be comparable to or shorter than the nuclear relaxation time of a bound water molecule, else it could not communi-

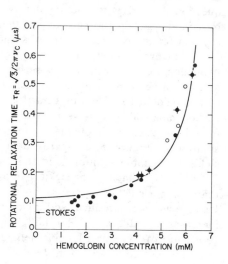

Figure 7. Variation with hemoglobin concentration of the rotational relaxation τ_R time for solutions of (●) liganded Hb A and (○) suspensions of oxygenated AA erythrocytes at 35°C, derived using Equation 3

Points with spokes are for oxy Hb A; the others are for carbonmonoxy Hb A. Data are for samples with differing values of pH and buffer composition, as indicated in Ref. 13. Solid line is from a least-squares comparison of the data with Equation 5, the model theory for microscopic viscosity developed in the text: $\tau_R/\tau_{R0} = (1 - (31/R - 31)^3)^{-1}$.

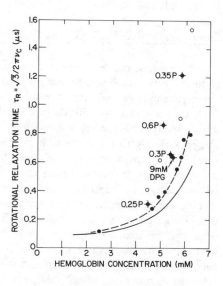

Figure 8. Variation with hemoglobin concentration of the rotational relaxation time τ_R for solutions of (●) oxy Hb S and (○) suspensions of oxygenated SS erythrocytes at 35°C, derived using Equation 3

Solid line represents analogous data for hemoglobin A from Figure 7. The points represent samples with differing values of pH and buffer composition, as indicated in Ref. 13. Dashed line indicates data points for solutions containing either zero or 0.1M phospate, and no 2,3-diphosphoglycerate (DPG); points representing solutions with greater phosphate content are indicated (P means potassium phosphate).

cate to the solvent what it sensed of the rotational motion. The lower bound for τ_M can be increased by using larger proteins and lower temperatures, whereas the upper bound can be decreased by looking first at ^1H, then ^2H, and finally ^{17}O relaxation rates since the latter is 10^3 faster than that of protons. These experiments have been performed (10). It was readily shown that there is no substantial distribution of τ_M values, since ^{17}O and ^2H relaxation rates kept pace with each other. Finally, by moving the two bounds on τ_M until they overlapped, τ_M was "bracketed" out of existence and the conclusion of the inapplicability of the concept of τ_M in the NMRD of protein solutions was reached (10). Moreover nothing in the temperature dependence of NMRD data to date suggests an activation energy that can be associated with τ_M (6, 7).

There is an additional argument against a two-site model, perhaps based more on esthetics than established fact, that was discussed a decade ago when the first NMRD data were published (6). This relates to the number of water molecules bound in the protein-associated sites; i.e., the fraction F of the water molecules in the bound sites at any one time or, equivalently, the fraction of time any one molecule spends in these sites. From the most general theoretical view of the A term (14), its value should equal $1/T_{1W}$ multiplied by the ratio $F\tau_R/\tau_{RW}$ (where $\tau_{RW} \simeq 5 \times 10^{-12}$ s is the τ_R of a free solvent molecule).

For a 10% (by weight) solution of a "typical" globular protein of, say, 50,000 Daltons (2 mM), one hydration layer corresponds to about 700 water molecules (or 25% additional mass), corresponding to F = 2%. From Figure 1 and Equation 3, $\tau_R \simeq 10^{-7}$ s, so that $F\tau_R/\tau_{RW} = 400$. This rough theory applied to protons, for which $1/T_{1W} = 0.3$ s^{-1}, gives A = 120 s^{-1} if all the hydration waters contribute, or a "relaxivity" R (where R is A ÷ protein molarity) of 60 (s mM)$^{-1}$. This is to be contrasted with an experimental value (cf. Figure 2) of about 0.5 (s mM)$^{-1}$, which is about two orders of magnitude less than the estimate based on a two-site model with irrotationally bound waters of hydration. Allowing for rotation, which could reduce the discrepancy by a factor of about five (6), the immediate inference is that only a few percent of the water molecules in the first hydration shell (about 7 molecules in the present case) contribute to the observed NMRD phenomena, assuming the validity of a two-site model. These molecules must all have roughly the same value of τ_M, and their number must be insensitive to pH. The remaining water molecules must exchange either much more slowly, and thereby not contribute to the NMRD at all, or exchange much more quickly so that they do not sense the rotational motion of the protein molecules.

Our view, based on our inability to find a τ_M value for these special water molecules, and on the unrealistic requirement that such special sites must be both few in number and universal in occurrence, independent of the nature of the surface and surface charge of the macromolecule, is that these specialized sites do not exist. Rather one must search for another mechanism to explain the A term of the NMRD phenomena in solutions of diamagnetic proteins.

Models and Mechanisms

The A Term. The mechanism underlying A is somewhat elusive. We have argued in some detail that any type of two site model is ruled out by the experimental results (10), and we still feel these arguments valid, suggestions to the contrary notwithstanding (17). Rather, it would appear that a hydrodynamic effect, simple to articulate but difficult (for us) to estimate, is responsible for A. We will call it the "slosh effect". Consider the rotational Brownian motion of a single protein molecule. In a time of the order of τ_R, the rotational correlation time, a fixed directrix in the molecule will have altered its direction by about one radian, meaning that the protein molecule for this time had an average angular velocity of τ_R^{-1} radians s^{-1} and an average angular momentum equal to this value multiplied by its moment of inertia. These are fluctuations of a vector quantity that, of course, must average to zero over the long term. They arise from interactions of the protein molecules with the water molecules because the system is at a non-zero temperature. These interactions all conserve angular momentum, and thus all the water (regarded as a continuum fluid) surrounding the protein molecule roughly midway to the neighboring protein molecules must have an equal and opposite fluctuation in its angular momentum. That is, as the orientation of the protein molecules fluctuates, the surrounding solvent must "slosh" about in an anticorrelated fashion to conserve the total angular momentum of the system. By virtue of solvent-solvent interactions, some of this sloshing motion will be converted from angular momentum of the continuum fluid to angular momentum of the individual solvent molecules (by the principle of equipartition). Thus the solvent molecules (all of them, as argued below) will have a small component of their thermal motion that follows the Brownian motion of the protein molecules.

It is this effect that we have suggested (7) is responsible for the observed NMRD spectra. It corresponds to an extremely small perturbation of the thermal motion of the solvent water molecules, and is therefore difficult to derive by the usual theoretical methods. However, it is in the nature of NMRD experiments that the effect of slower motions gets magnified (6); in the present case, the intrinsically small perturbation of the dynamics of the solvent molecules is increased by the ratio $\tau_R/\tau_{RW} \simeq 10^4$. (The area under the NMRD spectrum, proportional to $A\nu_c$, corresponds to the proportion of the thermal energy in that particular mode of motion; therefore, the less ν_c, the greater A for the same intrinsic effect, as is implicit in Figure 2.)

A remark is in order concerning the distance that the disturbance due to the fluctuating motion of a protein will propagate in solution. If a spherical protein molecule immersed in an infinite solvent is rotated at a steady rate, ultimately all the solvent will rotate with a velocity that falls off very slowly, essentially as the reciprocal of the distance d from the protein. If, however, the molecule is oscillated at a frequency ν_c, the range of the disturbance will be screened or shielded by the factor $\exp(-d/\lambda)$ because of viscosity, very much akin to the Debye screening of charges in solutions of electrolytes (18). The screening length $\lambda = (\pi \nu_c \rho / \eta)^{-1/2}$, where ρ is the solvent density, is 4000 Å for $\nu_c = 10^7 \text{ s}^{-1}$. For our typical protein solution, the protein radius r = 25 Å and the mean interprotein distance is 100 Å, much less than λ and d $\ll \lambda$. This is the argument for regarding all the solvent half-way to the neighboring protein molecules as being influenced by the motion of a single protein; the screening due to this "kinematic" viscosity is not important except for extremely dilute solutions.

It should be emphasized that, though we believe that we know the source of the A term, the causative mechanism has not been established unequivocally. Certainly, no quantitative theory is available (c.f. (7)). Therefore, the problem must be regarded as open. Nonetheless, the experimental procedures and the data themselves are extremely useful in investigations of water-protein and protein-protein interactions.

The D Term. We favor the view (6) that the water molecules in the hydration layer exchange very quickly, and are responsible for the D term. The magnitude of D is of order 10% of A, and the associated ν_c is 200-300 MHz compared to 2-3 MHz for the "typical" protein being considered (7). A very rough initial estimate for D would be that it is about equal to A if one assumes that the entire hydration layer contributes to the NMRD, but with a correlation time about two orders of magnitude or more shorter than that found for the A term. Thus the D term, the one for which we have the least information, is in our view attributable to all the water molecules in the hydration shell of the protein in solution with correlation times, using $\nu_c \simeq 250$ MHz, of about 10^{-9} s, or about two orders of magnitude slower than for solvent molecules ("sticky" water). These must be water molecules near surface polar groups, since their correlation times have been computed to be very long (19) compared to those for water molecules near uncharged residues, which are within factors of two of that of pure solvent (20). This consideration would reduce the estimate for D to a value close to that observed.

In a self-consistent way, we can argue that the slowing down of the water molecules in the hydration layer is due to the steric hindrance of the dynamics of the diffusion of solvent due to the presence of the charges on the protein surface itself. We can estimate the distance L that such a water molecule diffuses in a correlation time, using a diffusion constant 100 times slower than that of water: $L \simeq \sqrt{4D\tau_c} = (4 \times 5 \times 10^{-16})^{1/2} \simeq 5\text{Å}$. This distance corresponds to between one and two layers of water. Thus the correlation time is about the time a water molecule takes to diffuse out of range of the surface that is hindering its motion.

This view, that there are one or two layers of solvent water near the protein surface that behave very much like bulk water, but with a correlation time for many of its molecules greater by about 100 fold, is consistent with the results of measurements of dielectric dispersion of protein solution (21, 22). These data also show a continuing dispersion, in this case of the high frequency dielectric constant, in the region of several hundred MHz, indicating a slowing of the orientational relaxation time of a fraction of the solvent water. We associate this water with water molecules in the first (and perhaps second) hydration shells near polar groups, much the same water that appears in the X-ray data. The dynamics of this water is altered by the steric and hydrogen-bonding requirements imposed by the protein surface, those requirements that may be inferred from the X-ray results. The correlation time is essentially the exchange time, $\sim 10^{-9}$ s, and arises from the dynamics of diffusion of solvent near the surface. In particular, there is nothing to suggest that the exchange of water from these layers is any slower. If it were in the range $10^{-5} - 10^{-8}$ s, it would show in the A term, but it does not. It is hard to think of a type of bonding of water to the surface of a typical protein that would hold the water molecules to the surface for still longer times, and at the same time allow them to reorient freely. Moreover, it should be recalled that the orientational relaxation time of the interface water in frozen protein solutions is not much slower than this ($\sim 10^{-8}$ s) at $-35°$(2). Thus there is little margin for the existence of water molecules exchanging in times much slower than 10^{-9} s.

Cross Relaxation. Considerations of the data of Figure 5 have indicated that as solvent protons in protein solutions are progressively diluted with deuterons, the proton relaxation rate becomes determined more and more by the relaxation rate of the protein protons as transmitted through the mechanism that couples these two classes of protons at the proton-solvent interface. Though we have no clear view of the details of the coupling mechanism as yet, it can occur through exchange of magnetization without exchange of protons from a hydration shell. The relaxation of the protein protons can in principle be influenced by the presence of paramagnetic ions in the protein that, for example as in cyanomethemoglobin, may not influence directly the relaxation of solvent protons. Thus, though the question remains to be investigated, it does appear that NMRD of solvent protons can yield information about nuclear magnetic relaxation processes within protein molecules that contain buried paramagnetic ions, information that may be difficult to obtain by other techniques.

Another interesting implication of the cross-relaxation results is relevant to the question of compartmentalization of water within tissue. Because relaxation rates in the presence of cross-relaxation must be derived by solution of coupled differential equations, the observed relaxation rates are not simply the sum of the intramolecular and cross relaxation contributions to the relaxation. For a typical situation, the solvent relaxation may be influenced substantially by cross-relaxation, even though its time-dependence remains dominated by one eigenvalue, and so appears exponential. However, the relaxation of the protein protons (when observable), will almost invariably show two exponential contributions. Elimination of the cross term will eliminate one of these, alter the other and also alter the solvent proton relaxation rate; this of course can be accomplished by replacing most of the H_2O solvent by D_2O. The change in the relaxation behavior, from double to single exponential, could be readily misconstrued as the replacement of one of two classes of protons by deuterons. In tissue, the situation is more complex, but it is our belief (cf. 15) that much of the proton relaxation data in deuterated tissue samples should be reexamined in light of the recent cross-relaxation results; many published conclusions may be of limited validity.

Microscopic and Macroscopic Viscosity

The data in Figures 7 and 8 were among the first (12) to show a difference between the behavior of liganded normal adult and sickle hemoglobin; only a difference in the solubility limit had been demonstrated previously (23). As noted earlier, even the macroscopic viscosities as a function of hemoglobin concentration were found to be the same (16). In contrast, the NMRD results indicate that the miscrocopic viscosities are different, and indeed suggest that the sickle hemoglobin tetramers are aggregating into larger units of 2-4 tetramers. This inference has since been confirmed by electron spin resonance experiments on spin-labelled carbonmonoxy-sickle-hemoglobin (24). The question arises as to the fundamental differences between macroscopic and microscopic viscosity that make the former insensitive to aggregation and the latter depend on it.

We propose a single quantitative hydrodynamic model theory of microscopic viscosity that fits the data very well (the solid line, Figure 7) yet contains only two parameters. Once again, consider a spherical protein of radius r in solvent, this time enclosed by an outer concentric sphere of radius R as a boundary. It is known from classical hydrodynamics (18) that as R is reduced from ∞, the drag on the protein, if it is rotated, increases as though the viscosity η of the solvent increased from the free solvent value η_0 as

$$(\eta/\eta_0) = 1/(1-(r/R)^3) \ . \tag{4}$$

We now consider a protein solution, and regard the neighboring protein molecules around it as approximating a spherical cage of radius (D-r), where D is the average separation of the centers of the protein molecules. Reasoning from Equations 3,4 and the data of Figure 7, we assert that

$$\tau_c = \tau_0/(1-(r/D - r)^3) \tag{5}$$

should afford a reasonable description of the concentration dependence of the microscopic viscosity of nonaggregating spherical proteins. The solid line in

Figure 7 results from a least squares comparison of Equation 5 and the data. The values found for the two parameters are r = 31 Å and τ_0 = 64 ns. The actual radius of a hydrated (one monolayer) hemoglobin molecule (assumed spherical) is 30 Å. The value calculated for τ_0 using Stokes' Law is 37 ns.

It must be conceded that the concentration dependence given by this simple theory is very good, particularly at the higher concentrations. The excellent agreement of the numerical value of the derived and actual values of r indicate that the essence of the protein-protein interactions is contained in the theory; the protein-protein interaction is dominated by hydrodynamic effects, rather than electrostatic interactions, for example. The latter can fall off more slowly with distance and may be responsible for the deviations of the model fit at low concentrations of protein. Variations in ν_c with protein charge (i.e., pH) for fixed concentration have been observed by NMRD (12); the effective interactions are maximum near the isoelectric point. It is here that the fluctuations in interprotein separation at fixed protein concentration are greatest; protein molecules can approach more closely than when they are charged. Our simple model clearly does not include these second order effects.

Macroscopic viscosity, in contrast to microscopic viscosity, should not depend on the size of the protein molecules, but only on the volume fraction ϕ. This is well known in the limit of low concentration of protein where the Einstein relation for spherical particles (25) holds

$$(\eta - \eta_0)/\eta_0 = 2.5\,\phi \ . \tag{6}$$

But more generally, for all concentrations of spherical proteins, $(\eta - \eta_0)/\eta_0$ must only be a function of ϕ; the argument is straightforward, and can be said to rely on dimensional analysis or scaling theory. Classical hydrodynamics is a theory of continuum fluids in the strict mathematical sense; there is no intrinsic dimension that can be assigned to the fluid; there is no atomic scale. Therefore, the radius of a protein molecule, which must enter the theory of macroscopic viscosity in a comparative way, must be compared with the only other dimension in the problem: D, the intermolecular separation. But for fixed ϕ, the ratio (r/D) is independent of protein size. It doesn't matter whether a given amount of protein in a classical fluid is divided into many small molecules or a few large ones; the macroscopic viscosity will, indeed must, remain the same (so long as the protein molecules retain the same shape).

Thus, the aggregates of sickle hemoglobin tetramers must be reasonably spherical, on the basis of the measurements of macroscopic viscosity (16); and not too large, but nonetheless present, on the basis of the NMRD results.

Conclusions

The foregoing has been a presentation of a selected group of phenomena observable by NMRD experiments on solutions of diamagnetic proteins.

The point to emphasize is that the data reflect the influence that the presence of solute protein and suspended cells has on the averaged dynamic history of the solvent molecules. We infer from the data a view of hydration, solvent-protein interactions, and protein-protein interactions that is hydrodynamic on a scale comparable to the size of a protein molecule, and kinetic on atomic dimensions. Hydration, to the extent that it represents a specialized water layer at the protein surface, refers to water molecules that assume a geometry consistent with the hydrogen bonding possibilities to surface residues, but which exchange rapidly with the bulk solvent; any slowing of the motion of the solvent molecules is due to steric problems encountered when diffusing near the protein surface, particularly near polar groups. The time scales are of order 10^{-9} s which, though 100-fold slower than the time that characterizes the motions of solvent molecules in solution, is nonetheless fast compared to the corresponding relaxation times of the much larger protein molecules. We find no evidence of specialized binding sites with exchange times longer than 10^{-9} s.

Presently, the empirics is ahead of the theory; the model mechanisms that we propose we have only been able to quantitate in one instance, that of the dependence of protein reorientational relaxation time on protein concentration. Nonetheless, we have been able to clarify the distinction between macroscopic and microscopic viscosity; to measure protein-protein interactions within cells; and to demonstrate magnetization transfer from protein protons to solvent protons: all of this is consistent with the dynamics of water-protein interactions that we infer and have discussed.

Abstract

The presence of solute protein imposes on the rapid rotational Brownian motion of solvent water a small but readily measurable component that has the characteristics of the slower Brownian motion of the protein molecules. This phenomenon, known for somewhat over a decade, was first observed as an increase in the magnetic relaxation rate of solvent protons and has since been investigated in depth by studying the magnetic field-dependence of the relaxation of solvent protons and dueterons. The data are unexpectedly rich, and though the nature of the underlying solute-solvent interaction remains obscure, the results provide information on water-protein and protein-protein interactions in both solutions and cell suspensions. In addition to clarifying the concept of bound water, as is discussed, measurements extended to protein solutions in mixed H_2O/D_2O solvent indicate unexpected cross-relaxation interactions between solute and solvent protons. These last results imply that the interpretation of an increasing body of relaxation measurements of tissue samples needs to be reexamined and perhaps reinterpreted.

Literature Cited

1. Kuntz, Jr., I.D.; Kauzmann, W. Adv. in Protein Chem., 1974, 28, 239-345.
2. Kuntz, I.D.; Brassfield, T.S.; Law, G.D.; Purcell, G.V. Science, 1969, 163, 1329-1331.
3. Dubin, S.B.; Clark, N.A.; Benedek, J. J. Chem. Phys., 1971, 54, 5158-5164.
4. Agarwal, R.C. Acta Cryst., 1978, A34, 791-809.
5. Yonath, A. (private communication).
6. Koenig, S.H.; Schillinger, W.E. J. Biol. Chem., 1969, 244, 3283-3289.
7. Hallenga, K.; Koenig, S.H. Biochemistry, 1976, 15, 4255-4264.
8. Gupta, R.K.; Mildvan, A.S. J. Biol. Chem., 1975, 250, 246-253 (cf. footnote 3).
9. Cole, K.S.; Cole, R.H. J. Chem. Phys., 1941, 9, 341-351.
10. Koenig, S.H.; Hallenga, K.; Shporer, M. Proc. Nat. Acad. Sci. USA, 1975, 72, 2667-2671.
11. Fabry, M.E.; Koenig, S.H.; Schillinger, W.E. J. Biol. Chem., 1970, 245, 4256-4262.
12. Lindstrom, T.R.; Koenig, S.H. J. Magn. Reson., 1974, 15, 344-353.
13. Lindstrom, T.R.; Koenig, S.H.; Boussios, T.; Bertles, J.F. Biophysical J., 1976, 16, 679-689.
14. Abragam, A., "The Principles of Nuclear Magnetism"; Oxford University Press (Clarendon): London, 1961.
15. Koenig, S.H.; Bryant, R.G.; Hallenga, K; Jacob, G.S. Biochemistry, 1978, 17, 4348-4358.
16. Chien, S.; Usami, S.; Bertles, J.F. J. Clin. Invest., 1970, 49, 623-634.
17. Walmsley, R.H.; Shporer, M. J. Chem. Phys., 1978, 68, 2584-2590.
18. Lamb, H., "Hydrodynamics", 6th ed; Dover: New York, 1945.
19. Karplus, M. (private communication).
20. Karplus, M. (this Symposium).
21. Grant, E.H.; South, G.P. Advan. Mol. Relaxation Processes, 1972, 3, 355-367.
22. Pennock, B.E.; Schwan, H.P. J. Phys. Chem., 1969, 73, 2600-2610.
23. Benesch, R.; Benesch, R.E.; Yung, S. Biochem. Biophys. Res. Commun., 1973, 55, 261-265.
24. Johnson, M.E.; Danyluk, S.S. Biophysical J., 1978, 24, 517-524.
25. Einstein, A. Ann. d. Physik, 1906, 19, 289-306; 1911, 34, 591-595.

RECEIVED January 4, 1980.

Distribution of Water in Heterogeneous Food and Model Systems

P. J. LILLFORD, A. H. CLARK, and D. V. JONES

Unilever Central Resources, Colworth House, Sharnbrook, Bedford, England

In recent years, numerous examples of complex spin-spin relaxation have been reported in tissue. Most prevalent are studies on striated muscles from various sources (1). Belton et al. (2) proposed three fractions of water in frog muscle. Based on a graphical deconvolution of spin-spin experiments, T_2 values of 230 m.sec (15%), 40 m.sec (65%) and 10 m.sec (20%) were obtained. Derbyshire and Duff report biphasic spin-spin decays in porcine muscle with relative fractions dependent on the degree of rigor (3). Derbyshire and Woodhouse found complex relaxation in fish, porcine, frog and human muscle was confirmed by Hazelwood et al. (4) and Pintar et al. (5).

The explanation of these NMR phenomena are complicated by the existence of two theories for the behaviour of water in tissue. These are represented by the membrane theory (6) in which intracellular water is assumed to have the properties of ordinary liquid water, but that its solutes may be different from the extracellular solution due to the semi-permeable nature of the intact cell membrane. A later theory due to Ling (7) suggested that intracellular water may be extensively structured due to the high concentration of fixed charges on macromolecules.

The latter theory has been challenged by the measurement of diffusion of intracellular water; i.e. NMR methods have shown that the small difference in measured diffusion coefficients between intra and extracellular water can be accounted for by obstruction effects (8). Most authors now agree that the majority of intracellular water is "free" and account for the reduction in the tissue T_2 values by assuming that fast exchange occurs between a small "bound" fraction, and the larger "free" fraction. In the absence of structured water and slow exchange, the complex transverse relaxation is more difficult to explain. There is no doubt that the complex behaviour is within the water protons because of the size of resolvably different relaxation processes (2).

Several authors have discussed the origin of the multi-exponential transverse decay. Pearson et al (9) suggested that

0-8412-0559-0/80/47-127-177$05.00/0
© 1980 American Chemical Society

the onset of multiexponential decay during rigor was induced by
modification of exchange rates, or by the existence of two
physically separated water phases. The latter argument was
preferred. Belton et al (2) interpreted their three phases in
terms of protein bound, intracellular and extracellular water,
whereas Hazlewood et al, identifying a similar three fractions,
argued that the intracellular water was not exchange averaged (4).

Diegel and Pintar suggested that a separate distribution of
correlation times associated with exchange diffusion contributed
to the T_2 relaxation (5). Their analysis showed that this could
not be a complete explanation of the observed multiphasic
behaviour and that some compartmentalisation within the tissue
was necessary to explain the effects. They also argued that since
exchange within any one compartment was likely to be fast, the
assignment of a resolvable phase to the "bound" water fraction
was not necessarily valid, since each compartment contained
exchange averaged bound and free water.

Some difficulties in interpretation have been encountered
due to the existence of complex decay in transverse relaxation
but simple decay in longitudinal relaxation. More recent
experiments have shown that longitudinal relaxation processes are
also complex when accurate measurements are made (10).

All of the authors imply that separation of water phases
probably occurs at the cellular level. The semi-permeable nature
of the cell membrane towards ions and solutes which are capable
of relaxing water protons provides compartments in which
relaxation rates can be significantly different, even when water
transport across the membrane is very rapid. Indeed this property
of whole tissue has been used in the development of an NMR method
of determining water transport across erythrocyte membranes (11).

Complex relaxation is not however confined to biological
tissue. Following a freeze-thaw cycle, non-exponential decay is
observed for the water protons in an agarose gel (12); and is
also readily observed in meat models made from completely
synthetic man-made structures (13). In view of the absence of
membranes or any semi-permeable barriers in these wholly
fabricated materials, the general relevance of compartmentalisation to the observation of complex relaxation needs to be
re-examined.

Experimental Procedures and Results

A. Methods. T_2 measurements were made by the Carr-Purcell-
Meiboom-Gill method (14). τ spacings were: agarose 200 μsec;
soya fibres 300 μsec; raw and cooked/drained muscle 40 μsec; and
cooked muscle 100 μsec. 100 scans were accumulated in each case.

T_1 measurements were made using a 180° - τ - 90° sequence.
For each value of τ, 20 magnetisation values were accumulated.

All samples were thermostatted at 14°. Data were analysed
by multi-exponential fitting using a weighted least squares

analysis, or by deconvolution analysis (15).

B. Materials. Agarose was from Marine Colloids, Inc. (REX5468), and used without further purification. A sample of gel was prepared by pressure cooking a 4.8 wt% dispersion of agarose in distilled water for 10 min. Approximately 0.3 ml of the solution was placed in a 8 mm O.D. NMR tube and gelled by cooling. After relaxation measurements on the gel had been performed, the sample was frozen rapidly in liquid nitrogen, thawed and re-measured.

Soya protein fibres were supplied by Courtault Ltd. Their analysed composition is given in Table I. A small block of fibres (0.5 g) was cut from the fibre tow and placed in an NMR tube, with the fibres oriented parallel to the tube wall.

Table I. Analytical Composition of Courtauld Soya Fibres (Kesp)

Water %	Oil %	Protein $(N_2 \times 6.25)$ %	Ash %	NaCl %	pH
67.5	8.5	19.2	2.2	2.1	5.4

Muscle Tissue. A sample of muscle tissue was taken from the L-dorsi of a Lincolnshire Red Heifer post rigor. Approximately 0.5 g of tissue was placed in an NMR tube with the fibres oriented parallel to the tube wall, and the transverse proton relaxation measured immediately. The sample was then immersed in a water bath at $85°C$ for 5 min., cooled and remeasured. Finally, the excess liquor was drained from the sample, and the relaxation remeasured.

C. Results. Agarose:- In Figure 1, the effect of one freeze/thaw cycle on the spin-spin relaxation of agarose gel is shown. For the initial homogenous gel, the relaxation can be satisfactorily described by a single exponential process characterised by a T_2 value of 46 m.sec. After freeze/thaw, the decay was complex and required a number of discrete exponentials for its adequate description. The decay was fitted to 3 and 4 processes with the results given in Table II.

Soya protein fibres:- Non-exponential relaxation was detected in both T_1 and T_2 measurements (Figure 2). The T_2 decay was fitted to 3 and 4 processes with the results given in Table III.

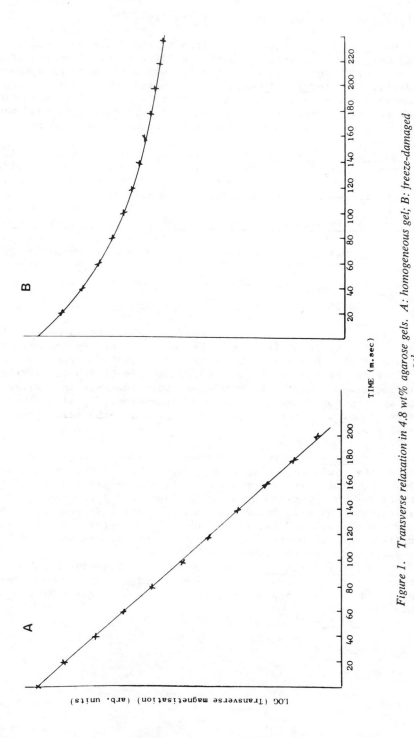

Figure 1. Transverse relaxation in 4.8 wt% agarose gels. A: homogeneous gel; B: freeze-damaged gel.

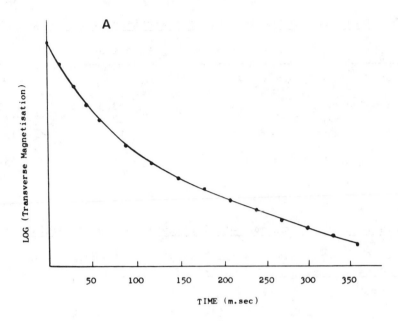

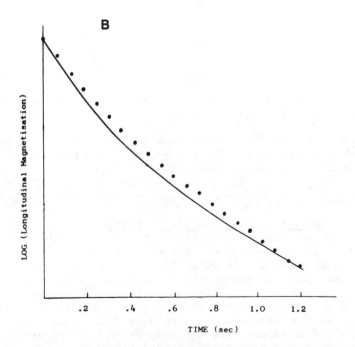

Figure 2. Water proton relaxation in soya fiber bundle. A: transverse relaxation; B: longitudinal relaxation. Curve B shows the experimental points in comparison with the theoretical curve calculated from the data in Table III.

Table II. Multiexponential Analysis of Agarose Gels

Gel State	No. of Processes	Amplitude %	T_2 (m.sec.)
Fresh	1	100	46
Frozen/Thawed	3	63.3	37
		22.5	157
		14.2	1140
Frozen/Thawed	4	11.5	18
		60.6	47
		16.0	240
		11.9	1290

Table III. Multiexponential Analysis of Courtauld Fibres

No. of Processes	Amplitude %	T_2 (m.sec.)	Protein Conc.%	T_1 (m.sec.)*
3	51	20		
	31	91		
	17	39		
4	38	16	50	100
	30	51	15	222
	24	209	3	500
	6	675	1	700

* T_1 data obtained from relaxation experiments on homogeneous gels of the appropriate concentration.

Muscle Tissue. Complex relaxation was detected in both the raw and cooked muscles. The transverse relaxation was fitted to both 3 and 4 processes (Table IV).

Discussion

A. Theories of Fast and Slow Exchange. In 1957, Zimmerman and Britten published a theoretical treatment of the relaxation of water protons absorbed on silica gel (16). In this system, both uni- and multiphasic relaxation decays were observed. The authors were able to account for the change in the number of observed phases by taking into account the relaxation times of the water protons in bound (absorbed) or free (non-absorbed) states, and the lifetimes of water molecules in each state. Two asymptotic expressions were derived, which have frequently been used in subsequent studies of water relaxation.

The first refers to fast exchange, i.e. when the reciprocal lifetime, or exchange rate of water in any phase is very large

Table IV.   Multiexponential Analysis of L. Dorsi (Heifer) Muscle

Muscle State	No. of Processes	Amplitude %	T_2 (m.sec.)
Raw	3	3.6	1.7
		90	46.7
		6.4	151
Cooked	3	48.4	22.6
		14.4	102
		37.3	1100
Cooked/Drained	3	4.6	1.2
		74.3	22.8
		21.1	71
Raw	4	2.6	0.23
		2.7	2.5
		88.9	47.2
		5.8	160.0
Cooked	4	32	4.5
		49.3	25.7
		11.0	129.0
		36.8	1120
Cooked/Drained	4	5.4	0.3
		11.8	10.3
		74.4	28.6
		8.3	119.7

compared to its relaxation time in that phase. The observed
relaxation is then uniphasic with relaxation rate given by the
weighted average of all the phases (Equation 1).

$$1/T_{av} = \sum_{i=1}^{n} P_i/T_i \qquad (1)$$

The second refers to slow exchange, when exchange rates are small
compared with relaxation rates. In this case, the relaxation is
multiphasic and the time dependence of the magnetisation becomes
the weighted average of effects taking place separately in each
phase (Equation 2).

$$\overline{M(t)} = \sum_{i=1}^{n} P_i \, e^{-t/T_i} \qquad (2)$$

In hydrated biological systems, protons are present in both
the aqueous solvent and in the dissolved or suspended substrates
(protein, fats, sugars, etc.). Clearly, the C-H protons of the
substrates represent a separate phase and since exchange is
extremely slow between CH protons and water, Equation 2 obtains
and multiphasic relaxation is to be expected. Fortunately, the
spin-spin relaxation times of substrate CH protons ($\sim$10 usec)
tend to be very short compared to water ($\sim$50 m.sec) so that
discrimination is easy. However, in many tissue systems, the
water proton relaxation itself exhibits multiphasic behaviour.
    Before attempts are made to explain the origin of the effects
reported here, other examples of complex relaxation must be
considered, to determine whether a generalised approach to water
proton relaxation in heterogeneous systems can be developed.

B. Experiments with Model Systems.

    1. Agarose. In previous work (17) it has been shown
that the observed relaxation in homogeneous gels is described by
the fast exchange approximation.

$$1/T_{av} = P_w/T_{2w} + P_b/T_{2b} \qquad (3)$$

Where $P_w + P_b$ represent the fractions of free and bound water
respectively, and T_{2b} is the spin-spin relaxation time of the
bound phase.
    At the temperature at which this sample was measured, the
rate equation is more accurately described by Equation 4.

$$1/_{T_{av}} = P_w/_{T_{2w}} + \sum_{i=1}^{n} P_{bi}/_{T_{2bi}} + P_c/(T_{2c} + \tau) \quad (4)$$

where the b phase represent motionally modified water molecules having a distribution of T_{2b} values, and the c phase represents a second, very motionally restricted set of protons probably the hydroxyl protons of the agarose molecule. These have an intermediate exchange rate characterised by τ, the exchange lifetime. Even when exchange between the c phase and other components is slow, a multiphasic decay is not normally observed, since the intrinsic relaxation time of the hydroxyl proton is in the usec range, comparable with that of the CH protons of the agarose skeleton (17). The observation of complex relaxation after freeze-thaw damage (Figure 1B) cannot therefore arise from the presence of non-exchangeable (CH) protons, or to a modification of exchange rate of agarose hydroxyls but is a property of the water protons within the structurally modified agarose network.

2. Spun protein fibres. The observed spin-spin relaxation decay for the fibre sample is shown in Figure 2A. Marked non-exponential behaviour is apparent (Table III). In experiments elsewhere on homogenous soya gels (18), simple exponential decays were observed. Relaxation of the non-exchangeable protein protons is not normally detected, being an order of magnitude faster, with relaxation times $\sim$ 25 μsec. Again the source of the non-exponential behaviour in fibres must be associated with the location of water protons within the fibre matrix.

3. Cooked meat. Spin-spin relaxation in raw and cooked meat also exhibited multi-exponential behaviour (Table IV). The degree of deviation from a single relaxation was more marked in the cooked sample than that reported in pre- or post-rigor muscle (2,3,4,5).

4. Summary. Gross deviation from single exponential relaxation was encountered in the T_2 experiments described here, yet only the raw muscle samples contain intact membrane systems. Neither is there any reason to believe that the agarose or protein gel systems contain barriers that are semi-permeable to either water or small solutes. The observed complex decays cannot therefore originate in compartmentalisation. The phenomenon was observed with carbohydrate and protein substrates; in man-made or natural structures; and with water contents from 95% to 70%. The only feature common to all the samples is that their structural arrangement involves a significant heterogeneity in distribution of mass. No existing theory is adequate to explain the general nature of this complex relaxation behaviour

and some new explanation must be sought.

C. Diffusion Distance Limited Exchange.

In deriving the expression for fast exchange, Zimmerman and Britton (16) assumed that every water molecule had the same probability of exchange between free and bound sites so that a single valued average T_1 or T_2 value is measured. Such a probability of exchange can only be averaged if every water molecule experienced the same environment. In samples with gross heterogeneity in distribution of the bound sites, which is maintained over the timescale of an NMR experiment, this cannot be fulfilled, and the observed magnetisation decay must then be described by the sum of all the relaxation decays for each environment, which in the limit corresponds to each water molecule (Equation 5).

$$\overline{M(t)} = a \, \exp\left[-t(P_w/T_w+P_b/T_b)\right] + b \, \exp\left[-t(P_w{}'/T_w+P_b{}'/T_b)\right] + \text{etc.}$$

$$= \sum_{i=1}^{n} N_i \, \exp\left[-t(P_{wi}/T_w + P_{bi}/T_{bi})\right] \tag{5}$$

Equation 5 is in exactly the same form as the slow exchange equation of Zimmerman and Britton so that complex relaxation decay is to be expected. The difference is that each phase has a weight averaged fast exchange relaxation time determined by the probability (P_{bi}) of the water molecule exchanging with a bound site with relaxation time T_{bi}. In the general case, where each water molecule has a probability of exchange with a number of different types of bound sites, the terms should be expanded to include this distribution, and the magnetisation decay is described by Equation 6.

$$\overline{M(t)} = \sum_{i=1}^{n} N_i \, \exp\left[-t(P_{wi}/T_w + \sum_{j=1}^{n} P_{bij}/T_{bij})\right] \tag{6}$$

The significance of Equation 6 in terms of the structures or structural changes giving rise to changes in the complex magnetisation decay can now be examined.

D. Conditions for the Observation of Complex Relaxation

To extract the significance of Equation 6 in these terms, it is easiest to consider its value in a limited form which may frequently be encountered,

i.e. $P_{bij} = P_{bi}$, $P_{wi} = P_w$, $T_{bij} = T_b < T_{2w}$ which implies that

water in the bound sites had equivalent relaxation times, much
shorter than that of free water. The probability of a water
molecule exchanging with any bound site is then determined by the
concentration of bound sites within the "diffusion distance" of
each water molecule.

In other words, <u>unless the concentration of bound sites is</u>
<u>the same within the volume of space sampled by any one water</u>
<u>molecule in its own intrinsic relaxation time, then, on average,</u>
<u>non-exponential decay must be observed.</u>

In its simplified form, Equation 6 becomes:

$$\overline{M(t)} = \sum_{i=1}^{n} N_i \exp\left((-t/T_b)P_{bi}\right) \tag{7}$$

If the decay curve is resolved into discrete processes, then
the relaxation rates

$$1/T_i \simeq P_{bi}/T_b$$

and since T_b is assumed to be constant for the system, the

$1/T_i = k$ (probability of collision) $= K$ (substrate density),
alternatively $T_i \propto$ (density)$^{-1} \propto$ pore size.

The amplitude of the ith process (N_i) represents the fraction
of water molecules in a domain of equivalent solute density or
equivalent pore size. A plot of N_i versus $1/T_i$ is then an
approximation to a spacial density distribution function for the
heterogeneous sample.

It is important to recognise that this theory does not
require that any complex decay should be necessarily described by
a few discrete relaxation processes. A distribution of times is
possible and may well be more appropriate.

E.   The Spacial Resolution of the Distribution Function

It is clear from the above, that more than one value of the
relaxation time will be obtained only if the concentration of the
protein or polysaccharide solute varies within dimensions of the
order of the "diffusion distances" for the water molecule. Since
the water molecule would relax with a time constant ~ 2 seconds
even in the absence of any solute, then the diffusion distance
relevant to these measurements is given by $\sim \sqrt{6Dt}$. Since D for
free water is $\sim 10^{-5}$ cm.$^{-2}$ sec.$^{-1}$ and t 2 sec., then the diffusion
distance ~ 100 μ. Since with careful experimentation, a deviation
of 10% from simple exponential relaxation is detectable, materials
which contain pores of ~ 10 μ or larger or heterogeneous density
over ~ 10 μ, distances can give rise to multi-component relaxation.

A similar analysis of complex relaxation has been indicated previously by Packer in studies of water in elastin (19). However, his suggestion that the measured intrafibre relaxation time (rather than the intrinsic relaxation time of free water) determines the space sampled by any water molecule appears to be unnecessary.

From the above discussion, it is clear that measurement of the change of the complex water proton decays can be related directly to structural changes within samples, provided the conditions of Equation 6 are met. Unfortunately, these conditions are rather strict and will rarely be completely fulfilled. Caution must be applied in interpreting the data with respect to the following complications.

(i)   The terms embodied in $\sum_{i=1}^{m} P_{bij}/T_{bij}$ in Equation 6 will not be comparable if different substrates are examined.

(ii)   These terms need not remain comparable within a single substrate if a structural change involves major changes in the conformation of the macromolecules, or shifts in pH and ionic strength, all of which change both P_b and T_b. Fortunately, the latter effects can usually be anticipated or allowed for by measuring the substrate system in a homogeneous state, when P_b/T_b can be measured directly.

(iii) If small molecules dissolve into the free water, then the relevant diffusion distance will be decreased by decreasing $\sqrt{6Dt}$. Again some correction for this effect can usually be applied by removing some of the interstitial fluid and obtaining an independent measurement of its relaxation time.

(iv)   It is assumed that all of the relaxation processes with apparent time constants of   10 m.sec. to 2 sec. derive from water protons. In the case of samples containing large amounts of liquid fat, dissolved oligosaccharides or amino acids, this may not be the case, since C-H proton relaxation rates fall within this range. Some other measurement (e.g. integration of the high resolution spectrum) may be necessary to correct for this potentially major source of error.

F.   Comparison of Theory with Results

Bearing in mind the constraints referred to above, it is now possible to check the validity of the theory with respect to the model systems investigated here.

1. Agarose. The action of freezing and thawing an

agarose gel causes a complex relaxation decay to be observed in the transverse relaxation. According to the theory developed above, we assign this change in the relaxation behaviour to the production of heterogeneous agarose distribution in the freeze damaged sample. The discrete, resolved relaxation times can therefore be interpreted in terms of domains within the gel having their fraction characterised by the fractional amplitudes of the relaxation processes, and their agarose concentration characterised by the separable relaxation times.

By assuming a linear dependence of $1/T_2$ vs agarose concentration (Figure 3), the apparent mass of agarose in the freeze damaged sample can be calculated. The results are expressed in Table V.

Table V. Mass Balance for Agarose Freeze/Thaw Experiment

	Fresh	Frozen/Thawed			Frozen/Thawed			
No. of processes	1	3			4			
T_2 (m.sec.)	46	37	157	1000	18	47	240	1290
Apparent concn.(%)	4.8	6	1.3	0.1	12.5	4.8	0.8	0.1
Amplitude (%)	100	63.3	22.5	14.2	1.11	60	16	11.9
Agarose wt.	4.8	3.78	0.293	0.014	1.38	2.88	0.12	0.14
Total agarose wt. at g/100 g sample	4.8							

It can be seen from Table V that the apparent mass of agarose present in the freeze damaged sample increased with the number of exponentials used in describing its relaxation. Since agarose mass was conserved in the freezing experiment, the higher order fitting appears to be more appropriate, which is supported by the improved agreement between calculated and observed NMR data when 4 processes are used. Increasing the number of processes beyond 4 is not warranted with respect to the agreement between calculated and observed relaxation, but it is quite possible that 4 processes represent only an approximation to the true distribution of relaxation processes in the freeze damaged sample.

2. Protein fibres. The transverse relaxation of the protein fibre samples were fitted to both 3 and 4 exponential processes (Table III). Again, the agreement between experimental and calculated data was better for the 4 process fit.

The theory developed above for the description of complex relaxation suggests that both transverse and longitudinal relaxation should be affected by the restrictions of diffusional averaging. Secondly, provided the terms embodied in $\sum_{j=1}^{m} P_{bij}/T_{bij}$

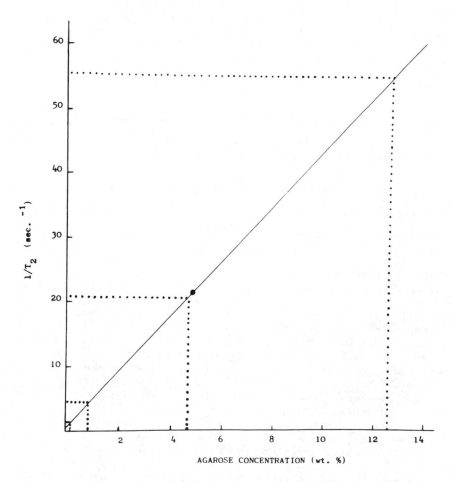

*Figure 3. Agarose concentration dependence of water proton transverse relaxa-
tion: (· · · ·) frozen/thawed gel (4-process fit).*

are known for the substrate, the form of the longitudinal
relaxation should be calculable from the transverse relaxation
and vice versa. In experiments performed elsewhere (18), the
transverse and longitudinal decays of homogeneous soya protein
gels have been measured as a function of protein concentration,
pH, ionic strength, etc. Using this transverse relaxation data,
the apparent concentration of the domains have been calculated
(Table III). From this information, and the concentration
dependence of the longitudinal relaxation times of the gels, the
expected longitudinal decay for the fibre sample was calculated,
and compared with experiments (Figure 2). Considering the
approximations made in the mathematical transformations, the
agreement between simulated and observed decays is very good;
suggesting that Equation 6 describes the dominant mechanism of
complex decay.

3. Muscle tissue.

a. Raw, post rigor muscle: Analysis by both 3 and
4 process fitting indicates that the relaxation is dominated
($\sim$ 90%) by a relaxation with $T_2 = 47$ m.sec (Table IV). Also,
approximately 6% of the decay is characterised by a process with
$T_2 \sim 150$ m.sec. The difference in the parameters derived from the
two fitting regimes lies in their analysis of the rapid initial
decay, which in a 3 process treatment is described by a single
exponential with $T_2 = 1.7$ m.sec, but in a 4 process treatment is
described by two discrete relaxations.

Examination of the difference between experiment and fitted
parameters indicated that the 4 process fit was a better representation of the data.

By analogy with previously reported results (2,4) the 6%
water relaxing with $T_2 \sim 150$ m.sec can be assigned to an extracellular component, the remaining 94% being intracellular.
Previous workers have suggested that the observation of discrete
relaxations for intra and extracellular space is a result of the
presence of the cell membrane which either restricts the rate of
exchange of extracellular water with intracellular bound sites,
or provide compartments in which the concentration of fast
relaxing sites is different. In the light of the results on model
systems reported above, the extent to which viable membranes
cause multiphasic relaxation must be reconsidered, since the
presence of extracellular spaces of ~ 50 μ dimension may itself
cause a significant degree of complex relaxation, whether or not
membranes are present or intact.

The intracellular water is represented by two or three
relaxation processes. There is no reason to suppose that the
faster components represent "bound" water, firstly because their
magnitude is too small to be equated with the 0.5 to 1 g/g protein
of water normally associated with protein hydration, and secondly,
because exchange between free and bound water in protein systems

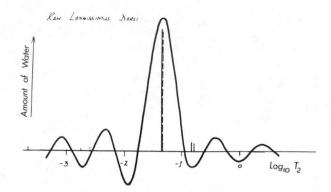

Figure 4. *Distribution of water proton relaxation times: (——) 3-process fit; (- - -)*
4-process fit.

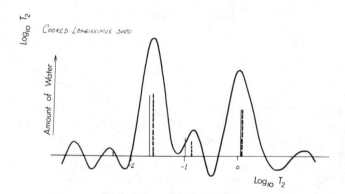

Figure 5. *Distribution of water proton relaxation times: (——) 3-process fit; (- - -)*
4-process fit.

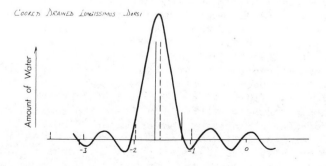

Figure 6. *Distribution of water proton relaxation times : (——) 3-process fit; (- - -)*
4-process fit.

has previously been shown to be fast.

b. Cooked, post rigor muscle: The action of heat produced a dramatic shift in the relaxation time spectrum (Table IV).   Firstly, ∼30% of the decay exhibited a T_2 of ∼1 sec. This component was easily identified, being visibly separate from the tissue mass, and removable by decanting.   The water remaining within the bulk of the tissue still exhibited complex relaxation which was dominated by a process with $T_2 = 25$ m.sec.   This decrease in relaxation time on cooking results from both the increase in concentration of protein, due to the tissue shrinkage, and the thermal denaturation of myofibrillar and connective tissue proteins.   The relaxation is, however, still significantly complex, despite the fact that cellular and intra cellular membranes have been destroyed.   This observation (together with previous microscopic studies of cooked meat, which clearly indicate the presence of spacings from 10 to 250 μ) suggests that diffusional distance restrictions to exchange, embodied in Equation 6, account for the complex relaxation in these cooked meat samples.

G.   Deconvolution of Complex Relaxation

It has been recognised above that the description of the relaxation in terms of a distribution of relaxation times may be more appropriate than resolution into a few discrete relaxation times.   A relaxation time distribution function can be derived directly from the decay curve, and mathematical treatments of the problem have been outlined previously, though with reference to stress relaxation or creep retardation rather than any electromagnetic decay process (20).   For comparison with the results outlined above, the data for the raw and cooked muscle samples were deconvoluted following the method of Roessler (20).   In Figure 4, 5 and 6, the results are shown, together with the results of multiexponential analysis.   The significance of the 3 and 4 process fitting results are more easily assessed when compared with the distribution analysis.   Clearly, the distribution is a better representation of the information content of the relaxation experiments.

Summary

In this paper, it has been demonstrated experimentally that complex water proton relaxation can occur in a wide variety of biopolymer systems.   The detection of complex relaxation cannot therefore be used as a criterion for the existence of separate domains into which water or small solutes are compartmentalised by permeable or semi-permeable membranes.   A theory has been developed which, to a first approximation, can explain the experimental observations and requires only that a heterogeneous distribution of the substrate be maintained over the timescale of

water proton relaxation (i.e. 2 seconds). The theory also
predicts that the water relaxation is sensitive to heterogeneous
mass distribution down to the dimensions of $\sim 10 \mu$; and that the
resolution of complex relaxation into a few discrete processes
represents a special case of what should better be considered a
continuous distribution of microheterogeneities.

This treatment represents a new approach towards the kind of
information that can be obtained from water proton relaxation.
In the past, the technique has been used to study water of
hydration, i.e. the small amount of water influenced in its motion
by close association with the substrate. Normally, relaxation in
such systems is, or can be arranged to be, described by a single
relaxation time.

This study of systems in which this is not the case has led
to the conclusion that while hydration water is important, it
serves as a means by which a water molecule 'recognises' the
presence of a surface. The significance of the relaxation times
becomes their ability to describe the distribution of distances
from water molecules to surfaces, or the local substrate density
in a microenvironment.

Literature Cited

1. James, T.L., Ed. "N.M.R. in Biochemistry", Academic Press,
    New York 1975.

2. Belton, P.S.; Jackson, R.R.; Packer, K.J. Biochim. Biophys.
    Acta, 1972, 286, 16.

3. Duff, I.D , Ph.D. Thesis, University of Nottingham, 1973.

4. Hazlewood, C.F.; Chang, D.C.; Nichols, B.L.; Woessner, D.E.
    Biophys. J., 1974, 14, 583.

5. Diegel, J.C.; Pintar, M.M. Biophys. J. 1975, 15, 855.

6. Toesteson, D.C. in "The Cell Function of Membrane Transplant",
    Prentice Hall, Englewood Cliffs, N.J. 1964.

7. Ling, G.N. "A Physical Theory of the Living State", Ginn
    (Blaisdell), Boston, Mass. 1962.

8. Wang, J.H. J.A.C.S., 1954, 76, 4755.

9. Pearson, R.T.; Ducc, I.D.; Derbyshire, W.; Blanshard, J.M.V.
    Biochim. Biophys. Acta, 1974, 362, 188.

10. Chang, D.C.; Hazlewood, C.F.; Woessner, D.E., Biochim.
    Biophys. Acta, 1976, 437, 253.

11.   Conlon, T.; Outred, R.   Biochim. Biophys. Acta, 1972, 288,
      354.

12.   Ablett, S.; Lillford, P.J.; Baghdadi, S.M.A.; Derbyshire, W.
      Magnetic Resonance in Colloid and Interface Science, A C.S.
      Symposium Series 34, 1976, 29, 344.

13.   Lillford, P.J , in Plant Proteins, G.Norton., Ed. Easter
      School in Agricultural Science, University of Nottingham,
      Butterworth and Co., 1976, p.289.

14.   Meiboom, S.; Gill, D., Rev. Sci. Instrum. 1958, 29, 688.

15.   Clarke, A.H.; Lillford, P.J., in preparation.

16.   Zimmerman, J.R.; Brittin, W.E.   J. Phys. Chem., 1957, 61,1328

17.   Ablett, S.; Lillford, P.J.; Baghdadi, S.M.A.; Derbyshire, W.
      Journal of Colloid and Interface Science, 1978, 57, 2, 355.

18.   Mahdi, A., Ph.D. Thesis, University of Nottingham, 1979.

19.   Ellis, G.E.; Packer, K.J.   Biopolymers, 1976, 15, 813.

20.   Roessler, F C., Proc. Phys. Soc. B., 1955, 68, 89.

RECEIVED January 4, 1980.

PROTEINS:
ORDERED WATER

Modeling Water–Protein Interactions in a Protein Crystal

JAN HERMANS and MICHELLE VACATELLO[1]

Department of Biochemistry, School of Medicine, University of North Carolina, Chapel Hill, NC 27514

Present knowledge of the details of the conformation of proteins is based almost exclusively on results of studies of protein crystals by x-ray diffraction. Protein crystals contain anywhere from 20 to 80% solvent (1) (dilute buffer, often containing a high molarity of salt or organic precipitant). While some solvent molecules can be discerned as discrete maxima of the electron density distribution calculated from the x-ray results, the majority of the solvent molecules cannot be located in this manner; most of the solvent appears to be very mobile and to have a fluctuating structure perhaps similar to that of liquid water. Many additional distinct locations near which a solvent molecule is present during much of the time have been identified in the course of crystallographic refinement of several small proteins (2,3,4,5, 6), but in all cases the description of solvent structure in the crystal is incomplete probably because only a statistical description is inherently appropriate.

One hopes eventually not only to be able to describe the structure of water near the surface of a protein, but also to understand this structure in terms of energies of protein-water and water-water interactions. The energy of a given configuration of protein and solvent molecules can be estimated in a straightforward manner. However, to obtain a representative sample of the statistical ensemble of possible configurations of water and protein requires calculation by Monte Carlo (7,8,9) or molecular dynamics methods (10,11,12,13). Because of the magnitude of the computation required to calculate an adequate sample by either one of these methods, we have developed a simpler, approximate calculation whose results provide insight into the ordering forces exerted by the protein in solvent space. Results obtained with this method are discussed first; discussion of calculation and properties of a statistical sample of protein-water configurations is presented in the second half of this paper.

[1] Current address: Instituto Chimico, University of Naples, Italy

0-8412-0559-0/80/47-127-199$05.00/0
© 1980 American Chemical Society

Energy map of solvent space near protein molecules.

We have made the following approximate calculation to estimate protein-water interactions by a less cumbersome procedure: it
is assumed that the protein molecule has a unique fixed structure
determined by x-ray crystallography and interactions are calculated between the protein and a single water molecule in the absence
of other solvent molecules. Using this simple system, one may
consider all positions and orientations of the single water molecule relative to the protein in a step-wise manner. We present
here the result of this calculation for the crystal of bovine
pancreatic trypsin inhibitor (BPTI). The calculated energy,
mapped in three dimensions, is a highly informative description of
the crystal's solvent space.

Calculations were performed for the asymmetric unit of the
crystal, which has a volume near 12,000 Å^3 and contains one protein molecule (MW=6700); one estimates that one third of the
crystal volume is filled with solvent.

All protein atoms except hydrogens bonded to carbon were
explicitly represented. Positions of polar hydrogens were calculated on the basis of coordinates of the heavier atoms (5) and
standard geometric constraints (14). The non-bonded energy for
the interaction of the water molecule and the protein was calculated as a sum of electrostatic and 6-12 Lennard-Jones attractive
and repulsive contributions for the three atoms of the water
molecule and all atoms of the protein within 6 Å of the water
oxygen's center. Empirical parameters were used for the 6-12
potentials (15,16,17), partial atomic charges were values obtained
with molecular orbital calculations (18); for hydrogen-bonded
interactions modified 6-12 parameters were used (18).

The water molecule's oxygen atom was successively centered at
all points of a simple rectangular grid with a mesh of approximately 1 Å^3. At each location a set of 276 different orientations
of the molecule was considered, obtained by rigid-body rotations
in steps of approximately 30°; the energy was subsequently minimized by smaller rotations with as starting point the coarse
orientation that had the lowest energy. The final minimum energy
for each grid point was retained as the parameter characteristic
of protein-water interaction. The entire calculation required
about 1 hour of c.p.u. time on an IBM 370 model 165. (The minimi-
zation step does not greatly alter the energy and could be omitted
for a considerable reduction of computation time.)

The calculated energy is most informative when represented as
a contour map in three dimensions together with the location of
crystallographically-determined water molecules. The positions of
47 water molecules identified in the crystal of BPTI (5) have been
marked in the map shown in Figure 1.

The first set of contours must be drawn at zero energy (dotted contour in Figure 1). If a lower energy is chosen for the
first contour, then any parts of the volume that are not within

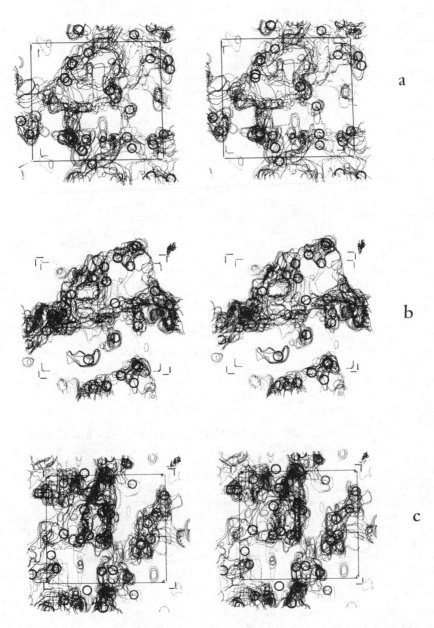

a

b

c

Figure 1.   Energy map of solvent space near protein molecules

the 6 Å cutoff distance from at least one atom of the protein are erroneously considered to be inaccessible to solvent. However, not all volume within the zero-energy contour is accessible to solvent. Non-bonded interactions of each water molecule in the crystal must be quite favorable if the water molecule is to prefer the crystal environment to that of the aqueous solution in equilibrium with the crystal. Therefore, a location for which the protein-water interactive energy is near zero can be occupied by a water molecule only if several favorable water-water interactions are possible for a water molecule at this location.

Isolated low-energy volumes large enough to accommodate a single water molecule are not expected actually to contain a water molecule, unless the protein-water energy is low, probably below -15 kcal/mole. There are several such low-energy pockets in the BPTI crystal. The energy of all but one of these pockets is greater than -15 kcal/mole, and these pockets are, therefore, not expected to contain water molecules. The energy of one pocket is circa -30 kcal/mole. (Figure 1b, below center). A water molecule placed in this pocket can form four hydrogen bonds with polar groups of a single BPTI molecule that entirely envelops this location and the x-ray crystallographic studies accurately place a water molecule in the pocket (5). The absence of crystallographic water molecules in the pockets with higher energy may not be entirely conclusive, since only one third of the crystal's solvent has been located as a result of the x-ray work. However, the small size of each of these pockets would strongly constrain the location of a water molecule contained in it and the electron density map would consequently show a sharp maximum, readily recognizable in a careful crystallographic analysis. It is, therefore, unlikely that any of these higher-energy pockets contains a water molecule.

The remainder of the volume enclosed by the zero-energy contour is a single highly interconnected network of channels and spaces. Within this volume, the contour at -5 kcal/mole encloses a single, less highly interconnected, network of channels. In two places the contours define a tube-shaped channel closed at one end; in both the energy goes well below -15 kcal/mole. (Figure 1b, lower right and slightly right of center.) The shorter of the two channels contains one crystallographic water molecule very near a local minimum of the energy. The longer channel contains a hydrogen-bonded chain of three crystallographic water molecules. As the one water molecule in the isolated low-energy volume, these three molecules are surrounded by a single BPTI molecule (5).

The next-most constraining part of the volume is a low-energy channel with a cross section that is narrow in one direction and somewhat wider in the other. This channel is filled with five crystallographic water molecules, so arranged that they can form five hydrogen bonds with each other. (Figure 1b and c, right of center.)

Besides these narrow spaces, the volume within the zero-energy contour surface contains wider spaces of various size connected by short and not very narrow constrictions. There are four such large spaces in the asymmetric unit. The largest of these is directly connected to two symmetry-related copies of itself and to one or two copies of each of the three smaller spaces. (Top half of Figures 1a and b). The crystallographic work places 4, 6, 9 and 17 water molecules, respectively, in these spaces.

This accounts for all 47 water molecules located by x-ray crystallography. The crystal contains an estimated 140 water molecules per asymmetric unit; thus the locations of 90 additional water molecules are uncertain. The map of Figure 1 suggests where these water molecules are to be found: more than half of them in the large cavity and the remainder in the three smaller cavities. Only very few of the missing water molecules are expected in parts of the narrow channels where no water molecule has been located crystallographically.

Overall features of map and crystallographically determined water structure may be summarized as follows. The distribution of lowest energies of protein-water interaction, within approximately 1 Å of the crystallographic water locations, is centered between -15 and -10 kcal/mole. At 4 locations the energy is below -20 kcal/mole, at 3 between -5 and 0 and at a single location the energy is greater than 0 kcal/mole. The location of 17 of the 47 molecules can be clearly identified with a local minimum of the energy. A further 13 molecules are connected, directly or indirectly, to at least one of these 17 molecules, (presumably) via hydrogen bonds. It appears, therefore, that interactions with protein molecules and a first set of ordered water molecules combine to create an additional set of low-energy locations for water molecules.

We have been struck by the observation that of 10 separate energy minima below -20 kcal/mole, only 4 are associated with a crystallographic water molecule. Similarly, many minima between -15 and -20 kcal/mole do not have a crystal water nearby. Most of these minima occur near the ends of charged and other hydrophilic amino acid side chains, that are expected to be specifically hydrated. Intense thermal motion and/or disorder of long hydrophilic side chains is evident from the results of crystallographic refinement (5): terminal atoms of long side chains either have high thermal parameters or else have not even been located by crystallographic methods, but instead have been assigned positions consistent with those of other atoms and standard stereochemistry. Water molecules bound to these side chains will generally be subject to even more intense thermal motion, and thus will not correspond to easily identifiable maxima of electron density.

This analysis of water structure in the crystal of a protein, based on consideration of crystallographically-determined water locations and of a map of the energy of interaction of a single water molecule with the protein, demonstrates what we believe to

be the general usefulness of such a map. Our experience suggests
that these maps are useful because they simplify the interactions
between water and surrounding protein in two important ways: (1)
the complex of interactions with many atoms of the protein is
reduced to a single scalar, whose value defines the space occupied
by solvent and is significant only within this space, and, (2) the
result is a three-dimensional scalar function that can be mapped
and visualized using familiar graphic techniques. Contour surfaces drawn at negative energies delimit volumes that may contain
protein-bound water molecules. A contour surface drawn at a
positive energy outlines shape and volume of the protein molecule;
such a surface is similar (but not identical) to molecular surfaces defined according to purely geometric criteria (19,20,21).

Monte Carlo Simulation of the Crystal

1.   Principle of Simulation.   The preceding description of protein-solvent interactions by means of energy map of solvent space
near protein molecules can be considered only as a first, rough,
approximation to the complete problem. While shape and principal
features of solvent space near the protein surface are clearly
outlined in these maps, it is easy to see that inclusion in the
calculation of the strong and highly directed interactions between
water molecules will result in a different characterization of
solvent space, especially as concerns hydration shells beyond the
first. Because of the apparently disordered state of much of the
solvent, a statistical description is required, i.e., a large
number of different configurations of the system of protein and
solvent must be considered, each of which contributes to the
average properties of the system in proportion to a Boltzmann
factor.

For such large systems, the number of possible configurations
is so large that these must be sampled. Two conceptually different but computationally related methods for sampling configuration space in proportion to actual population at a given temperature have been successfully applied in recent years to a variety
of systems; these are the molecular dynamics and the Monte Carlo
methods. The system to be simulated is initialized in some suitably chosen configuration. For molecular dynamics, the force
experienced by each particle according to the given force field is
calculated and the equations of motion are integrated numerically.
A molecular dynamics simulation follows the evolution of the
system in real time; the time interval between succeeding configurations must be small enough to follow the most rapid fluctuations, with frequencies of the order of 10^{15}. As a consequence,
the total real time that can be covered in a molecular dynamics
simulation is at best of the order of 100 psec.

In order to obtain the Monte Carlo sample successive configurations of the system are generated by small random changes; the
probability of accepting a new configuration into the sample is

equal to the Boltzmann factor, exp(-ΔE/RT), if the total energy, E, increases, and 1 if it decreases; if the new configuration is not accepted, the old one is repeated in the sample and is considered again as starting point for a new random displacement (22).

With both methods, a large number of configurations is generated in order to bring the system near equilibrium, before actual generation of the sample is begun. Equilibrium properties can be obtained as simple averages over the sample, since the Boltzmann factor has been implicitly included in the probability of selecting each new configuration.

2.   Problems of Sampling Configuration Space.   If molecular dynamics and small-displacement Monte Carlo methods are to be used to generate, in a reasonable length of time, samples that are representative of configuration space, then two requirements must be met. The first is that low-energy configurations, i.e., configurations that should be heavily represented in the sample, must be adjacent in configuration space, Γ; the second is that the starting configuration must be in or near the portion of Γ that is to be sampled. This can be realized easily if the important region of Γ corresponds to a single energy minimum.

On the other hand, if Γ contains several different minima of the energy separated by high-energy (or low-entropy) barriers, then the sample may well include only configurations from one of these regions, selected by choice of the initial configuration. In that case it is imperative to ensure that subsequent simulation will take place in the most populated low-energy area of Γ.

This aspect of initialization does not seem to be a problem for a liquid such as water (11). However, redistribution of solvent between several poorly communicating solvent-filled cavities in the BPTI crystal is highly improbable during ordinary small-step simulation. The 'equilibrium' distribution of solvent molecules among these cavities depends only on the chosen starting configuration, unless a mechanism is provided for transfer of solvent molecules among cavities.

In a molecular dynamics simulation jumps among different parts of configuration space are excluded, since each configuration is the mechanical consequence of the preceding one. In principle, a Monte Carlo calculation is not thus limited. In practice, the trajectory in configuration space followed in a Monte Carlo simulation usually consists of a series of closely spaced points; this is so because large changes of configuration generally lead to unacceptably large increases of energy. Special methods are required that enhance the frequency of generating steps to other low-energy parts of configuration space.

We describe in this paper preliminary results of a Monte Carlo simulation of protein-solvent interactions in the BPTI crystal during which water molecules are rapidly redistributed troughout the available space between protein molecules. The likelihood that large displacement of a water molecule is acceptable energetically is enhanced by not attempting these large displacements to positions selected entirely at random. Instead, large jumps are attempted only to positions that are energetically relatively favorable. A list of such positions is maintained in the form of the coordinates of a set of dummy water molecules that occupy changing, energetically favorable positions in the system, but that do not, in turn, influence system configuration. The dummies' positions are adjusted both by small and by large displacements; positions of real waters are adjusted by small displacements and by occasional exchange of positions of a real and a dummy water.

Undoubtedly, the realism of any simulation of a protein-solvent system will be further enhanced if all parts of the system experience thermal motion. Because water molecules interact most strongly with polar side chains and because the conformation of long polar side chains may be subject to considerable thermal disorder, flexibility of side chains, in particular, is a major aspect of protein-solvent interaction. Thermal motion of side chains could be provided for rather simply by rigid-body internal rotation of side chains about side-chain single bonds as part of the Monte Carlo process.

3. Application to the BPTI Crystal. The system to be simulated consisted of the protein atoms of one BPTI molecule (5) and 140 water molecules. The required number of water molecules could be calculated both from the volume of the crystal for which protein-water energy is zero or negative (solvent space) (9) and from unit cell volume and density of protein and water. Protein-water interactions were calculated as in the first part of this article, protein-protein interactions as described elsewhere (23). Interactions between water molecules were calculated using the ST2 model, introduced by Rahman and Stillinger in a molecular dynamics simulation of liquid water (11).

Initial water positions were determined as follows: using the energy map (Figure 1) as a reference, a set of 140 water molecules was randomly inserted into the solvent space of the asymmetric unit; the molecules were treated at this stage as non-overlapping rigid spheres of 1.0 Å radius and water hydrogens were assigned random initial positions consistent with fixed water geometry. Another 1000 dummy water molecules were randomly inserted into solvent space in a similar manner, except that these could overlap with each other.

The configuration could change as a result of two types of small displacement, i.e., internal rotation of side chains about single bonds and changes of location and orientation of real waters. (Translations were limited to 0.2 Å along each coordinate axis, rotations to 18° about bonds or about each coordinate axis.) Acceptance of a change depended on the usual stochastic test on the Boltzmann factor. In calculations of non-bonded energy a considerable part of the computation is devoted to 'bookkeeping' operations, required to determine which pairs of non-bonded atoms contribute significantly to the energy. For this reason, unsuccessful motions were immediately reattempted (up to 19 times) with other randomly selected translation and rotation, with reference to a parameter table created during the first attempt. In what follows, a step represents selection of a different element to be moved, regardless of the number of attempts required to obtain an acceptable move (or to abandon the move). The number of attempts was circa 4 times the number of steps. A portion of the calculation was devoted to maintenance of the 1000 dummy waters in energetically favorable positions relative to protein and real waters, by application, both of small displacements, and of displacements to anywhere within the asymmetric unit. (Adjustment of dummy positions did not alter system configuration or increase the number of steps of the simulation.)

System configuration could also change as the result of a large displacement, or jump, of a real water. Such a jump could be performed only to the position of a dummy water. Acceptance of such a jump again depended on a stochastic test on the Boltzmann factor. If the jump was successful, then the real water's prior coordinates were saved as the new coordinates of the dummy water. As initial equilibration of the system progressed, it was found that acceptance of a reasonable fraction of all attempted exchanges required prior thermal equilibration of the position of each selected dummy water with respect to protein and real waters. Execution of a sufficient number of attempted exchanges of positions of dummy and real waters caused a considerable increase of computer time. However, use of positions of dummy waters instead of positions selected randomly within the asymmetric unit caused an increase of the probability of exchange and a decrease of the length of the thermal equilibration step of dummy waters selected for exchange, and limited the increase of computer time.

4. Results. In Figure 2 are shown changes of the average energy per water molecule and of average water-protein and water-water energies during the simulation. The energy decreased during the first 30,000 steps; the first 30,000 configurations were consequently discarded. The subsequent set of 50,000 configurations was used as a statistical sample. The percentage of successful jumps of water molecules decreased during the first part of the simulation and reached about 2% at equilibrium. The 80,000 steps required about 10 hours of c.p.u. time on an IBM 370/165.

The distribution of energies of water molecules is shown in Figure 3; (presumably trapped) high-energy waters that occured in an earlier Monte Carlo simulation (7) and molecular dynamics study (24), have apparently been eliminated as a result of interchange of dummy and real water molecules. The distribution of water-protein energies suggests that there may be a separation into classes of waters with 0, 1, 2, 3 and 4 protein-water hydrogen bonds, respectively, each hydrogen bond contributing circa -6 kcal/mole to the energy. However, all trace of this separation is absent from the distribution of the total energy which is much narrower than that of either of its two parts.

Results of simulation of side-chain motion are shown in Figure 4, as average r.m.s. displacement of atoms as a function of distance (expressed as number of bonds) from the first single bond in the side chain used as pivot axis. Average r.m.s. displacements of these atoms calculated from crystallographic thermal parameters (5) are only somewhat higher than r.m.s. displacements during Monte Carlo simulation. Use of a static protein backbone during the simulation presumably causes at least part of this difference. We find no readily apparent correlation between crystallographic thermal parameters and r.m.s. displacements of individual atoms.

An average solvent density has been calculated for the simulated sample, with assumption of a density distribution $\exp(-x^2/0.25)$ for the water molecule, x being distance from molecular center in Å. A contour map of the calculated density is shown in Figure 5; minor differences aside, the layout of this Figure is the same as that of the potential map of Figure 1.

The density map is characterized by a large number of pronounced maxima. The map contains 158 maxima, 18 more than the number of water molecules. At circa 120 maxima relative density is greater than 30%. This value of the central density corresponds to displacement from an equilibrium position according to a Gaussian probability with r.m.s. displacement of 0.5 Å (and a crystallographic thermal parameter, B, of 19). Distribution of maxima according to height is shown in Figure 6. One may consider peak height to be determined by two factors: fraction of the configurations in which a molecule is present in the volume of the peak (occupancy) and probability of displacement of the molecule from a central low-energy position. Neglecting occurence of peaks with occupancies below 1, and assuming a Gaussian probability for displacement from the center, we could calculate r.m.s. displacement, and hence crystallographic B-value, for any given peak height; these are shown as additional scales at top of Figure 6.

A single volume containing relatively disordered water can be found in the large solvent cavity (top half of parts a and b of Figures 1 and 5). Everywhere else, i.e., along the surface of the large cavity and in smaller cavities and channels, one finds regularly spaced maxima that correspond to a highly ordered solvent structure.

Positions of 47 crystallographically-located waters have been drawn in Figure 5 as circles of 1 Å radius. Of these molecules, 20 are within 1 Å of a pronounced maximum of the density; a further 17 lie between two closely spaced pronounced maxima; only 4 are not within 1 Å of a volume with simulated density greater than the average for liquid water.

Thermal parameters, B, of the 47 crystallographic waters of BPTI range from 7 to 60; all but 7 values are between 12 and 44

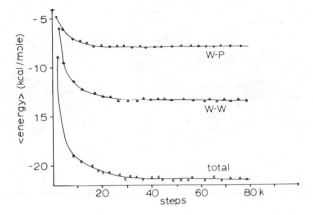

Figure 2.   Changes of the average energy per water molecule, average water–protein energy, and average water–water energy during simulation

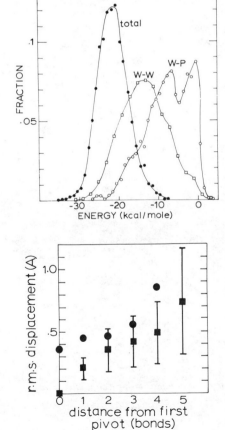

Figure 3.   Distribution of energies of water molecules

Figure 4.   Results of simulation of side-chain motion; average displacement of atoms as a function of distance from the first single bond in the side chain

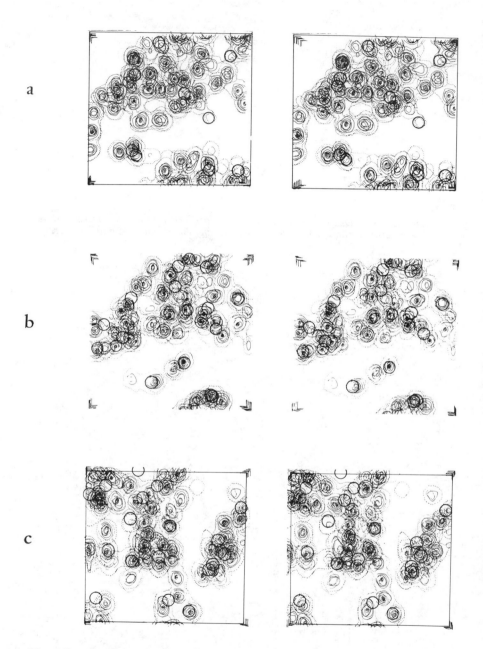

a

b

c

Figure 5.   Contour map of calculated average solvent density for a simulated sample

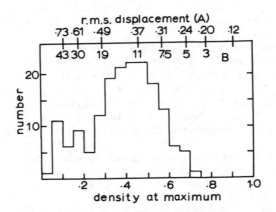

Figure 6.   Distribution of maxima (see Figure 5) according to peak height

(5). As a result of use of a stationary protein backbone, sharpness of maxima of the simulated density is exaggerated; therefore, the ranges of thermal motion of crystallographic and simulated waters are in reasonable agreement. The single water molecule entirely surrounded by protein has both lowest thermal parameter and narrowest distribution of simulated density.

According to the crystallographic results, ordered water in the BPTI crystal consists of a small set of molecules hydrogen-bonded to protein; according to the results of Monte Carlo simulation, more than twice as many water molecules are ordered as a result of water-water and protein-water interactions. The difference between experimental and simulated solvent structure may be due in part to the presence of inorganic ions in the crystals studied by x-ray crystallography; the concentration of salt inside the crystal is unknown, but the asymmetric unit might contain as many as 5 phosphate and 10 potassium ions. The presence of these ions in varying positions will decrease apparent solvent order, to an extent that we cannot presently assess. We discern in addition possible reasons why these descriptions disagree, related to problems of technique both of crystallography and of simulation.

The total number of crystallographic waters of BPTI is small compared with the number of 127 water molecules in the refined crystal structure of rubredoxin, a protein of similar size (6). Water molecules in the rubredoxin crystal are hydrogen-bonded to groups on the protein and to one another. Crystallographic solvent structure in rubredoxin is qualitatively similar to simulated solvent structure in the BPTI crystal; however, the range of crystallographic thermal parameters of water molecules in rubredoxin (from 15 to 60) corresponds to a less ordered structure than that of the simulated solvent in BPTI. (For simple practical reasons, simulation of the rubredoxin crystal has not yet been feasible.) Since initial crystallographic placement of water molecules and hence the total number of water molecules placed and refined, is based on a, to some extent subjective, interpretation of electron density maps, it is perhaps understandable why two such different results have been obtained in different laboratories: refinement of rubredoxin with many ordered water molecules can be considered to have been a test of the hypothesis that much ordered water exists in the crystal of a protein, whereas this hypothesis presumably has not influenced refinement of BPTI. Positions and heights of maxima of a simulated water distribution provide a unique alternative starting point for initialization of crystallographic refinement of solvent structure in the BPTI crystal. We believe that results of such a refinement will provide a critical test of validity of results of the Monte Carlo simulation.

Various needed improvements of the simulation will tend to cause the resulting average structure to be more disordered, and thereby will enhance agreement with crystallographic results: extension of the duration of the simulation and parallel simula-

tions (with different initialization of the random-number genera-
tor) will be required to address possible statistical insufficien-
cies; conformational freedom of the entire protein must, of course,
be a feature of a definitive calculation.

Acknowledgements

We thank J. Deisenhofer for refined coordinates and thermal
parameters of BPTI and C. Carter for helpful comments. This work
was initiated at a CECAM workshop on molecular dynamics, held in
Orsay, France, June-August 1978 and organized by C. Moser and H.
J. C. Berendsen. Further support was received from the Consiglio
Nazionale delle Ricerche and from the National Science Foundation
(Grant PCM-76-22723).

Abstract

We have used the refined crystal structure of a small protein
(trypsin inhibitor) as a system on which to test methods of analysis of solvent structure near protein surfaces in terms of 6-12
and electrostatic potentials.
(1)   The energy of interaction between a unique conformation of
the protein and a single water molecule was found to be a valuable
description of solvent space, when presented as a contour map in
three dimensions. The zero-energy contour surface separates
protein and solvent space and may be considered to effectively
define protein surface. Solvent space within the zero-energy
contour consists of a single network of channels and spaces of
various sizes and shapes plus one isolated low-energy volume that
contains one water molecule. All but one of the 47 crystallographically-located water molecules are within the zero-energy contour,
and many are in volumes of quite low energy.
(2)   A Monte Carlo simulation was performed in order to sample the
equilibrium ensemble of protein-solvent configurations. Transla-
tion/rotation of solvent and internal rotation of side-chains provided motion. Provision was made for small motions, to obtain
rapid local equilibration, and for large motions, to obtain proper
distribution of solvent. Simulated solvent structure is found to
be highly ordered: All water molecules in a first, and part of
those in a second layer at the protein surface maintain a unique
hydrogen-bonded network, that may be considered 'anchored' to the
protein at low-energy positions of solvent space. The network
contains twice as many ordered water molecules as have been located by x-ray crystallography. Possible reasons for the difference
include high molarity of salt in the crystals, incomplete freedom
of motion in the simulated crystal, insufficient size of the
simulated statistical sample, and incomplete crystallographic
refinement of solvent structure.

Literature Cited

1. Matthews, B. Ann. Rev. Phys. Chem., 1976, 27, 493-523.
2. Watenpaugh, K. D.; Sieker, L. C.; Herriott, J. R.;
    Jensen, L.H. Acta Cryst., 1973, B29, 943-956.
3. Carter, C. W.; Kraut, J.; Freer, S. T.; Zuong, W.; Alden, R.
    A.; Bartsch, R. G. J. Biol. Chem., 1974, 249, 4212-4225.
4. Huber, R. G.; Kukla, D.; Ruhlmann, A.; Epp, O.; Formanek, H.
    Naturwissenschaften, 1970, 57, 389-392.
5. Deisenhofer, J.; Steigemann, W. Acta Cryst., 1975, B31,
    238-250.
6. Watenpaugh, K. D.; Margulis, T. N.; Sieker, L. C.;
    Jensen, L.H. J. Mol. Biol., 1978, 122, 175-190.
7. Hagler, A. T.; Moult, J. Nature, 1978, 272, 222-226.
8. Owicki, J. C.; Scheraga, H. A. J. Am. Chem. Soc., 1977, 99,
    7403-7412.
9. Clementi, E.; Corongiu, G.; Jönsson, B.; Romano, S. FEBS
    Letters 1979, 100, 313-317.
10. McCammon, J. A.; Gelin, B. R.; Karplus, M. Nature, 1977,
    267, 585-590.
11. Rahman, A.; Stillinger, F. H. J. Chem. Phys., 1971, 55,
    3336-3359.
12. Stillinger, F. J.; Rahman, A. J. Chem. Phys., 1972, 57,
    1282-1292.
13. Stillinger, F. J.; Rahman, A. J. Chem. Phys., 1974, 60,
    1545-1557.
14. Scheraga, H. A. Adv. Phys. Org. Chem. 1968, 6, 103-184.
15. Ferro, D. R.; Hermans, J. in "Liquid Crystals and Ordered
    Fluids", Johnson, J. F.; Porter, R., eds. Plenum, New York,
    1970; pp. 259-275.
16. Nelson, D. J.; Hermans, J. Biopolymers, 1973, 12, 1269-1284.
17. Gelin, B. R. Ph.D. thesis, 1976, Harvard University.
18. Poland, D.; Scheraga, H. A. Biochemistry, 1967, 6, 3791-
    3800.
19. Richards, F. M. J. Mol. Biol., 1974, 82, 1-14.
20. Richards, F. M. Ann. Rev. Biophys. Bioeng., 1977, 6,
    151-176.
21. Greer, J.; Bush, B. L. Proc. Natl. Acad. Sci. U.S.A., 1978,
    75, 303-307.
22. Metropolis, N.; Rosenbluth, A. W.; Rosenbluth, M. N.;
    Teller, A. H.; Teller, E. J. Chem. Phys., 1953, 21, 1087.
23. Hermans, J.; Ferro, D. R.; McQueen, J. E.; Wei, S. C. in
    "Environmental Effects on Molecular Structure and Properties",
    Pullman, B., ed., Reidel, Dordrecht, Holland, 1976; pp.
    459-483.
24. Rahman, A.; Hermans, J. in "Report on Cecam Workshop: Models
    for Protein Dynamics", Cecam, Orsay, France, 1976, pp. 155-
    158.

RECEIVED January 21, 1980.

The Determination of Structural Water by Neutron Protein Crystallography

An Analysis of the Carbon Monoxide Myoglobin Water Structure

BENNO P. SCHOENBORN and JONATHAN C. HANSON

Biology Department, Brookhaven National Laboratory, Upton, NY 11973

Neutron crystallography (1) is an ideal technique for studying the water structure of proteins, particularly in cases where H_2O is replaced by D_2O. The scattering factor of hydrogen (Table I), although negative, is large enough for clear depiction of H atoms in high resolution Fourier maps, but hydrogen is often replaced by deuterons in order to reduce the large incoherent scattering from hydrogen atoms.

Table I

Scattering Length $(10^{-12}$ cm)

Element	Neutrons	X Rays
H	−0.374	0.28
D	.667	0.28
C	.665	1.7
N	.940	2.0
O	.580	2.3

This isotopic exchange not only brings about a large reduction in background but also distinguishes exchangeable from non-exchangeable hydrogens. The scattering factor of deuterium is nearly twice as large as that of hydrogen, and increases the visibility of water molecules in Fourier maps, so that in a neutron map a water molecule appears about three times as strong as in the equivalent electron density map. Furthermore, the absence of any radiation damage by neutrons eliminates scaling errors, since only one crystal is used to collect a complete data set. The use of only one crystal ensures that the derived structure is not an average of many different structures that might exist in different crystals.

The analysis of water structure can be further enhanced by mixing H_2O and D_2O which like isomorphous heavy atom replacement,

0-8412-0559-0/80/47-127-215$05.00/0
© 1980 American Chemical Society

HIS 5EF H-BONDS TO ASP 17H AND D₂O

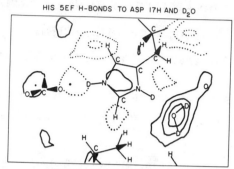

NEUTRON DIFFERENCE DENSITY MAP
TO LOCATE D_2O ($0.25F/A^3$)

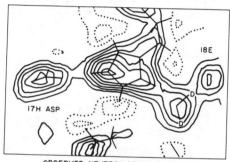

OBSERVED NEUTRON DENSITY MAP
D_2O INCLUDED IN PHASES ($0.5F/A^3$)

Figure 1.   Neutron density map

Top: section of neutron differences density map ($F_o - F_c$, α_c); the calculated structure factors (F_c, α_c) do not include any water molecules. The peak at the bottom right depicts a molecular feature not included in the structure factor calculation. The shape of that peak suggests a D_2O molecule. Middle: the same section of the neutron density map (F_o) with phases (α_c) for all atoms depicted including the D_2O molecule. Bottom: the same section of the neutron density map (F_o) with phases (α_c) for all atoms depicted. In this case, the phases were calculated for H_2O not D_2O. The Fourier feature obviously does not agree with the difference map; the correct solution is shown in the middle of the figure with a D_2O molecule D bonded to N_{δ_1} of histidine (5EF) and D bonded to O of isoleucine (18E).

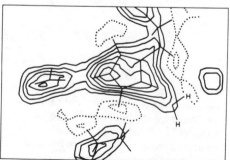

CALCULATED NEUTRON DENSITY MAP
H_2O INCLUDED IN PHASES ($0.5F/A^3$)

produces a controlled change in scattering density. The degree
of isomorphism depends, however, on the structural equivalence
between H_2O and D_2O. The resultant peak heights of water
molecules can therefore be changed in a controlled way and can be
used to distinguish water peaks from other constituents.
Furthermore, the characteristic peak shapes of H_2O and D_2O
molecules (Figure 1) permit the elucidation of their orientation
and therefore of their hydrogen bonding arrangements.

     Results from an early unrefined met myoglobin neutron
analysis (2) based on the x ray structure by Watson and Kendrew
(3) showed a total of 106 water peaks. In that analysis, only
peaks that had a peak height equivalent to one-half D_2O molecule
were considered. More recently, Takano (4), in a refined high
resolution x-ray analysis of met myoglobin, found 72 water
molecules, and in the present refined neutron analysis of CO-Mb,
only 40 water molecules were found. These water molecules were
identified in a difference map where the features depicting the
protein had been subtracted. The protein structure itself had
been refined by successive real-space refinement (5,6,7) of eight
different maps comprising more than thirty refinement cycles
(7,8). The difference map was then searched for features
equivalent to peak heights of at least one-half water molecule.
Note, that, in such difference maps, calculated without the
phases of the water molecules, a water molecule will appear at
half height (9). A careful study of these "unphased" difference
maps is very informative because they contain little bias;
whereas phased maps tend to reproduce features even if there is
no contribution to the structure factor magnitude, i.e., if water
molecules are introduced into the set of phases, they will
produce Fourier features even if they are not present. The
orientations of the D_2O molecules were examined with the aid of
computer graphics (10). The two deuterons and the oxygen atom
were placed to best fit the neutron difference density while
assuming ideal D_2O geometry and to accommodate D bond donors and
acceptors on the protein's surface. The occupancy of the
observed water molecules was determined by the real-space program
from the difference map. The neutron density of a fully occupied
D_2O in such an unphased difference map would be $0.96 \cdot 10^{-12}$ cm,
and the calculated error of these difference maps is only
$.1 \cdot 10^{-12}$ cm. The actual peak height is, however, reduced by
thermal parameters and by the fact that only about 80% of the
water of crystallization is D_2O, as shown by NMR analysis (11) of
the Mb-CO crystals (pH 5.6 in 80% ammonium sulphate).

     Inspection and analysis of the Mb-CO difference map resulted
in a finding of 40 water molecules per asymmetric unit (one
myoglobin molecule). No peak with a weight of less than 0.4
water molecule was included in this number. Several features in
the remaining difference map might represent water molecules, but
at a low level of occupancy (<0.4). Refinement of the water
locations and weight was performed by two cycles of real-space

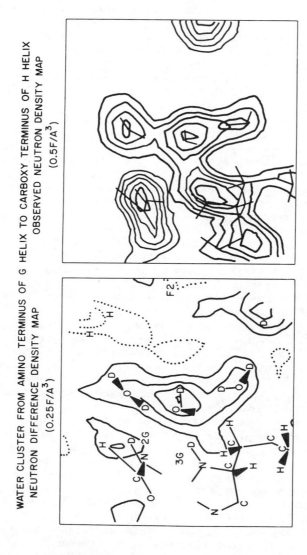

Figure 2. Section of the neutron difference density map ($F_o - F_c$, α_c); the calculated structure factors do not include any water molecules. Peaks in the center are interpreted as three water (D_2O) molecules. On the right-hand side the neutron density map (F_o) for the same section with all atoms included in the phase calculation (α_c).

Table II. Water Molecules Found

RESIDUE NUMBER AND TYPE	C_α	N	CO	POLAR GROUP	RESIDUE NUMBER
1 Val NA1	H	D	o	D_2O------H17'	1
2 Leu NA2	H	D	o	D_2O	2
3 Ser A1	H	D	o		3
4 Glu A2	H	D→	o←D_2O---D_2O	C_δ—O_{ϵ_1}—D_2O	4
5 Gly A3	H	D	o	D_2O------H21'	5
6 Glu A4	H	D	o	C_δ—O_{ϵ_2}—D_2O H9	6
7 Try A5	H	D	o		7
8 Gln A6	H	D	o	C_δ—N_{ϵ_1}----D_2O	8
9 Leu A7	H	D	o		9
10 Val A8	H	D	o		10
11 Leu A9	H	D	o		11
12 His A10	H	D	o	N_{ϵ_2}	12
13 Val A11	H	D	o	N_{γ_1}----D_2O D_2O---B15'	13
14 Try A12	H	D	o	N_ϵ	14
15 Ala A13	H	D	o--D_2O--------D_2O--D_2O--D_2O---B14'	15
16 Lys A14	H	D	o--D_2O--E2'	N_ζ D_2O---C2'	16
17 Val A15	H	D	o		17
18 Glu A16	H	D	o	C_δ—O_{ϵ_1}—O_{ϵ_2}	18
19 Ala AB1	H	D	o		19
20 Asp B1	H	D	o	C_γ—O_{δ_1}—O_{δ_2}	20
21 Val B2	H	D	o		21
22 Ala B3	H	D	o		22
23 Gly B4	H	D	o		23
24 His B5	H	D	o	N_{ϵ_2} N_{δ_1}	24
25 Gly B6	H	D	o		25
26 Gln B7	H	D	o	C_δ—O_{ϵ_2}—D_2O N_{ϵ_1}	26
27 Asp B8	H	D	o	C_γ—O_{δ_1}—O_{δ_2}	27
28 Ile B9	H	D	o		28
29 Leu B10	H	D	o		29
30 Ile B11	H	D	o	N_ϵ	30
31 Arg B12	H	D	o	C_ζ—N_{η_1} N_{η_2}	31
32 Leu B13	H	D	o		32
33 Phe B14	H	D	o--A13		33
34 Lys B15	H	D	o--A13	N_ζ----D_2O	34
35 Ser B16	H	D	o	O_γ	35
36 His C1	H	D	o	N_{ϵ_2} N_{δ_1}----D_2O	36
37 Pro C2	H		o--A13	D_2O	37
38 Glu C3	H	D	o--D_2O---┐	C_δ—O_{ϵ_2}—O_{ϵ_1}	38
39 Thr C4	H	D	o	O_γ	39
40 Leu C5	H	D	o		40

Table II. Continued

RESIDUE NUMBER AND TYPE		C_α	N	CO	POLAR GROUP	RESIDUE NUMBER
41	Glu C6	H	D	o	$C_\delta{<}^{O_{\epsilon_2}}_{O_{\epsilon_1}}$ — $\boxed{D_2O}$ — D_2O D_2O ⌐	41
42	Lys C7	H	D	o	N_ζ	42
43	Phe CD1	H	D	o		43
44	Asp CD2	H	D	o $\boxed{D_2O}$	$C_\gamma{<}^{O_{\delta_2}}_{O_{\delta_1}}$ $H2'$	44
45	Arg CD3	H	D	o $- D_2O$	$C_\zeta{<}^{N_{\eta_1}}_{N_\epsilon}{<}_{N_{\eta_2}}$	45
46	Phe CD4	H	D	o		46
47	Lys CD5	H	D	o	N_ζ	47
48	His CD6	H	D	o	N_{δ_1}	48
49	Leu CD7	H	D	o	N_{ϵ_2}	49
50	Lys CD8	H	D	o		50
51	Thr D1	H	D	o	N_ζ	51
52	Glu D2	H	D	o		52
53	Ala D3	H	D	o	O_γ	53
54	Glu D4	H	D	o	$C_\delta{<}^{O_{\epsilon_1}}_{O_{\epsilon_2}}$	54
55	Met D5	H	D	o	$C_\delta{<}^{O_{\epsilon_1}}_{O_{\epsilon_2}}$	55
56	Lys D6	H	D	o	N_ζ	56
57	Ala D7	H	D	o	$A14'$	57
58	Ser E1	H	D	o	O_γ	58
59	Glu E2	H	D	o	$C_\delta{<}^{O_{\epsilon_2}}_{O_{\epsilon_1}}$	59
60	Asp E3	H	D	o	$C_\gamma{<}^{O_{\delta_2}}_{O_{\delta_1}}$	60
61	Leu E4	H	D	o		61
62	Lys E5	H	D	o	N_ζ	62
63	Lys E6	H	D	o	N_ζ	63
64	His E7	H	D	o	N_{δ_1}	64
65	Gly E8	H	D	o	N_{ϵ_2}	65
66	Val E9	H	D	o		66
67	Thr E10	H	D	o	O_γ	67
68	Val E11	H	D	o		68
69	Leu E12	H	D	o		69
70	Thr E13	H	D	o	O_γ	70
71	Ala E14	H	D	o $\boxed{D_2O}$ $\boxed{D_2O}$		71
72	Leu E15	H	D	o		72
73	Gly E16	H	D	o		73
74	Ala E17	H	D	o		74
75	Ile E18	H	D	o		75
76	Leu E19	H	D	o		76
77	Lys E20	H	D	o	N_ζ	77
78	Lys EF1	H	D	o	N_ζ	78
79	Lys EF2	H	D	o	N_ζ	79
80	Gly EF3	H	D	o	N_ζ	80

Table II. Continued

RESIDUE NUMBER AND TYPE	C_α	N	CO	POLAR GROUP	RESIDUE NUMBER
81 His EF4	H	D	o		81
82 His EF5	H	D	o		82
83 Glu EF6	H	D	o		83
84 Ala EF7	H	D	o		84
85 Glu EF8	H	D	o		85
86 Leu F1	H	D	o		86
87 Lys F2	H	D	o		87
88 Pro F3	H		o		88
89 Leu F4	H	D	o		89
90 Ala F5	–	D	o	H24	90
91 Gln F6	H	D	o		91
92 Ser F7	H	D	o		92
93 His F8	H	D	o		93
94 Ala F9	H	D	o		94
95 Thr FG1	H	D	o		95
96 Lys FG2	H	D	o		96
97 His FG3	H	D	o		97
98 Lys FG4	H	D	o		98
99 Ile FG5	H	D	o		99
100 Pro G1	H		o		100
101 Ile G2	H	D	o		101
102 Lys G3	H	D →	o		102
103 Tyr G4	H	D	o		103
104 Leu G5	H	D	o		104
105 Glu G6	H	D	o		105
106 Phe G7	H	D	o		106
107 Ile G8	H	D	o		107
108 Ser G9	H	D	o		108
109 Glu G10	H	D	o		109
110 Ala G11	H	D	o		110
111 Ile G12	H	D	o		111
112 Ile G13	–	D	o		112
113 His G14	H	D	o		113
114 Val G15	H	D	o		114
115 Leu G16	H	D	o		115
116 His G17	H	D	o		116
117 Ser G18	H	D	o		117
118 Arg G19	H	D	o		118
119 His GH1	H	D	o		119
120 Pro GH2	H		o		120

Table II. Continued

RESIDUE NUMBER AND TYPE		C_α	N	CO	POLAR GROUP	RESIDUE NUMBER
121 Gly	GH3	H	−	o		121
122 Asn	GH4	H	D	o	C_γ $\begin{smallmatrix} N_{\delta_1} \\ O_{\delta_2} \end{smallmatrix}$	122
123 Phe	GH5	H	D	o		123
124 Gly	GH6	H	D	o	$C6'$	124
125 Ala	H1	H	D	o		125
126 Asp	H2	H	D	o	C_γ $\begin{smallmatrix} O_{\delta_2} \\ O_{\delta_1} \end{smallmatrix}$ $--D_2O$	126
127 Ala	H3	H	D	o		127
128 Gln	H4	H	D	o	C_δ $\begin{smallmatrix} O_{\epsilon_2} \\ N_{\epsilon_1} \end{smallmatrix}$ $--D_2O$	128
129 Gly	H5	H	D	o		129
130 Ala	H6	H	D	o		130
131 Met	H7	H	D	o		131
132 Asn	H8	H	D	o	C_γ $\begin{smallmatrix} N_{\delta_1} \\ O_{\delta_2} \end{smallmatrix}$ $-----$ A4	132
133 Lys	H9	H	D	o	N_ζ $-----$	133
134 Ala	H10	H	D	o		134
135 Leu	H11	H	D	o		135
136 Glu	H12	H	D	o	C_γ $\begin{smallmatrix} O_{\epsilon_1} \\ O_{\epsilon_2} \end{smallmatrix}$ $---D_2O$	136
137 Leu	H13	H	D	o	A2′	137
138 Phe	H14	H	D	o		138
139 Arg	H15	H	D	o	C_ζ $\begin{smallmatrix} N_{\eta_1} \\ N_{\eta_2} \\ N_\epsilon \end{smallmatrix}$	139
140 Lys	H16	H	D	o	N_ζ $-----$	140
141 Asp	H17	H	D	o	C_γ $\begin{smallmatrix} O_{\delta_1} \\ O_{\delta_2} \end{smallmatrix}$	141
142 He	H18	H	D	o		142
143 Ala	H19	H	D	o		143
144 Ala	H20	H	D	o	A2′	144
145 Lys	H21	H	D	o	N_ζ $-----$	145
146 Tyr	H22	H	D	o	O_η	146
147 Lys	H23	H	D	o	N_ζ $-----$ F2	147
148 Glu	H24	H	D	o	C_δ $\begin{smallmatrix} O_{\epsilon_2} \\ O_{\epsilon_1} \end{smallmatrix}$	148
149 Leu	HC1	H	D	o		149
150 Gly	HC2	H	D	o		150
151 Tyr	HC3	H	D	o		151
152 Gln	HC4	H	D	o	O_η C_δ $\begin{smallmatrix} N_{\epsilon_1} \\ O_{\epsilon_2} \end{smallmatrix}$	152
153 Gly	HC5	H	D	o		153
HEME				o	C $\begin{smallmatrix} O---D_2O \\ O \end{smallmatrix}$	

refinement reducing the rms shift to 0.13 A.   Figure 1 is a
typical section of the difference map, showing a water molecule
lying between several protein atoms, and Figure 2 is another
section showing a small water cluster that bridges two protein
molecules.   The water molecules found are listed in Table II,
which is similar to the summary of the original analysis (2) and
can be directly compared.

Refinement of the protein structure reduced the number of
water molecules from the 108 originally found to less than 50,
with only 18 at full occupancy.   Of the 40 waters found, only 25
are present in Takano's structure.

The effect of salt and pH on the role of bound water (12) is
still unclear and might explain some of the large differences
that have been observed in protein crystallography particularly
in hemoglobin and myoglobin.   Some of the smaller still
unassigned peaks might be due to other water molecules, to noise,
or to Fourier termination errors.   Data to higher resolution
(1.5 A) have been obtained with recently developed faster data
collection methods utilizing a two-dimensional position-sensitive
counter, and are being analyzed.   Such a higher resolution map
will permit the analysis of fractional bound water to an
occupancy of less than 0.2.   Data collection in different H_2O/D_2O
mixtures will further increase the reliability of the chemical
identity of those Fourier peaks.

Apart from identifying individual bound water molecules on
the surface of the protein, neutron maps can be used to determine
the average solvent density in the regions between protein
molecules.   This is done by averaging the featureless background
density over small volumes and expressing it in terms of
scattering length/A^3.   The calculated volumetric scattering
density is $6.3 \cdot 10^{-14}$ cm/A^3 for D_2O and $-.6 \cdot 10^{-14}$ cm/A^3 for
H_2O.   The present analysis shows a density fluctuation of 15%,
the D_2O density ranging from 85% to 70%.   With a solvent
composition of 80% D_2O and 20% H_2O, this would give a density
from 1.06 to 0.88.   In this case, the Fourier scale was
calculated so that the average observed densities of the
different atoms in the protein corresponded to the expected
values (Table I).   In order to obtain a more reliable absolute
scale, data must, however, be collected in H_2O and in D_2O.   In
addition to proper scaling, it is also necessary to subtract the
proper contribution of the protein structure.   In order to do
this accurately, precise atomic location, weight, and temperature
factors are needed.   The real space refinement procedure can
carry out this subtraction well, but even with the present
refined 1.8-A map, groups that are disordered and have high
temperature factors are difficult to subtract properly.

Literature Cited

1. Schoenborn, B. P. and Nunes, A. C.   1972.   Ann. Rev.
    Biophys. Bioeng. 1, 529-552.
2. Schoenborn, B. P.   1971.   Cold Spring Harbor Symp. Quant.
    Biol. 36, 569-575.
3. Watson, H. C.   1968.   In Progress in Stereochemistry, Vol.
    4, pp. 299-333, Butterworth, London.
4. Takano, T.   1977.   J. Mol. Biol. 533-568.
5. Diamond, R.   1971.   Acta Cryst. A27, 436-452.
6. Schoenborn, B. P. and Diamond, R.   1975.   Brookhaven Symp.
    Biol. 27, Sect. II, 3-11.
7. Norvell, J. C. and Schoenborn, B. P.   1979.   Brookhaven
    Symp. Biol. 27, Sect. II, 12-23.
8. Hanson, J. and Schoenborn, B. P.   1979.   (in preparation).
9. Luzzati, V.   1953.   Acta Cryst. 6, 142-148.
10. Hanson, J., Ringle, W.M. and Love, W. E.   1973.   Am. Cryst.
    Assoc. Abst. 5, 2.
11. Moore, P.   1977.   Anl. Biochem. 82, 101-108.
12. Perutz, M. G.   1977.   Biosystems 8, 261-263.

Research supported by the U.S. Department of Energy.

RECEIVED January 4, 1980.

Density of Elastin–Water System

MARIASTELLA SCANDOLA and GIOVANNI PEZZIN

Centro di Studio per la Fisica delle Macromolecole del C.N.R., Via Selmi 2, 40126 Bologna, Italy

A variety of different experimental techniques has been used to investigate the hydration of fibrous proteins: among them calorimetric (1,2,3), dielectric (4) and dynamic-mechanical measurements (5,6), equilibrium sorption data (7,8,9), infrared spectroscopy (10) and Nuclear Magnetic Resonance (11,12).

Little information is however available in the literature on the specific volume of hydrated fibrous proteins. To the best of the authors' knowledge, the only set of data published to date refers to hydrated keratin, was published by King (13), and later discussed by Rosenbaum (14).

Volumetric data can be essential in the thermodynamic treatment of the "polymer-solvent" interaction process. The lack of them for many important fibrous proteins is due to the difficulty of measuring the density, at controlled temperature and hydration degree, for these systems. As far as elastin is concerned, it has been reported that when completely hydrated this protein has a negative and very large coefficient of thermal expansion (15), a result which has been interpreted as evidence of a hydrophobic character of the protein (16).

Elastin, which is substantially amorphous but fibrous at all levels of investigation (starting from the largest filaments which are about 6 μm in diameter and down to about 10 nm (17,18)), is a fragile, glassy substance when dry and has a glass-to-rubber transition temperature at about 200°C (19); upon hydration or solvation with appropriate solvents, it becomes a rubbery system with the glass transition below room temperature (20).

Calorimetric data have shown that only half of the total water sorbed by elastin (about 0.6 g water / g dry protein) is really "bound", the remaining water being freezable (1). The volumetric data reported in the literature (15,16) refer therefore to an essentially heterophase system, so that the negative and very large coefficient of thermal expansion of the fully hydrated protein does not appear to be suitable for the interpretation of the thermoelastic data and calculation of the

0-8412-0559-0/80/47-127-225$05.00/0
© 1980 American Chemical Society

protein-water interaction parameter.

It is the purpose of the present work to collect reliable volumetric data on the system elastin-water and to compare them with the scarce literature data on similar natural polymer systems or on synthetic polymeric systems. The measurements have been purposely limited to the region of low water contents where the system can be considered homogeneous.

Experimental

Samples of native ox ligamentum nuchae elastin and of collagen-free elastin were obtained as reported previously (19).

In order to carry out the density measurements in a wide range of temperatures, a suitable glass pycnometer was used: the pycnometer capillary is placed laterally with respect to the chamber containing the sample and the filling liquid, and is provided with an upper (expansion) and a lower (contraction) spherical cavity. A low molecular weight silicon oil (Dow Corning 200/20) was used as the filling liquid.

Dry samples (about 3 g of elastin strips less than 1 mm thick) were stored under vacuum over P_2O_5 for several days, carefully weighed on an automatic balance, placed in the pycnometer, covered with silicon oil and degassed by means of a rotative pump until no gas release by the sample was observed. The pycnometer was then closed, immersed in a bath maintained at the measurement temperature and, when the thermal equilibrium was reached, the liquid level in the lateral capillary was adjusted with a small syringe.

Weighing of the pycnometer, after careful drying, was carried out at room temperature and the density of the protein sample ϱ was calculated from the equation:

$$\frac{m}{\varrho} = V_p - \frac{m_t - m_p - m}{\varrho_s} \tag{1}$$

where: V_p is the volume of the pycnometer
m^p is the weight of the sample
m_t is the total weight of the pycnometer containing sample and filling liquid
m_p is the weight of the empty pycnometer
ϱ_s^p is the density of the silicon oil

The density of samples of purified elastin containing appropriate amounts of water was determined with the same procedure. The hydration levels chosen are those corresponding to water weight fractions of 0.05, 0.14, 0.18 and 0.29. Due to the difficulties in reproducing the same water weight fraction, only one measurement of density was carried out at each water content. However, the experimental scatter is small, as seen from the linearity of the data plotted in Figures 1 and 2.

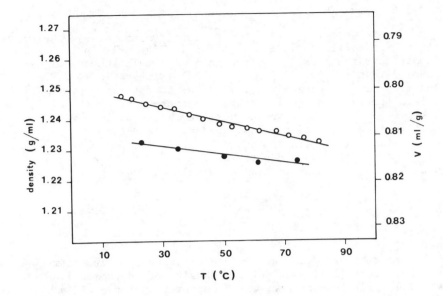

Figure 1. Temperature dependencies of the density and specific volume of dry elastin: (○) native; (●) purified.

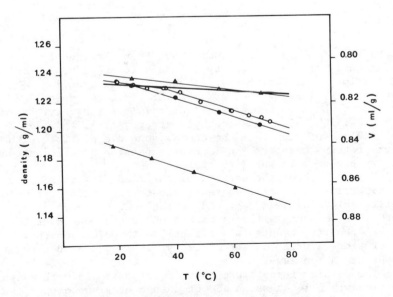

Figure 2. Density and specific volume of dry and hydrated purified elastin samples as a function of temperature at water weight fractions of: (—) 0; (△) 0.05; (○) 0.14; (●) 0.18; (▲) 0.29

Results and Discussion

The temperature dependence of the density of dry samples of native and purified elastin is shown in Figure 1. In the temperature range explored, the native protein shows a higher density than the purified one, typical values being 1.245 g/ml and 1.232 g/ml, respectively, at 25°C. In a first approximation, this difference in density can be accounted for by the different composition of the native and purified protein. If native elastin is considered a two phase composite material (approximately 80% elastin and 20% collagen), disregarding other minor components whose density data are not available, the density of the composite can be calculated from the equation:

$$\frac{1}{\varrho} = \frac{w_E}{\varrho_E} + \frac{w_C}{\varrho_C} \tag{2}$$

Assuming $w_E = 0.80$, $w_C = 0.20$ and taking $\varrho_E = 1.232$ g/ml and $\varrho_C = 1.32$ g/ml (21), a calculated density of the composite $\varrho = 1.247$ g/ml is obtained, which compares satisfactorily with the experimental density of native elastin.

The volumetric expansion coefficients, calculated from the equation:

$$\alpha = -\frac{d \ln\varrho}{d T} = \frac{d V}{V d T} \tag{3}$$

(where V is the specific volume) have been obtained by least square analysis of the density data and found to be 1.9×10^{-4} °C^{-1} and 1.1×10^{-4} °C^{-1} for the native and purified protein respectively. These values are similar to those typical of glassy amorphous or semicrystalline polymers (polystyrene: $\alpha = 1.7 - 2.1 \times 10^{-4}$ °C^{-1}; poly(methyl methacrylate): $\alpha = 2.2 - 2.7 \times 10^{-4}$ °C^{-1} (22)). Unfortunately, no comparisons are possible with other proteins owing to the lack of expansion coefficient data for dry proteins in the literature.

Figure 2 shows the density of purified elastin as a function of temperature and water content. The corresponding expansion coefficient values are reported in Table I. At temperatures lower than 30°C and with the exception of the most hydrated sample, density changes only slightly with water content; at higher temperatures, on the contrary, density depends strongly on the hydration level.

The density-temperature curve for the sample with weight fraction $w_1 = 0.14$ shows an abrupt change of slope at a temperature of about 38°C; the volume expansion coefficient increases from 2.5×10^{-4} °C^{-1} to 5.4×10^{-4} °C^{-1} at this temperature. The latter value, like those for samples with

higher water content (see Table I) is typical of amorphous
polymers above their glass transition temperature T_g
(polystyrene: $\alpha = 5.1-6.0 \times 10^{-4}$ °C^{-1}; poly(methyl methacrylate)):
$\alpha = 5.6-5.8 \times 10^{-4}$ °C^{-1} (22)).

Table I - Expansion coefficient and volume contraction of
hydrated purified elastin

w_1	$\alpha \times 10^4$ ($°C^{-1}$)	$(\Delta V_m/V_o) \times 10^2$		
		30 °C	45 °C	60 °C
0	1.1			
0.05	2.1	- 1.77	- 1.66	- 1.54
0.14	2.5 (T<T_g) 5.4 (T>T_g)	- 3.15	- 2.83	- 2.28
0.18	5.4	- 4.01	- 3.47	- 2.96
0.29	5.9	- 2.49	- 1.91	- 1.37

A glass transition temperature of about 40°C for the hydrated
elastin sample ($w_1 = 0.14$) fits fairly well into the glass
transition temperature–water content curve calorimetrically
determined by Kakivaya and Hoeve (23) for hydrated elastin.
    It is known that Simha and Boyer (24), treating the glass
transition of polymers as an iso-free volume state, have derived
the equation:

$$(\alpha_L - \alpha_G) \cdot T_g = K_1 \qquad (4)$$

where α_L and α_G are the volume expansion coefficients above and
below T_g and K_1 is a constant that for a variety of polymers
assumes the value of about 0.11. The application of this
equation to the elastin–water system (i.e. to the sample with
$w_1 = 0.14$, $T_g = 38$°C and the α values of Table I) gives $K_1 = 0.09$.
This result seems reasonable, since it has been suggested (24)
that in a polymer–solvent system the solvent contributes free
volume below T_g so that α_G becomes larger than for the
unplasticized polymer and, consequently, K_1 assumes values lower
than 0.11. Furthermore, experimental results on PVC plasticized
with dioctyl phthalate by Heydemann and Guicking (25) show that
$\Delta\alpha$ at T_g is lower for a 10% plasticized sample than for pure
PVC; this fact, and the simultaneous decrease of T_g, will result
in a lowering of K_1 for solvated polymers.
    It has to be pointed out that, in the temperature range
explored in the present work, the volume expansion coefficient
of hydrated elastin is positive and constant on both sides

of T_g. The only literature data of volume expansion coefficient for elastin refer to the protein in swelling equilibrium with water (15,16); the reported coefficients are strongly negative and increase with increasing temperature in the range 20-60 °C. Typical quoted values are $\alpha = -1.4 \times 10^{-2}$ °C^{-1} at 20°C, $\alpha = -4 \times 10^{-3}$ °C^{-1} at 40°C and $\alpha = 0$ °C^{-1} at 62°C (16). The strongly temperature dependent volume changes of highly hydrated elastin (in swelling equilibrium conditions) have been interpreted by Gosline (16) in terms of changes in hydrophobic interactions. In the relatively low hydration range explored in the present study, no evidence of such a mechanism has been found; in this respect, hydrated elastin behaves in the same way as many synthetic polymer-solvent systems.

As far as the important problem of the polymer-solvent volume changes on mixing is concerned, it is seen from Figure 2 that at each temperature the density of the elastin-water system remains practically constant in the hydration range from the dry protein to the water content corresponding to the onset of the glass transition. This implies that the glassy system "elastin plus water" has a higher density than that expected from volume additivity calculations. This is clearly visualized in Figure 3 where the experimental specific volumes (obtained from Figure 2 by suitable interpolations at three selected temperatures) are compared with the "ideal additive volumes" of the system, represented by the lines connecting the pure liquid water and dry glassy elastin volumes.

It is well known that volume changes on mixing are likely to occur in polymer-solvent systems, and both theoretical and experimental studies have been devoted to this subject (26,27,28). When the volumes of the components are not strictly additive, both theoretical and phenomenological approaches include an excess volume term in the thermodynamic equations for the system. These treatments, however, concern polymeric solutions in equilibrium, i.e. the mixture is liquid or rubbery, not glassy. For solutions in the glassy state, no theoretical descriptions of the excess volume of mixing seems to be available up to date.

According to Rehage (29), when the polymeric mixture is glassy, other volume changes that he calls "freeze-in volume" superimpose themselves on the volume changes of mixing caused by molecular interactions. No practical way of calculating an approximate value of the "freeze-in volume" is available for our system, since volumetric data for glassy water at room temperature are needed (in principle, they could be obtained by suitable extrapolation from low temperature volumetric data on glassy water).

One of the parameters most commonly used to quantify volume changes on mixing is the fractional volume change $\Delta V_m/V_o$ (where V_o is the "ideal" volume of the mixture assuming additivity and ΔV_m is the difference between the experimental volume of the mixture and the "ideal" volume V_o). Values of the fractional

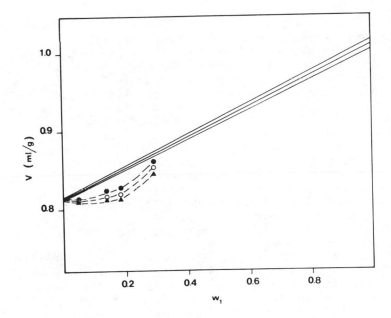

Figure 3. Experimental specific volume of hydrated elastin as a function of concentration at three selected temperatures: (▲) 30°C; (○) 45°C; (●) 60°C; straight lines are the "ideal additive" volumes (lower: 30°C; middle: 45°C; upper: 60°C)

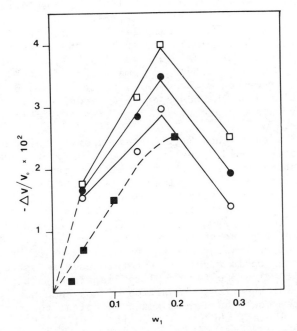

Figure 4. Fractional volume change as a function of water content for elastin at three temperatures: (□) 30°C; (●) 45°C; (○) 60°C; (■) data calculated from Ref. 31 on the system polycarbonate–dibutyl phthalate are included for comparison

volume change $\triangle V_m/V_o$ calculated at three selected temperatures
are reported in Table I. The volume contractions are relatively
large and sometimes exceed the volume contractions reported for
typical water-alcohol systems (30), which are known to be the
largest found in binary equimolecular organic mixtures.

The fractional volume changes of Table I are shown in
Figure 4 together with some values for polycarbonate
plasticized with dibutyl phthalate, calculated from the data
of Wyzgoski and Yeh (31). The volume contraction for the
elastin-water system is remarkable even at low levels of
hydration for the three temperatures considered. It increases
with increasing water content, reaching a peak value at a water
weight fraction of about 0.2. The maximum expected and observed
volume changes on mixing for equilibrium polymeric solutions
occur near the composition $w_1 = 0.5$ (28); the occurrence of a
peak at $w_1 = 0.2$ for our system could be tentatively explained
as related to the limited solubility of water in the protein (1)
and/or to the occurrence of the glass transition.

The data for the system polycarbonate-dibutyl phthalate,
which are included in Figure 4 for the sake of comparison, show
a behavior similar to that described for the elastin-water
system. The datum for the most highly plasticized sample (DBP
content: 20%), which at the measurement temperature (25°C) is
probably very near the onset of the glass transition, seems
to indicate a less effective volume contraction than at lower
plasticizer content.

The only other set of data available in the literature on
density as a function of water content for a fibrous protein
appears in the work of A. T. King (13) on hydrated wool at 25°C.
The hydration range is similar to that of the present work, and
the overall results are analogous: at low levels of hydration
the density of wool was found to remain approximately constant;
at higher water contents the density decreased approaching the
values calculated on the basis of volume additivity. The maximum
fractional volume change $\triangle V_m/V_o$ from King's data is about 5%,
i.e. larger than ours.

According to Rosenbaum (14), the experimentally determined
dry wool density could actually be the density of a polymeric
material containing "small static voids" originally occupied by
the water molecules in the native state of the hydrated protein;
for wool, Rosenbaum's "specific volume without voids" is 7%
smaller than the experimental specific volume.

In our view, it is quite difficult to accept a similar
interpretation for the system elastin-water (and probably even
for the system wool-water) since the filling up of "small static
voids" by water would result in a heterogeneous system, i.e. a
composite with liquid water droplets dispersed in the rigid
glassy protein. The glass transition of any such heterogeneous
polymer system would be constant and identical to that of the
dry protein, whereas the glass transition of elastin is strongly

depressed by interaction with water, the decrease being marked especially at low levels of hydration (at $w_1 = 0.06$, T_g is already 100°C lower than that of the dry protein (23)). This result is possible only if the two components interact at the molecular level.

In conclusion, the most interesting results of the present work are: first, that the volumetric temperature coefficients of elastin at low and moderate levels of hydration are positive and small (rather than negative and large, as reported for the fully hydrated but heterophase protein) and, second, that the interaction with water is accompanied, as reported previously for wool, by large volume changes, which appear to be indicative of strong interactions at the molecular level.

Literature Cited

1. Ceccorulli, G.; Scandola, M.; Pezzin, G. Biopolymers, 1977, 16, 1505.
2. Haly, A. R.; Snaith, J. W. Appl. Polym. Symp., 1971, 18, 823.
3. Dehl, R.E. Science, 1970, 170, 738.
4. Algie, J. E.; Gamble R. A. Kolloid Z. Z. Polym., 1973, 251, 554.
5. Scandola, M.; Pezzin, G. Biopolymers, 1978, 17, 213.
6. Baer, E.; Kohn, R.; Papir, Y. S. J. Macromol. Sci-Phys., 1972, B6, 761.
7. Bull, H. B. J. Amer. Chem. Soc., 1944, 66, 1499.
8. Zimm, B. H.; Lundberg, J. L. J. Phys. Chem., 1956, 60, 425.
9. Simha, R.; Rowen, J. W. J. Amer. Chem. Soc., 1948, 70, 1663.
10. Bendit, E. G. Biopolymers, 1966, 4, 539.
11. Ellis, G. E.; Packer, K. J. Biopolymers, 1976, 15, 813.
12. Lyerla, J. R.; Torchia, D. A. Biochemistry, 1975, 14, 5175.
13. King, A. T. J. Text. Inst., 1926, 17, T 53.
14. Rosenbaum, S. J. Polym. Sci., 1970, C 31, 45.
15. Mistrali, F.; Volpin, D.; Garibaldo, G. B.; Ciferri, A. J. Phys. Chem., 1971, 75, 142.
16. Gosline, J. M. Biopolymers, 1978, 17, 697.
17. Gotte, L.; Mammi, M.; Pezzin, G. Connective Tissue Res., 1972, 1, 61.
18. Gotte, L. Proceedings of the Third John Innes Symp., 1976, p. 39.
19. Pezzin. G.; Scandola, M.; Gotte, L. Biopolymers, 1976, 15, 283.
20. Gotte, L.; Pezzin, G.; Stella, G. D. in "Biochimie et Physiologie du Tissu Conjonctif", Comte, P., Ed., Centre Tech. Cuir, Lyon, 1966, p. 145.
21. Astbury, W. T. J. Int. Soc. Leather Trad. Chem., 1940, 24, 69.
22. Brandrup.J.; Immergut, E. H. Editors, "Polymer Handbook", 2nd Ed., Wiley, New York, 1975.
23. Kakivaya, S. R.; Hoeve, C. A. J. Proc. Natl. Acad. Sci. USA, 1975, 72, 3505.

24. Simha, R.; Boyer, R. F. J. Chem. Phys., 1962, 37, 1003.
25. Heydemann, P.; Guicking, H. D. Kolloid Z. Z. Polym., 1963, 193, 16.
26. Hildebrand, J.; Scott, R. L. "The Solubility of Nonelectrolites", 3rd Ed., Reinhold, New York, 1950.
27. Scatchard, G. Trans. Faraday Soc., 1937, 33, 160.
28. Sanchez, I. C.; Lacombe, R. H. Macromolecules, 1978, 11, 1145.
29. Rehage, G. ; Borchard, W. in "The Physics of Glassy Polymers", Haward, R. N., Ed., Applied Science, London, 1973.
30. Weissberger, A., Editor, "Technique of Organic Chemistry", 3rd Ed., Physical Methods (Part I), Interscience, New York, 1959, p. 144.
31. Wyzgoski, M. G.; Yeh, G. S. Polymer J., 1973, 4, 29.

RECEIVED January 4, 1980.

Comparison of Weight and Energy Changes in the Absorption of Water by Collagen and Keratin

M. ESCOUBES
Université Claude Bernard, 69621 Villeurbanne, France

M. PINERI
Centre d'Études Nucléaires de Grenoble, 38041 Grenoble Cédex, France

The collagen and keratin constitute the two most important protective fibrous proteins in living systems. The role of water in connection with these polymers in biological processes has been the object of a number of studies for more than a century but remains controversial. We have shown, in preceeding article concerning collagen (1), that the analysis of data for weight absorption and kinetics of absorption of water as studied by torsion pendulum experiment of hydrated collagen permit an improvement on literature data and suggests a hydration scheme involving five stages.

This article concerns collagen fibers from rat tail tendon. The object is to confirm, improve or modify the proposed model by studying different collagen samples and secondly fibrous proteins such as the α keratin of human hair which exhibits similar hydration properties.

1. Review of results on collagen fiber from rat tail tendon

This type of collagen has the advantage of illustrating the most recent structural data. It is necessary to review briefly this data to describe the proposed hydration model, and to aid in characterizing other samples of fibrous protein discussed later.

The molecule of collagen has the form of a rigid rod (2) with a length of 2900 Å and a diameter of 12.5 Å with a molecular weight of 300,000. This rod consists of three helical polypeptide chains with three parallel axes separated by 4.5 Å (triple helix or tropocollagen).

Each sequence of three residues (taken as a weight reference) possesses one glycine residue ($-NH-CH_2-CO-$) and on average one proline or hydroxyproline ($-N-CH-CO-$) and has a molecular weight of about 270 gm. The three chain complex is stabilized essentially by hydrogen bonding between the CO and NH units. Models have been proposed (3,4) involving one and two hydrogen bonds for each group of three residues. In each model the stabilization is completed by two bridged water molecules. The destruc-

0-8412-0559-0/80/47-127-235$05.00/0
© 1980 American Chemical Society

tion of hydrogen bonds by heating represents the majority of the
enthalpy involved in the helix-coil or collagen-gelatin transitions.

- Collagen molecules spontaneously form aggregates in the form
of microfibrils. Electron microscopy data (5) suggest an assembly
of five molecules in which each two adjacent molecules are shifted
by a quarter of their length (allignment "quarter stagger") on a
distance of 670 Å. The succession of the two molecules assures an
interval of 350 Å. This gives a free volume of the order of 10 %
and a periodicity of 670 Å in the dehydrated state.

- Finally the juxtaposition of microfibrils in a tetragonal
arrangement limited to the sublattices and largely associated with
the amorphous region results in a fiber diameter between 1000 and
5000 Å.

Molecular cohesion in the microfibril as well as in the amorphous regions is due to two types of bonding. Firstly weak covalent
bonds (of the type ester, imide, peptide) between lateral groups
of certain residues as partic acid, glutamic acid, lysine) and secondly from very strong linkages located only at the non helical
end areas of the molecule (telopeptidique areas) which cause the
eventual condensation of the two aldehyde derivatives of the
lysine and hydroxylysine forms. These two types of linkages can
also be present between chains of the same molecule. The latter
is also largely responsible for the insolubility in acid media of
old collagen.

Fibers of rat-tail tendon contain few linkages in the terminal telopeptidique areas. We have shown (1) that these fibers retain 1 % of water under a vacuum of 10^{-4} torr at 20°C and that the
hydration in the vapor phase at 20°C shows 3 relatively sharp steps
in terms of the energetic and mechanical data, before the appearance of free water near 90 % relative humidity.

- The first step between 0 and 10 % with a energy of 17 kcal/
mole (corresponding to the formation of double hydrogen bond
without breaking other preexistant bonds) is attributed to intramolecular water and corresponds quantitatively to two water molecules per reference unit.

- The second step between 10 and 25 % with an energy of 13 kcal/
mole may correspond to bridging of water between the collagen molecules occurring with a diminution of the cohesive energy of the
microfibril.

- The third step between 25 and 50 % with an energy of 10 kcal/
mole does not correspond to the presence of free water according
to most of the data in the literature.

The latter occurs near 50 %. This step is associated with a
decrease in the coefficient of diffusion. We have attributed this
in part to the formation of water clusters within the periodically
arranged free volume zones in the micofibril.

Thus the hydration of rat-tail tendon collagen fibers appears to be connected with different structural levels of the protein. It is interesting to confirm this pattern by studying other fibrous proteins.

2. Sample characterization

2.1 - Collagen samples

_ Turkey leg tendon at 22 months.

This tissue illustrates the calcification of collagen produced with time. The mineral-protein interaction has been largely studied by small angle X-ray studies (10), electron microscopy (11) and neutron diffraction (12). It seems to be established that calcium phosphate (mostly apatite crystals) is distributed along the length of fiber with same periodicity of 670 Å which characterizes the shift of two adjacent collagen molecules. Despite the absence of information concerning the lateral arrangement, it seems reasonable to speculate that the mineral is developed not only at the level of the fiber but also in the holes of the microfibril as much as the size of the crystals calculated from diffraction spots agree with the size of the holes. The highly calcified apatite in the tissues is found between the fibrils without any periodicity and is not observed by diffraction.

- Films of reconstituted collagen at 20°C and 50°C

The remarkable properties of collagen have led to many diverse application : solutions, gels, films, tubes, sponges. In particular the film is used in application as diverses as the fabrication of sausage skins and the treatment of burns or blood dialysis and artificial kidneys. It is thus especially interesting to understand the behaviour with reference to the water present in reconstituted collagen. The samples analysed here have been prepared by the method developed by the "Centre Technique de Cuir" at Lyon (France) from tannery products (Brevet Français n° 1596790 of Nov. 27.1968) : the protective skin of veal is ground, washed with a phosphate pad and exchanged with water to obtain at first fiber in a 0.1M solution of acetic acid, reprecipitated by addition of NaCl, dialysed with exchanged water and lyophilic agents. The lyophylic acid soluble product is dissolved in a 0.1 M acetic acid solution at a concentration of 3 %, contrifuged, deaerated and finally films are formed by evaporation at different temperatures in special frames.

The biochemical and biophysical studies and mechanical properties of the film have been compared with those of the lyophylic acid soluble products (11, 12) :

. a partial insolubility, or an increase in intra and intermoleculaire bonds is present in the end regions of the fiber ;

. a more irregular placement of molecules is present in the interior of the network of fiber which can be interpreted by a disappearance of the "quarter stagger" allignment ;

. an increase in the denaturation temperature (for the collagen
  gelatin transformation) which implies an increase in bonds sta-
  bilizing the triple helix ;

. the elastic properties attributed to the triple helix are maintained for film preparation temperatures less than 46℃ confirming the retention of triple helix structure below this temperature.

It should be added that the two films analysed are especially
cross-linked and do not contain the "quarter stagger" shift and
that the film formed at 50℃ no longer retain the triple helix
structure.

2.2 - Keratin samples

- α-keratin from hair

Human hair occurs in the form of a fiber with a central core
containing the keratin within an external cuticule having the form
of scales which contains 10 % of its mass.

Chemically, keratin is formed from polypeptide chains formed
from 20 different amino acids. The relative proportion of which
vary for different keratins. Some possess an acid side group (12 %
of which are glutamic acid) others a amine function (30 % of which
are lysine), others a hydroxyl (10 % of which are serine). But the
keratin is above all characterized by an important quantity of
sulfur due to the presence of cystine units which form a disulfide
bridge between two chains greatly contributing to the stability of
the proteins. In conclusion, the chains are connected by a number
of diverse interactions including hydrogen bonds, salt bonds and
covalent bonds.

Keratin has a crystallinity of about 30 %. Since 1950, X-ray
diffraction and electron microscopy have led to a model in which
the structural elements are (13,14) : the α-helix with a diameter
of 10 Å, the elementary fibril containing two to three helicoïdal
chains of 20 Å diameter, the microfibril or association of ten
elementary fibrils through the amorphous regions (of 80 Å diameter) and finally the fibrils or structures of several microfibrils
within an amorphous matrix.

The structure of the amorphous region in unknown but it is
known to be rich in sulfur with a number of disulfide bridges and
absorb more water than the microfibrils.

The absorption of water by keratin has been studied by N.M.R.,
I.R., dielectric and calorimetric methods (15-20). Keratin exhibits
a single line spectrum by N.M.R. with line-width depending on orientation. This is in contrast to collagen which shows a three line
spectrum. Lynch and Haly (15) have studied the influence of orientation on the spin-spin relaxation time of absorbed water in keratin
from rhinoceros horn. A priviledged rotation of the water is observed around an axis approximately parallel to the fiber direction.
From I.R. results Bendit proposes a weak association of water with
carbonyl groups in the crystalline phase. I.R. experiment on deute-

rated keratin also shows evidence for hydrogen–deuterium exchange processes (17). Leveque (18) has interpreted a maximum observed in depolarisation experiments as being associated with a reorientation of bound water molecules. Water molecules absorbed by keratin from wood have been classified as free or bound (21,22). A further subdivision of the bound water molecules has been proposed (23). Feughelman (24) has proposed a model implying different types of water molecules with one to four degrees of association. The relative percentage of these different types changes with the level of water absorption.

No clear experimental proof has yet been given for the existence of these several types of water molecules.

– Keratin of reduced hair and dyed hair

Natural hair can be greatly changed in terms of the cohesive bonds present.

By treatment with thioglycolic acid in ammonia, 33 % of the cystine ($HOOC - \underset{\underset{NH_2}{|}}{CH} - CH_2 - S - S - CH_2 - \underset{\underset{NH_2}{|}}{CH} - COOH$) is transformed to the form ($HOOC - \underset{\underset{NH_2}{|}}{CH} - CH_2 - SH$) and 22 % to the form lanthionine ($HOOC - \underset{\underset{NH_2}{|}}{CH} - CH_2 - S - CH_2 - \underset{\underset{NH_2}{|}}{CH} - COOH$). There is a decrease in the number of disulfide linkages making the hair fiber temporarily plastic or inelastically deformable. This reduction is used in hairdressing to create a permanent set in which the reduced hair is wound on a hair curler and fixed while the fiber reconstitute its initial chemical texture.

– The dying of hair is accomplished by the action of H_2O_2, the persulfate of sodium or ammonia which oxidize the cystine. An analysis of this treatment reveals the presence of several intermediate oxidation products and the final transformation of 32 % of the cystine to cysteic acid ($HOOC - \underset{\underset{NH_2}{|}}{CH} - CH - SO_3H$).

The decrease in crosslinking is comparable to that of reduced hair.

We have in this study tried to analyse the effect of these chemical modifications on the hydration properties of hair.

3. Experimental results

3.1 – Comparison of different collagen samples

The absorption isotherms for water at 20°C are shown in figure 1. We note :

1) that the calcified collagen absorbs less water than that of rat–tail tendon. The change is particularly important in the range of 20–80 % relative humidity ;

2) that the reconstituted collagen film exhibits the same hydra-

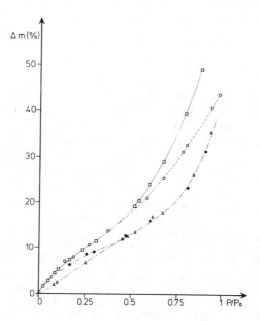

Figure 1.   Water sorption isotherms of different collagen forms: (□) rat tail tendon
collagen; (○) reconstituted collagen (1); (△) reconstituted collagen (2); (●) calcified turkey leg tendon

tion as the rat-tail tendon at low humidities but absorbs less
water than the rat-tail tendon above humidities of 50 % ;

3) that the reconstituted collagen film at 50°C absorbs less water
over the entire humidity range. The isotherm is similar to that
of the mineralized collagen near P/P_0 of 0.5.

The energy absorption data are shown in figure 2. The different steps of absorption observed for rat-tail tendon are not seen
for the other samples. Especially unusual is that for the calcified
collagen the energy of absorption is clearly less than 10 kcal/mole
as soon as the weight of absorbed water exceeds 10 %. For the two
films of reconstituted collagen the energy of absorption goes
through a weak maximum near a weight of 10 % ($P/P_0 \sim 0.4 - 0.5$)
and then decreases regularly.

3.2 - Comparison of different keratin samples

The absorption isotherms of keratin from human hair shows a
sigmoidal form as did those for collagen which are classified as
type II with a hysteresis over the entire humidity range as shown
in figure 3. A difference exists in the kinetics of absorption
when observed by successive increments. Longer times are required
to attain equilibrium for the keratin in the first increment than
for those above $P/P_0 = 0.6$ (figure 4).

A shift of the absorption isotherm is observed toward higher
water levels for reduced or dyed hair (figure 5).

The absorption energy profile for natural hair or dyed hair
shows in all cases a slightly increasing slope (below $P/P_0 = 0.6$)
with the molar heat below that for the liquefaction of water, followed by a part clearly decreasing up to relatively low values
(figure 6).

4. Discussion

The above consideration lead essentially to a comparison with
the model proposed for collagen fiber from rat-tail tendon. In
effect, if it is possible to relate the hydration properties to
structural or textural chemical differences in the samples, one can
hope to explain the collagen-water interaction on a molecular basis.

Collagen

For calcified collagen the hypothesis can be made that the
decrease in hydration as compared to that for the rat-tail tendon
corresponds to the filling of holes in the microfibril by apatite
crystals. This would be evident in the third step defined above.
But a shift is observed at the low levels. The size of the shift
is much more important than would be expected on the basis of the
volume of the holes in question since it extends to relative humidities near 50 %. It is also accompanied by a radical change in
the energy absorption profile and an overall decrease in the diffusion site.

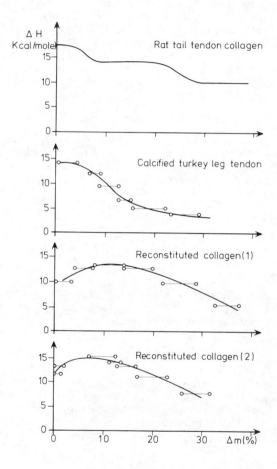

Figure 2.   Energy of hydration vs. the amount of water absorbed for different collagen forms

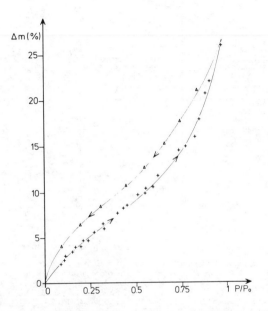

Figure 3. Human hair keratin: (+) water sorption isotherm and (△) water desorption isotherm

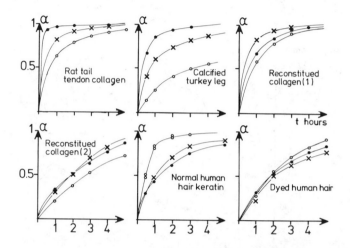

Figure 4.   Kinetics of hydration for different specimens.

α *is the ratio of the amount of water absorbed after time* t *vs. that absorbed at equilibrium. Symbols are defined as follows. Rat tail tendon collagen:* (●) $0 < P/P_o < 0.33$; (×) $0.43 < P/P_o < 0.58$; (○) $0.66 < P/P_o < 0.81$. *Calcified turkey leg tendon:* (●) $0 < P/P_o < 0.22$; (×) $0.33 < P/P_o < 0.48$; (○) $0.65 < P/P_o < 0.85$. *Reconstituted collagen (1):* (●) $0 < P/P_o < 0.20$; (×) $0.37 < P/P_o < 0.59$; (○) $0.78 < P/P_o < 0.95$. *Reconstituted collegan (2):* (●) $0 < P/P_o < 0.17$; (○) $0.26 < P/P_o < 0.48$; (×) $0.62 < P/P_o < 0.82$. Normal human hair keratin:* (●) $0 < P/P_o < 0.21$; (○) $0.2 < P/P_o < 0.33$; (×) $0.62 < P/P_o < 0.87$. *Dyed human hair keratin:* (●) $0 < P/P_o < 0.15$; (×) $0.4 < P/P_o < 0.6$; (○) $0.75 < P/P_o < 0.9$.*

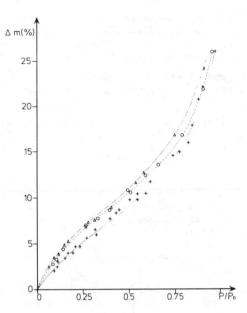

Figure 5. Water sorption isotherms for different keratin forms: (+) normal human hair keratin; (○) reduced human hair keratin; (△) dyed human hair keratin

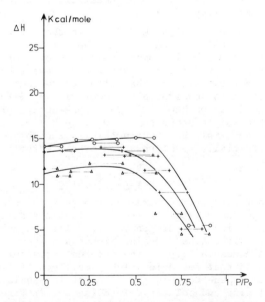

Figure 6. Molar heat of water sorption for different keratin forms: (+) normal human hair keratin; (○) reduced human hair keratin; (△) dyed human hair keratin

All of these changes can be explained if one supposes that the localized apatite crystals cause a strong crosslinking with an increase in the cohesive energy between molecule and a decrease in the chain mobility.

The energy and kinetic characteristics of swelling are changed except for step I corresponding to the absorption of intermolecular water.

For reconstituted collagen the hydration behaviour is also explained relatively well in terms of the model if the structural modifications described above (paragraph II) are taken into account. For the film prepared at 20°C we note :

. the increase in intramolecular bonding which increases the stability of the triple helix (increase in the denaturation temperature) implies a large decrease in the initial molar energy of hydration. The first intramolecular water molecules are required to overcome an increased cohesive energy while the following molecules do less and less to change the structure. This explains the progressive increase in the balance of energy in the course of the first absorption step.

. the disappearing of the "quarter stagger" structure and the increase in the number of the intermolecular crosslinking in the telopeptidic zones can explain the changes which are observed with this 20°C reconstituted collagen : decrease in the amount of absorbed water, decrease in the values of the energy and finally decrease of the diffusion coefficients which correspond to the accessibility of the sites.

The film prepared at 50°C differs in that it no longer contains the triple helix structure. If the helix-coil (or collagen gelatin) transformation is complete, the absorption should no longer be related to the different structural levels but instead controlled by the tortuosity of the diffusion paths. The rate of water absorption is considerably slower with a decrease in the weight absorbed at low humidities. But the energy profile remains similar to the sample prepared at 20°C. We believe that the disappearance of the triple helix structure is incomplete and that the water tends initially to be absorbed by the structures still present.

Hydration of different keratin samples

Keratin fibers from natural hair have structures comparable to collagen in the crystalline region. But the structure is restricted in addition by hydrogen bonds, polar and charged interactions of the main chains and side groups and above all by covalent bridges. The accessibility of the internal structure is reduced with respect to that for acid soluble collagen fibers. Nevertheless water can still penetrate in the fiber at low humidities because of its small size and polarity. In doing so, it fixes to the most easily available polar groups, shields the associated electric fields, and reduces the cohesive forces. This eases chain movements through an induced plasticization.

The weight and energy absorption data are compatible with these ideas. In effect :

. The energy absorption profile increases slightly in the first region which implies, as seen above for the case of reconstituted collagen, a plasticization of the medium by the first molecules absorbed.

. The absorption isotherm does not follow the BET expression in the initial part (20) :

$$\frac{P/P_0}{N(1 - P/P_0)} = \frac{1}{MC} + \frac{C-1}{MC} \frac{P}{P_0}$$

This equation is based on the assumption that the absorbed species (N in units of gm/gm) are immobilized at sites with concentration M (in units of gm/gm) at first in a monomolecular form and then subsequently in the form of multilayers. The BET transformation is convex with respect to the axis P/P_0 for keratin but linear for the case of rat-tail collagen **(figure 7)**.

This means that the concentration of absorption sites changes progressively in keratin. New absorption sites are created by the absorption of water. These are produced by the separation of molecular surfaces initially in contact.

. The diffusion coefficient increases in the course of successive absorption increments. This shows again that the accessibility of sites is increased by plasticization.

. The elastic modulus decreases regularly with the weight of water absorbed at temperatures above 250 K.

It must also be concluded that water is already absorbed at the polar sites even at low partial pressures. The energy of plasticization does not fully account for the water-polymer interaction. The large values of 13-14 kcal/mole suggest that water forms hydrogen bonds between chains and that the association is relatively strong. We note in this regard that keratin retains 2-5 % by weight of water under a vacuum of 10^{-4} torr at ambiant temperature.

For relative humidities above 0.6, the mechanism of absorption changes suddenly. In particular the molar energy of absorption decreases and becomes less than 10 kcal/mole. This may imply the presence of particularly weak interactions with the protein but another interpretation is also possible. Throughout the course of this step there is a replacement of the strong bonds previously formed in favor of interactions between water molecules for example. This conclusion is supported by the results from NMR, IR, and depolarisation current experiments (17,18,24) which seem to show that the relative proportions of different types of water polymer associations varies with the degree of absorption.

Reduced and dyed hair are characterized by the scission of disulfide links between chains. This scission involves in the two cases about 30 % of the cystine present. Therefore there is a

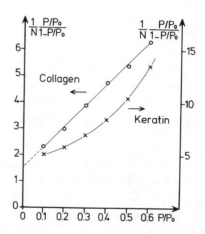

Figure 7.   BET representations for (○)
collagen samples and (×) keratin samples

decrease in the cohesive energy between chains and eventually the formation of new absorption sites.

In fact the absorption results are not greatly different : there is a slight increase in the amount of water absorbed (+ 10 % on average) and a simple shift of the energy profile (toward higher values in the case of reduced hair and toward lower values in the case of dyed hair). In total the chemical modifications involving the cystine do not lead to profound changes in the mechanism of water absorption at a molecular level for keratin.

Conclusion

These comparative studies show that only the hydration of acid-soluble collagen fibers exhibits well defined energetic steps which can be attributed respectively to intramolecular water, water in microfibrils and "free" water giving rise to an overall swelling of the fiber.

The intramolecular and intermolecular bonds which arise in calcified collagen and reconstituted collagen are the origin of the difference in the weight and kinetics of water absorption. The water acts as a plasticizer and brings into play other possible interactions as a result of the diffusion process.

The keratin fiber differs from that of collagen in particular due to the importance and variety of polar, salt and covalent interactions between the main chains and side groups. The water is expected to moderate the interactions in question by concentrating around the polar sites. There should subsequently be in the course of the swelling process or redistribution of the water molecules.

We have shown that the disulfide links between cystine groups of the chains are only one of several bonds involved in hair. The scission of these links by reduction or oxidation does not modify profoundly the hydration behaviour.

In conclusion, the large differences between natural collagen and keratin must be emphasized both in terms of the isotherm absorption and the nature of fixation sites of the water. We have shown that the influence of chemical modifications gives results which can be explained in terms of the model we have proposed.

Acknowledgements

We thank Dr. A. Miller, Dr. J.L. Leveque and Dr. Erbage for helpfull discussion and for providing different samples.

Literature Cited

1. Pineri M.H. ; Escoubez M. ; Roche G. Biopolymers, 1978, 17 n° 12, 2799
2. Ramachandran G.N. ; Kartha G. Nature, 1955, 176, 1062
                                   Nature, 1969, 174, 269
3. Rich A. ; Crick F.H.C. Nature, 1955, 176, 915-916
4. Ramachandran G.N. ; Sasisekharan V. ; Thathachari Y.T., in

Ramanathan, ed., Collagen. Interscience Publishers, Inc. New York, 1962, p. 81

5. Smith R.W. Nature, 1968, 219, 157-158
6. Fessler J.H. ; Bailey A.J. Biochim. Biophys. Acta, 1966, 117, 368-378
7. Bornstein P. Fed. Proc., 1966, 25, 1004-1009
8. Fernandez-Moran J. ; Engstrom A. Biochim. Biophys. Acta, 1957, 23, 260-264
9. Robinson R.A. ; Watson M.L. Ann. N.Y. Acad. Sci. 1955, 60, 596-628
10. White S.W. ; Hulmes D.J.S. ; Miller A. Nature, 1977, 266 n° 5601, 421-425.
11. Bernengo J.C. ; Roux B. ; Herbage D. Biopolymers, 1974, 13, 641-647
12. Journot M. Mémoire du C.N.A.M. Lyon Université, 1979
13. Crick F.H.C. Nature, 1952, 170, 882
14. Pauling L. ; Corey R.B. J. Am. Chem. Soc., 1950, 72, 5349
15. Lynch L.J. ; Haly A.R. Kolloid Zeitschuft und Zeitschouft five polymere, Band 239, Heft 1, 1970
16. Bendit E.G. Biopolymers, 1966, 4, 539
17. Bendit E.G. Biopolymers, 1966, 4, 561
18. Levêque J.L. ; Garson J.C. ; Boudouris G. Biopolymers, 1977, 16, 1725
19. Lynch L.J. ; Marsden K.H. J. Chem. Phys. 1970, 61, 7, 349.
20. Berendsen H.J.C., Mighelsen C., Ann. N.Y. Acad. Sci., 1965, 125, 365
21. Cassie A.B.D., Trans. Farad. Soc., 1945, 41, 458
22. Cooper D.N.E., Ashpole D.K., J. Textile Inst., 1956, 50, T 223
23. Windle J.J., J. Polym. Sci., 1956, 21, 103
24. Fenghelman M., Haly A.R., Textile Research Journal, 1966, 966
25. Zahn H., Berthsent T. et al, J. Soc. Cosm. Chem., 1963, 14, 529-543
26. Bore P., Arnaud J.C., Bull. ITF, 1974, 1, 75-105

RECEIVED January 17, 1980.

POLYSACCHARIDE INTERACTIONS
WITH WATER

New Insights into the Crystal Structure Hydration of Polysaccharides

T. BLUHM, Y. DESLANDES, R. H. MARCHESSAULT, and P. R. SUNDARARAJAN

Xerox Research Centre of Canada, Mississauga, Ontario L5L 1J9, Canada

With the exception of cellulose and chitin, plant polysaccharides are usually hydrated. Hydration often occurs in the crystalline regions as well as in the amorphous areas. When water of hydration is found in the crystallites, it may or may not affect the conformation of the polysaccharide backbone and in most cases, it affects the unit-cell dimensions, while in a few cases, the water appears to have no effect on unit-cell dimensions. The structures of six hydrated neutral polysaccharides will be examined with regards to the state of water of hydration in the structure. It will be seen that water may occur as columns or as sheets in these structures. The structures that will be discussed are (1→4)-β-Ḏ-xylan, nigeran, amylose, galactomannan, (1→3)-β-Ḏ-glucan and (1→3)-β-Ḏ-xylan. The chemical structures of these polysaccharides are shown in Figure 1.

The evidence to be used is based on x-ray diffraction analysis and not on swelling or moisture regain studies. This allows rather specific conclusions about the nature of the polysaccharide-water interaction but eliminates macromolecular aspects related to the osmotic phenomenon of polymer-water interaction, as in hydrogels. In polysaccharides, both aspects are important, for example: in seed germination, crystalline hydration is a first step in the process, while gelatinization of starch involves a hydrogel phenomenon with hydrate crystallites as pseudo-crosslinks.(1)

In lignocellulosics, the extreme affinity of dry wood for water is notorious. The Egyptians took advantage of the phenomenon to split large rocks by wedging one end of a beam in a cleft and placing the other end in contact with water. In a thermodynamic sense, this takes advantage of the swelling pressure (2) which is certainly related to the hydration of the

0-8412-0559-0/80/47-127-253$05.00/0
© 1980 American Chemical Society

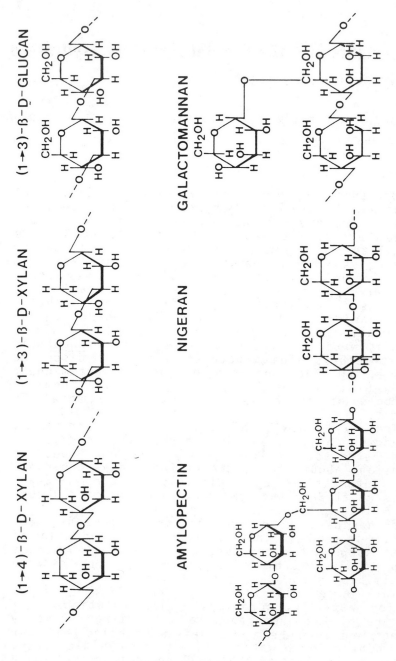

Figure 1.   Chemical structures of (1 → 4)-β-D-xylan, (1 → 3)-β-D-xylan, (1 → 3)-β-D-glucan, amylopectin, nigeran, and galactomannan

hemicelluloses. In fact, the same affinity could be involved in the familiar problem of water transport in trees, whereby water columns of hundreds of feet are stable.

Xylan [Poly (1→4)-β-D-XYLP]

Probably the second most abundant polysaccharide in the world, xylan (3) is found in association with cellulose in hardwoods and softwoods as well as in grasses. It is seldom found as a homoglycan, but rather with a substituted backbone 4-0-methyl-D-glucuronic acid and arabino-furanose being the common glycosidic substituents. O-Acetyl groups frequently occur along the chain as natural ester groups. Often, xylans are found with only arabinose substituents and these materials are referred to as arabino-xylans. They are important components of cereal grains and plant corms from which they can be extracted to be used as gums.

Xylans have always been classified with the hemicelluloses, the non-cellulosic polysaccharides of woody materials. In the paper-making process, they are assigned a hydration role, i.e., without them, strong interfiber bonds do not form unless extensive mechanical stock refining is used. Accordingly, they are classified as hydrocolloids or gums and are thought to act as a glue encouraging the sticking together of fibers in paper.

Although it was not realized at first, the hydrating properties of xylan extend to its crystalline state. The unit-cell is hexagonal (4) with:

$$a = b = 9.16\overset{\circ}{A} \text{ and } c(\text{fiber repeat}) = 14.9\overset{\circ}{A}$$

The base plan unit-cell projection (Figure 2) shows the positions of the water molecules in the unit-cell as deduced from an analysis of the equatorial x-ray diffracted intensities for a fiber diagram recorded at about 50% relative humidity. The water molecules cluster in distinct areas and form a helical column whose symmetry matches that of the xylan chains. In fact, the water of hydration may dictate the symmetry of the xylan chains. The energetically (theoretical) most stable conformation of the xylan chain involves two-fold symmetry, whereas in the hydrated crystalline environment, as deduced from x-ray diffraction, the xylan chains possess three-fold symmetry. The water molecules stabilize the three-fold structure by the formation of hydrogen bonds. This structure is an example of columnar hydration which allows a symmetric

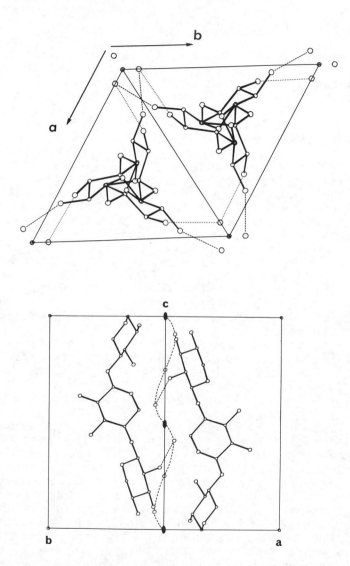

Figure 2. Top: projection in the ab *plane of (1 → 4)-β-D-xylan unit cell. Bottom: projection in the 110 plane of (1 → 4)-β-D-xylan unit cell*

expansion of the unit-cell as the humidity and corresponding degree of hydration increase. At 58% relative humidity, the unit-cell contains one water molecule per xylose unit while at 90% relative humidity there are two water molecules per xylose unit. At very low relative humidities, the crystallinity is poor. Table I shows the unit-cell dimensions of the various polymorphs.

Table I

Unit-Cell Parameters of (1→4)-β-D-Xylan Polymorphs

	"Dry"	"Hydrate"	"Dihydrate"
a = b (Å)	8.8	9.16	9.64
c (Å)	14.85	14.85	14.95
γ (deg.)	120	120	120

The influence of the arabinose substituent on hydration is illustrated by the unit-cell recorded for the gummy polysaccharide from corm sacs of Watsonia pyramidata where there seems to be two arabino-furanose units per backbone xylose unit. In this case, the hexagonal base plane expands to (5):

$$a = b = 14.0\text{Å} \quad c(\text{fiber repeat}) = 14.9\text{Å}$$

Since this sample loses crystallinity on drying, it is clear that water is involved in maintenance of the organized structure.

Nigeran [Poly (1→4)-α-D-GLCP-(1→3)-α-D-GLCP]

One of the polysaccharides most easily extracted from fungal cell walls is nigeran. It is soluble in warm water (∼60°C) and was shown to be an alternating co-polysaccharide composed of (1→3)-α and (1→4)-α linked D-glucose units.(6,7) The polymer is highly crystalline in the cell wall and was first isolated (6) in 1914. Its role in cell walls or mycelia is not clear at this time and by suitably adjusting the growth medium, one can obtain mycelia with as much as 40% by weight of nigeran.(8) Other polysaccharide constituents of the fungal cell wall are typically: chitin, (1→3)-α-D-glucan, and (1→3)-β-D-glucan, all of which are water insoluble.

The detection of crystalline nigeran in fungal cell walls can be accomplished readily by x-ray diffraction. Even for low percentages of nigeran (3-6%), the crystallinity is high and the reflections are easily recognized and accurately measured.(9) By working with purified material, it was shown that two distinct polymorphs can be identified; the "dry" and "hydrate" forms whose orthorhombic unit-cell dimensions are given in Table II.

Table II

Unit-Cell Dimension of "Dry" and "Hydrate" Nigeran

	"Dry"	"Hydrate"
a ($\overset{o}{A}$)	17.76	17.6
b ($\overset{o}{A}$)	6.0	7.35
c(fiber repeat) ($\overset{o}{A}$)	14.62	13.4

The major dimensional changes in the unit-cell are in the b and c dimensions. Since the latter is related to the chain conformation, its interpretation is important. Recent studies (10,11) have shown that there is no change in the 2_1 helix symmetry on going from "hydrate" to "dry" form. The conclusion is that a slight extension of the helix takes place while the contraction in the b dimension indicates a major structural transformation. Since only one dimension of the base plane is changing, one is tempted to describe the water of hydration in nigeran as sheet-like, with the sheets running parallel to the a axis.

Figure 3 shows electron micrographs of nigeran single crystals. These are lamellae grown from dilute solutions and do not necessarily bear any relationship to the crystalline morphology found in the fungal cell wall. Nevertheless, they have been invaluable in deriving the crystalline parameters of nigeran and are visible proof of the high chemical regularity of this material. Furthermore, the systems of parallel marks which are clearly visible in the "dry" form are macroscopic evidence of the stresses that occur in one direction of the unit-cell base plane as dehydration takes place on the grid of the electron microscope. When the water is removed by solvent exchange to methanol, the smoother uncracked surface is seen.

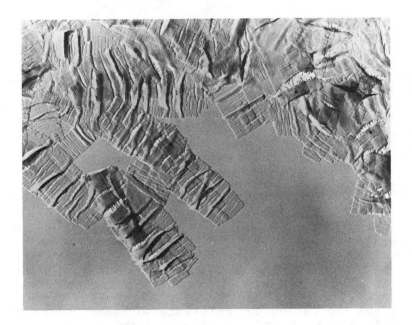

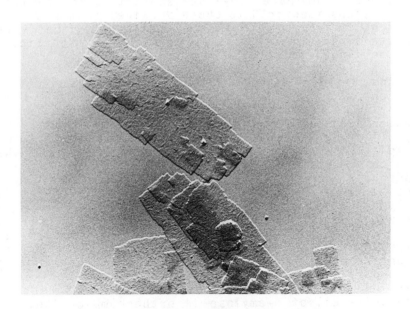

Figure 3. Top: nigeran single crystals grown from ether. Bottom: nigeran single crystals after solvent exchange with methanol.

A simple but descriptive picture of this hydrate structure (12) is as a sandwich with water of hydration in between layers of polysaccharides (Figure 4). This is clearly over-simplified, but in view of the observed morphological changes in the single crystal lamellae, it helps to understand the large scale dimensional changes which occur due to the additive effect of the small changes in b in each of the unit-cells which make up a crystal. The water of hydration also has an effect on the chain conformation, causing a contraction, even though the symmetry of the chain does not change.

Amylose [Poly (1→4)-α-D-GLCP]

Amylose is the linear homopolysaccharide of (1→ 4)-α-D-glucose which when associated with the branched homopolysaccharide amylopectin (Figure 1) [(1→ 4)-α-D-glucan with (1→ 6)-α-D-glucose branch points] forms the commonly occurring polysaccharide, starch. Starch occurs as crystalline granules in nature which give rise to three types of x-ray diffraction diagrams indicating the existance of three polymorphic structures; A, B and C. Since x-ray diffractograms obtained from pure amylose exhibit characteristic reflections identical to those obtained from A, B and C starch granules, it is considered that the crystallinity in starch is due to the amylose portion or to the linear branches of amylopectin.

Native A-starch occurs predominantly in cereal grains, B-starch is found in certain tuberous plants, C-starch is a rare form found in some plants and may actually be a combination of A- and B-starches. Partial conversion of B-starch to A-starch can be accomplished by adjusting temperature and humidity conditions, but complete conversion has never been achieved. When starch is dissolved in hot water, it spontaneously undergoes gelation and subsequent crystallization in the well-known process of retrogradation.(13) The polymorphic form of retrograded starch is B-starch, irrespective of the form of starch initially dissolved.

Recent crystal structure proposals for A- and B-amylose (14,15) consist of parallel-stranded, right-handed double helices (Figure 5) packed in an anti-parallel fashion. Each strand is comprised of six α-D-glucose units in the stable $4C_1$ conformation. The unit-cell of A-amylose is orthorhombic with:

a = 11.90Å, b = 17.70Å and c(fiber repeat) = 10.52Å

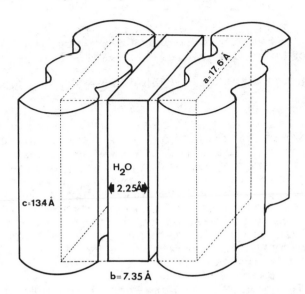

Figure 4. Schematic of nigeran unit cell, showing approximate location of water of hydration

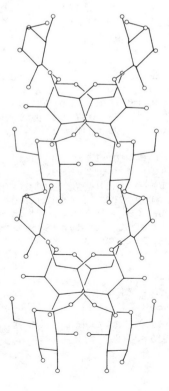

Figure 5. Double helical conformation of amylose

The fiber repeat of 10.52$\overset{o}{A}$ for the double helix indicates that each single chain has a repeat of 21.04$\overset{o}{A}$. The unit-cell of B-amylose is hexagonal with:

$$a = b = 18.50\overset{o}{A} \text{ and } c(\text{fiber repeat}) = 10.40\overset{o}{A}$$

In this case the single chain repeat is 20.8$\overset{o}{A}$, practically identical with the chain in A-amylose. The double helical conformation of A- and B-amylose are nearly identical, however, the two polymorphs differ significantly in degree and type of hydration. The water content of crystalline B-amylose varies from 5% to 27% depending on the relative humidity of the surroundings while A-amylose contains a nearly constant 6% water at various relative humidities. In neither A-nor B-amylose, do the unit-cell constants change with degree of hydration, however, the intensities of diffracted x-rays change with various degrees of hydration in B-amylose.

Upon inspection of the crystal structures of A- and B-amylose, shown schematically in Figure 6, one sees that the amylose double helices pack in hexagonal arrays in both structures; the major difference in the structures being that in B-amylose the centre of the hexagonal array is occupied by water of hydration, whereas in A-amylose this lattice site is filled with another amylose double helix. The water in B-amylose is labile, moving in and out of the structure with great ease. This water may in fact be transported in tuberous plants. In A-amylose, the water of hydration is more sheet-like and more tightly bound to the surrounding amylose double helices. Hence, the A-amylose structure is not as sensitive to the surrounding humidity as is the B-amylose structure.

Conversion of B- to A-amylose on the molecular level occurs with the loss of significant amounts of water followed by a movement of amylose chains into the lattice site vacated by the columnar water of hydration. Starch polymorphism in plants may be a result of the environment in which synthesis occurs. Synthesis and subsequent crystallization may occur as follows: amylose single strands are synthesized first, the strands then intertwine about each other forming the amylose double helix. Crystallization then occurs in either the A or B polymorphic form depending on the amount of water in the environment. This mechanism probably implies low degree of crystallinity in the final material, which is generally the case.

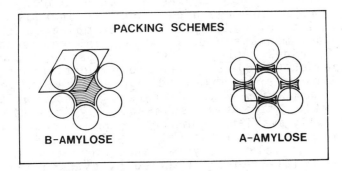

Figure 6. *Schematic of A- and B-amylose unit cell in the* ab *plane; amylose double helices are shown as circles.*

Galactomannan

Ivory nut mannan, poly $(1\rightarrow 4)$-β-D-mannose, is a well-known reserve polysaccharide which is water insoluble (16) and highly crystalline. It occurs also in date seeds where it was shown to disappear upon germination, a characteristic of seed endosperms. The native unit-cell, mannan I, is orthorhombic (17) with:

$$a = 7.21\overset{o}{A}, \quad b = 8.82\overset{o}{A} \text{ and } c(\text{fiber repeat}) = 10.3\overset{o}{A}$$

Intermolecular hydrogen bonds hold the adjacent chains tightly together, indeed, the density of native mannan generally surpasses that of native cellulose. Because of this high cohesion, mannan I is water insoluble and unhydrated. Substitution along the polysaccharide backbone increases water solubility since bulky groups disrupt the fit and regular hydrogen bonding scheme between adjacent molecules, thereby increasing the accessibility of hydroxyls to water molecules.

Galactomannans are a family of seed endosperm polysaccharides (16) with a mannan backbone and appended $(1\rightarrow 6)$-α-D-galactose substituents which render them water soluble (see Figure 1). A wide variety of galactomannans with different mannose/galactose (M/G) ratios have been studied. Typical of these commercially available gums are: Guar, Locust and Tara galactomannans where the ratios M/G are 1.9, 3.2 and 3.7, respectively. Crystalline, oriented films of these materials can be obtained by evaporation of an aqueous solution of these polysaccharides.(18,19)

Table III

Orthorhombic Unit-Cell Parameters (in Å) from Fiber Diagrams of Galactomannans

Sample	a	b	c
at 0% RH			
Guar	13.5	8.7	10.4
Locust	11.6	8.7	10.4
Tara	----	---	----
at 58% RH			
Guar	24.0	8.9	10.4
Locust*	24.0	8.9	10.4
Tara	24.2	9.0	10.4
at 78% RH			
Guar	33.2	9.0	10.4
Locust	30.6	9.0	10.4
Tara	28.3	9.1	10.4

X-ray fiber diagrams recorded at different relative humidities showed three important features (Table III):

o   New diffraction spots develop above 20% relative humidity necessitating a doubling of the unit-cell (a dimension). It seems that water induces a variation of the unit-cell by modifying the relative orientation of the chains.

o   The d-spacings of the most intense equatorial reflection (related to the a dimension only) increase as relative humidity increases; the b and c dimensions do not change with relative humidity; in fact, those values are the same as found in pure mannan I, where galactose is totally absent.

o   Finally, the a dimension of these three samples generally increases with the degree of galactose substitution with b and c remaining constant.

These three observations lead to the conclusion that a good crystalline model consists of a sheet-like arrangement of chains parallel to the unit-cell b axis as first proposed by Palmer and Ballantyne.(19)

By increasing the galactose content or by increasing the relative humidity, a "repulsive force" between the chains in the a direction is felt and the unit-cell expands accordingly. The packing forces in the b direction are the same no matter what the level of galactose substitution and this force seems responsible for the sheet-like hydration mechanism.

Since crystalline mannan I is not hydrated, it seems clear that the role of the $(1 \rightarrow 6)$-α-D-galactose substituent is to encourage hydration and plasticity. In wood cells where the galactoglucomannan (glucose present in the backbone) is a matrix substance, plasticization is probably the desired property. In seeds where galactomannan is a constitutent of an endosperm, controlled hydration to facilitate attack by some enzyme is probably the important feature. The facility of hydration can control the specific time of germination of seeds especially in a desert environment.

As is expected with this type of hydrated structure, the best x-ray patterns (best ordered sample) are obtained when the relative humidity is in the middle range (40% to 80%). At 98% relative humidity, the amount of water begins to solubilize the chains, thus destroying the crystallinity. At 0% relative humidity, the lack of water limits the mobility of the polysaccharide elements and the chains have difficulty finding the regular arrangement of the crystalline state.

(1→3)-β-D-Glucan and Xylan

Another polysaccharide which displays interesting hydration phenomena is (1→3)-β-D-glucan, often called paramylon (20), curdlan (21) or laminaran (16). The molecular crystalline arrangement of this polysaccharide consists of a triple helix formed by three intertwining 6_1 helices.(22,23) Two polymorphs (22) are observed when x-ray diagrams of well-crystallized fibers are recorded at different relative humidities. If the sample is placed under vacuum, the "dry" polymorph is obtained. The "hydrated" polymorph is found at 75% relative humidity and contains two water molecules per glucose residue. The transformation between the two forms is reversible, the critical relative humidity being around 20% depending on sample history.
        The unit-cell of both polymorphs is hexagonal with the parameters shown in Table IV. The striking variation in c can be easily explained in terms of a loss of symmetry in the crystalline structure when going from the "dry" form to the "hydrate" form. Since no physical modification of the fibers is observed, it is unlikely that the increase in fiber repeat is caused by an actual physical stretching of the chain. The triple helical structure is composed of three equivalent strands related by a three-fold symmetry operation. The repeat of such a structure is 1/3 of the repeat of a single chain (Figure 7). However, if the strands are not identical, the three-fold symmetry is lost and the fiber repeat of the whole structure is that of the single strand, i.e., that found in the "hydrate".

Table IV

Unit-Cell Parameters of the Polymorphs of (1→3)-β-D-Glucan and (1→3)-β-D-Xylan

	(1→3)-β-D-Glucan		(1→3)-β-D-Xylan	
	Dry	Hydrated	Dry	Hydrated
a = b (Å)	14.6	15.6	13.7	15.4
c (Å)	5.8	18.6	5.88	6.12
Helix Pitch (Å)	17.34	18.6	17.64	18.36

All of the unit-cell dimensions increase when converting from the "dry" form to the "hydrate". The a b plane expands equally in both directions and it is believed that the water of hydration is disposed between the triple helices in a columnar fashion (Figure 8). Interhelix hydrogen bonds are probably broken and/or replaced by new bonds involving water molecules. This mechanism could lead to the loss of three-fold symmetry. The three strands are no longer equivalent, possibly due to different hydroxymethyl group rotameric positions.

(1→3)-β-D-xylan behaves similarly to the glucan even though its C(5) carbon lacks the CH_2OH group. Nevertheless, the crystal structures of these two polysaccharides are very similar, consisting of triple helices.(24) The xylan unit-cell is hexagonal with the parameters given in Table IV. The effect of hydration is about the same as in the corresponding glucan structure. However, in the case of the xylan, the three-fold symmetry is not lost in the hydrated form. The correlation of this phenomenon with the absence of the hydroxymethyl group seems obvious. The presentation of three-fold symmetry is an indication that no water molecules can be accommodated in the middle of the triple helix. It is impossible to introduce three coplanar water molecules and retain the three-fold symmetry in a cavity of approximately 3A diameter.

In conclusion, it is very likely that the hydration of both (1→3)-β-D-glucan and (1→3)-β-D-xylan is in the form of columns between the triple helices which bring about an increase in the unit-cell dimensions of the hydrate.

Discussion

The observations discussed above were deliberately restricted to neutral glycans in order to avoid hydrating phenomena related to polyelectrolyte behaviour. The latter effect is found in the mucopolysaccharides (25) and sulfated algal polysaccharides.(26) The glycans of this review are from the cell walls of flowering plants, algae or fungi. Their roles are clearly structural, matricial or reserve (Table V). Present understanding of the short-range non-bonded interaction allows one to predict general features of single chain conformation, but details such as multiple helix formation or hydration and its effect on comformation are still beyond theoretical prediction.

In general, hydration of polysaccharides is an element of structural adaptation. Structures that hydrate will show stress relaxation under tension and

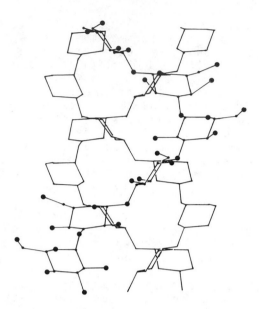

Figure 7. *Triple helical conformation of (1 → 3)-β-D-glucan*

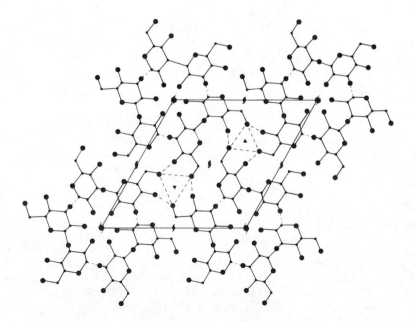

Figure 8. ab *projection of (1 → 3)-β-D-glucan anhydrous unit cell*

Table V
_____

Role in Nature, Hydrate Form,
Unit-Cell Type and Conformational
Symmetry of Various Polysaccharides

Polysaccharide	Role	Hydration	Unit-Cell	Conformational Symmetry
$(1\rightarrow3)$-β-D-Glucan	Structural, Reserve	Columnar	Hexagonal	6_1
$(1\rightarrow3)$-β-D-Xylan	Structural, Reserve	Columnar	Hexagonal	6_1
$(1\rightarrow4)$-β-D-Galactomannan	Matrix	Sheet-Like	Orthorhombic	2_1
$(1\rightarrow4)$-β-D-Xylan	Matrix	Columnar	Hexagonal	3_1
Nigeran	?	Sheet-Like	Orthorhombic	2_1
Amylose	Reserve	Columnar, Sheet-Like	Hexagonal, Orthorhombic	6_1

the rate will be a function of relative water content. Of the two categories discussed, columnar hydration is more reversible and less morphologically damaging. It is the only kind expected in a structural material. Sheet-like hydration causes uneven expansion or contraction of the material and thereby destroys the structural cohesion. The effect can be gauged by the appearance of the vacuum dried single crystals of nigeran shown in Figure 2a. The system of parallel cracks created by the drying stresses are irreversible and are of a size to allow enzymes to access the inner surfaces of a cell wall. Thus, mere drying of a fungal wall containing nigeran increases its susceptibility to enzyme attack.(27)

The strain effects due to dehydration are nowhere better illustrated than in the precautions that must be taken to dry and season wood before its use. This problem relates to water removal from capillaries but certainly the dehydration of the hemicelluloses which are in a paracrystalline order at the surface of the microfibrils (28) must play a role. Xylan (4) in hardwoods and galactoglucomannans (18) in softwoods have the characteristics of columnar and sheet hydration, respectively.

Seed germination is a phenomenon which requires moisture, hence hydration of the polysaccharide in the endosperm. Galactomannans seem tailored to adapt to the environmental requirements of plants located in tropical areas where moisture is seasonal. By comparison, starch hydration is probably more gradual and reversible, a situation more in keeping with a temperate climate.

From a thermodynamic point of view, one might expect that polysaccharide crystallites would display distinct hydrates and not show continuous variation in water content as a function of relative humidity. So far, the continuous variation in unit-cell parameters as a function of relative humidity seems to be the rule. However, the variation of cell parameters with relative humidity seems to follow the shape of the moisture sorption curve.(18) In all probability, the fine structure factor introduces localized strain effects which prevents detection of a unique hydrate at a given relative humidity.

Literature Cited

(1)   Flory, P. J., Faraday Discussions of the Chemical Society, 1974, 57, 7.

(2)   Freundlich,   H.,   "Kapillarchemie",   V2, Akademische Verlagsgesellschaft, M8H, Leipsig, Germany, 1930, p. 568.

(3)   Timell, T. E., Adv. Carbohydrate Chem., 1964, 19, 247.

(4)   Nieduszynski, I. A.; Marchessault, R. H., Biopolymers, 1972, 11, 1335.

(5)   Lelliott, C.; Atkins, E. D. T.; Juritz, J. W.; Stephen, A. M., Polymer, 1978, 19, 363.

(6)   Barker, S. A.; Carrington, T. R., J. Chem. Soc., 1953, 3588.

(7)   Barker, S. A.; Bourne, E. J.; O'Mant, D. M.; Stacey, M. A., J. Chem. Soc., 1957, 2448.

(8)   Dox, A. W.; Niedig, R. E., J. Biol. Chem., 1914, 18, 167.

(9)   Reese, E. T.; Mandels, M., Can. J. Microbiol., 1964, 10, 103.

(10)   Sundararajan, P. R.; Marchessault, R. H.; Quigley, G. J.; Sarko, A., J. Am. Chem. Soc., 1973, 95, 2001.

(11)   Perez, S., Roux, M.; Revol, J. F.; Marchessault, R. H., J. Mol. Biol., 1979, 129, 113.

(12)   Taylor, K. J., Ph.D. Thesis, Universite de Montreal, 1976.

(13)   Whistler, R. L.; Smart, C. L., "Polysaccharide Chemistry", Academic Press, New York, 1953.

(14)   Wu, H-C. H.; Sarko, A., Carbohyd. Res., 1978, 61, 7.

(15)   ibid, 26.

(16) Aspinal, G. O., "Polysaccharides", Pergamon Press, Elmford, New York, 1970.

(17) Nieduszynski, I.; Marchessault, R. H., Can. J. Chem., 1972, 50, 2130.

(18) Marchessault, R. H.; Buleon, A.; Deslandes, Y.; Goto, T., J. Colloid and Interface Sci., in press.

(19) Palmer, K. J.; Ballantyne, J. Am. Chem. Soc., 1950, 72, 736.

(20) Barras, D. R.; Stone, B. A., "The Biology of Euglena", Academic Press, New York, 1969.

(21) Harada, T., Process Biochem., 1974, 9, 21.

(22) Marchessault, R. H.; Deslandes, Y.; Ogawa, K.; Sundararajan, P. R., Can. J. Chem., 1977, 55, 300.

(23) Bluhm, T. L.; Sarko, A., Can. J. Chem., 1977, 55, 293.

(24) Atkins, E. D. T.; Parker, K. D., J. Polym. Sci., Part C, 1968, 28, 69.

(25) Winter, W. T.; Smith, P. J. C.; Arnott, S., J. Mol. Biol., 1975, 99, 219.

(26) Arnott, S.; Scott, W. E.; Rees, D. A.; McNab, C. G. A., J. Mol. Biol., 1974, 90, 253.

(27) Nordin, J.; Bobbitt, T., unpublished result.

(28) Marchessault, R. H.; Liang, C. Y., J. Polym. Sci., 1962, 59, 357.

RECEIVED January 4, 1980.

Measurement of Bound (Nonfreezing) Water by Differential Scanning Calorimetry

SUBHASH DEODHAR

Department of Paper Science and Engineering, University of Wisconsin, Stevens Point, WI 54481

PHILIP LUNER

College of Environmental Science and Forestry, State University of New York, Syracuse, NY 13210

Various terms have been used to characterize the water associated with cellulose fibers. Bound water, imbibed water, water of constitution, adsorbed water, fiber saturation point are some of the terms that have been used to describe the water in pulps and papers. The origin of each term can be traced to either theoretical considerations or to the experimental method of measurement. Bound water has been the most popular term used to describe the associated water. Bulk water or free water is that portion of water not associated (or not bound) with the fibers. Two measurement techniques may not yield identical values of bound water. The non-existence of a sharp boundary between bound water and free water is one reason for such discrepancies. In addition, two methods may be measuring different physical phenomena.

Measurement of Bound Water

NMR (Nuclear Magnetic Resonance) (1,2,3,4). This technique detects the mobility of protons in various energy states. The hydrogen atoms in bound water are at different energy levels than the hydrogen atoms in free water. These energy levels are measured and recorded in the form of NMR spectra. The bound water can be calculated from the NMR spectrum. NMR measurements may be done at any temperature. While NMR may be the most basic method for the measurement of bound water, it requires expensive equipment, trained personnel, and considerable preparation for each experiment. These requirements are not frequently available to the researcher in the paper industry.

DSC (Differential Scanning Calorimeter (4,5,6). When a wet pulp sample is cooled well below 0°C, the free water freezes but the bound water remains in the non-frozen state. When the frozen sample is heated in a calorimeter, the heat required to melt the frozen water can be measured. Non-frozen water, which is defined as the bound water, is the difference between the total water and the frozen water. The freezing of free water and non-freezing of

0-8412-0559-0/80/47-127-273$05.00/0
© 1980 American Chemical Society

bound water are thermodynamic phenomena, so the measurement is absolute, but as the bound water is calculated by difference, the measurement is indirect. The bound water is measured only at the freezing point. The experimental procedure is simpler than NMR.

Solute Exclusion Method (7,8). The bound water is defined as that water in a swollen fiber structure which does not act as a solvent for a critical size solute molecule. Essentially it measures the volume of all the pores smaller than the critical pore size. The critical size for the pores or the solute molecules is arbitrarily set. As the thermodynamics of bound water does not enter, this method is not absolute. The solute exclusion method is a relatively simple technique, which does not require expensive equipment but does require experimental precision.

Thermogravimetric or the Drying Rate Method (9,10). A wet fiber sample is dried under controlled conditions to obtain the drying rate curve. The moisture content at the boundary of constant-rate drying period and falling rate drying period is defined as bound water. The drying rates are highly dependent upon the diffusion rates and the geometry of the sample. These measurements may be made at any temperature. The bound water data on papermaking fibers can be used in estimating the rate of drying in the production of paper. The drying rate method is not a high powered analytical technique, but it does provide direct measurement of bound water as it influences drying of paper.

WRV (Water Retention Value). The water retained when fibers are subjected to external force is known as the Water Retention Value. Water retained by surface tension forces in addition to adsorbed water may be involved in this determination. The methods used in determining this value are centrifuging (water retained under standard centrifuging) (11,12), hydrostatic tension (water retained under standard tension) (13), and suction (water retained under suction or vacuum) (14). All these methods can be employed at any temperature and the experimental techniques are quite simple.

Experimental

DSC Measurement. The non-freezing water of wet pulp and paper samples was determined by Differential Scanning Calorimetry. A wet sample was hermetically sealed in a sample pan. The empty sample pan and the sealed pan were weighed. The sealed pan was quickly frozen inside the DSC chamber to -40°C and several minutes were allowed for the system to come to equilibrium. The sample holder assembly was then heated at a rate of 5°C/min. A scanning speed of 5°C/min was found to give the optimum values of peak height and peak spread. This minimized the errors in experimental measurements. After the DSC measurement, the pan was punctured and heated in a vacuum oven at 105°C to obtain the dry weight.

A strip chart recorder connected to the Perkin-Elmer DSC - 1B recorded the melting behavior. A straight base line was obtained when no physical change took place in the sample. As soon as the frozen water started melting, it absorbed heat and lowered the temperature. The instrument supplied heat to the sample so that the sample temperature was maintained equal to the reference temperature. The amount of heat supplied to equalize the temperatures was recorded on the strip chart recorder. The total area under the DSC curve given by the recorder is proportional to the total heat supplied to equalize the temperatures, i.e., the heat required to melt the frozen sample. An integrator performed the integration at the same time the curve was being traced. The integrator was calibrated using distilled water as a sample. Thus, from the integrator reading, the heat required to melt the frozen sample was calculated knowing the calibration factor.

Each wet pulp or paper specimen was centrifuged at 3000 rpm for 30 minutes. The centrifugal force was equivalent to 900g's. The centrifugal method of water retention value (WRV) under conditions of 900g and 30 minutes has been suggested by Scallan and Carles (12). Thus the water content of the pulps was reduced to the level of WRV before determining the non-freezing water. For each specimen, minimum of 5 runs were made and the average of five samples is reported as non-freezing water. (The complete set of results including the statistical analysis is available from the authors).

Pulp Preparation and Characteristics. Several papermaking variables were selected for studying the effects on the non-freezing water. These were beating, drying, pressing, removal of fines, and addition of salts. A hardwood bleached pulp was used for the once-dried pulp while for the never-dried pulp, a 50% yield spruce bleached kraft was used. The pulps were beaten in a Valley beater according to TAPPI standard T-200 and handsheets made. The non-freezing water values were determined for both the pulp and the rewetted handsheets. The density, breaking length, and elastic modulus of the handsheets were also measured by standard TAPPI procedures.

The effect of fines was studied for unbleached, never-dried spruce sulfite pulp (yield 57.6%). The fines were removed using a 200-mesh screen. The non-freezing water was measured for the pulp and the rewetted handsheets. To prepare a sheet from fines, the fines suspension was poured into a flat dish and the water allowed to evaporate under ambient conditions. The sheet did not disintegrate on rewetting.

To study the effect of applied pressure on non-freezing water, handsheets were made in a TAPPI standard sheet mold and manually pressed at various pressures up to 1000 psi. The handsheets were pressed between a blotter and a steel plate. When pressed at high pressures, the wet handsheets adhered strongly to the blotter

and to the steel plate. To prevent the adhesion and for easy separation of the sheet after pressing, a filter paper was placed between the sheet and the steel plate. For the sake of uniformity,
this procedure was followed for all pressures, even though it was
not necessary at low pressures. One group of handsheets was replaced in the water immediately after pressing so the fibers in
this group were never allowed to dry. The remaining handsheets
were dried in the humidity room under standard conditions. The
bound water was determined for these sheets after they were soaked
in water overnight. The density, breaking length, and elastic
modulus of the dried sheets were determined by TAPPI standard
methods. Non-freezing water measurements were made on sulfite,
kraft and mechanical pulps at several levels of pressure. The
sulfite pulp was pulped from spruce to a 57.6% yield. It was unbleached, unbeaten and never dried. The freeness of the dispersed
pulp was 635 ml CSF. The kraft pulp (50% yield) was pulped from
the same spruce as the sulfite pulp. It was unbleached, unbeaten
and never dried. The freeness of the dispersed pulp was 675 ml
CSF. The groundwood was in the dry lap form and when dispersed,
it had a freeness value of 465 ml CSF.

Effect of Salts on Non-freezing Water. The non-freezing water of pulps in the presence of salts was also determined. The
salts added to the pulp were KNO_3, CsCl, KI, $MgSO_4$, LiCl, Li_2SO_4,
and $Al_2(SO_4)_3$. The first three are considered structure breakers
while the last four are structure makers. The pulp used was a
57.6% yield spruce sulfite pulp. It was unbleached, unbeaten and
never dried. The pulp was dispersed in 1 M salt solution. To
study the effect of concentration, 0.5 M and 0.1 M salt solutions
were also used for dispersion in the case of KNO_3 and Li_2SO_4. The
pH of the dispersion before and after salt addition was measured.
The pH was not adjusted. The dispersion was stirred gently for a
few hours by a magnetic stirrer bar. The dispersion was centrifuged at 3000 rpm for 30 minutes (900g's) as had done for other
pulp samples before the non-freezing water determination. No attempt was made to wash the salts from the pulps.

Capillary Condensation - Freezing Point Depression

As a consequence of surface tension, there is a balancing
pressure difference across any curved interface. Thus, the vapor
pressure over a concave liquid surface will be smaller than that
over a corresponding flat surface. This vapor pressure difference
can be calculated from the Kelvin's equation:

$$RT \ln P_r/P_o = \frac{2\sigma V_m \cos\theta}{r}$$

Where P_r is the vapor pressure in the capillary of radius r, P_o is
the vapor pressure of free water; σ, V_m and θ are surface tension,
molar volume and contact angle of the water, respectively.

If it is assumed that an absorbed layer of water exists before capillary condensation takes place and that this layer consists of ordered or oriented water molecules, then the contact
angle in Kelvin's equation should be very close to zero. With
zero contact angle, vapor pressures in the capillaries are calculated from Kelvin's equation for capillaries from 10 Å to 200 Å,
Table I. The vapor pressure of water below 0 C (15) is compared
with the vapor pressures in the capillaries to obtain the freezing
points. Figure 1 shows the relation between the freezing point
depression and the capillary radius.

Table I: Capillary radii, vapor pressures and freezing point
of water from Kelvin's equation.

Capillary Radius, Å	Vapor Pressure, mm of Hg	Freezing Point, ^oC
∞	4.579	0
1000	4.526	−0.15
200	4.322	−0.8
150	4.258	−1.0
100	4.08	−1.6
75	3.925	−2.11
50	3.634	−3.15
40	3.41	−4.0
30	3.07	−5.4
20	2.368	−8.8
10	1.443	−15.0

The DSC curves of the non-freezing water for all pulp and
paper samples gave a minimum melting point of −4°C. A freezing
point depression of 4°C corresponds to a 40 Å radius capillary.
Since freezing or melting was not observed below −4°C, it may be
concluded that the water in capillaries smaller than 40 Å did not
freeze. A critical pore size can be defined as the largest pore
that can carry 100% non-freezing water. This value is chosen as
40 Å. The choice is subjective to some extent.

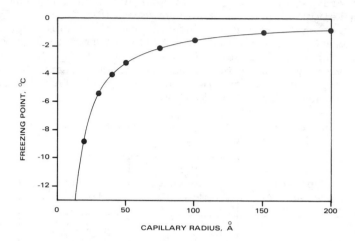

Figure 1.   Freezing point of water in capillaries (from Kelvin's equation)

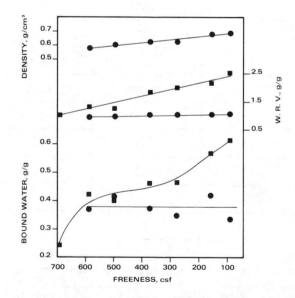

Figure 2.   Effect of beating on once-dried Kraft pulp: (■) pulp; (●) rewetted handsheets

Results and Discussion

Effect of Beating on Non-freezing Water. It is well known
that beating pulp results in an increased water retention value.
It was therefore of interest to study the effect of beating on the
non-freezing water on several types of pulps. Figure 2 shows the
changes in the non-freezing water and WRV of a dried hardwood
kraft (commercial) pulp and handsheets made from this pulp which
was dried and rewetted. Both the non-freezing water and water retention value for the pulp increased on beating. Two zones were
observed, one from 700 to 600 CSF and one from 300 CSF to lower
freeness values. However, both non-freezing water and WRV for the
rewetted samples remained constant.

Figure 3 shows the same measurements for a bleached, never-
dried kraft (Spruce) pulp. In contrast to Figure 2, beating did
not alter significantly the bound water values for the pulp or the
rewetted handsheet. However, the WRV for the pulp increased with
the initial beating and was higher than the hardwood kraft pulp
(Figure 2). It is interesting to note that the non-freezing water
of the rewetted papers of the two pulps was constant and similar,
and also that the beaten, once-dried hardwood pulp had a higher
non-freezing water value than the beaten never-dried softwood
pulp. The higher non-freezing water values of the fines in the
hardwood pulp may be responsible for these results. In fact, as
shown in Table II, the non-freezing water of fines may be four
times the value of fibers. Thus, pulp fines may contribute significantly to the non-freezing water values, but once dried into
sheets, they become part of the fiber and do not contribute to the
non-freezing water on rewetting.

Table II: Non-freezing water for fines, fines-free pulp and handsheets, the never-dried sulfite pulp beaten in valley beater.

Sample	Non-freezing water, g/g		W.R.V., g/g	
	av.	st.dev.	av.	st.dev.
whole pulp	0.568	0.03	2.92	0.55
fines-free pulp	0.538	0.13	2.17	0.30
Handsheets from whole pulp	0.443	0.03	1.26	0.11
fines-free pulp	0.465	0.12	1.26	0.17
fines	1.85	–	13.4	–
once dried fines	0.611	0.07	2.68	0.4

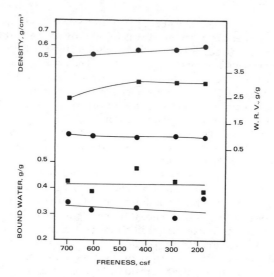

Figure 3.   *Effect of beating on never-dried Kraft pulp: (■) pulp; (●) rewetted handsheets*

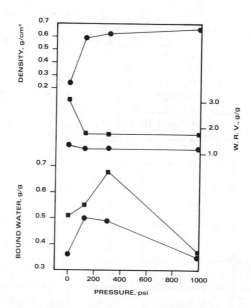

Figure 4.   *Effect of pressing on sulfite sheet: (■) sheets pressed and rewetted; (●) pressed, dried, and rewetted*

Effect of Pressing on Non-freezing Water. Figures 4, 5, and 6 show that the non-freezing water initially increases on pressed rewetted sheets and on pressed, dried, and rewetted handsheets for sulfite, kraft, and groundwood pulps. However, at higher pressures, the non-freezing water values decrease. This effect varies with the type of pulp.

At this point, the following picture can be developed to qualitatively account for these results. The non-freezing water in both pulps and paper sheets is the result of the strong water-cellulose surface interaction. The extent of this interaction depends on the surface characteristics of the fiber. Thus, hemicelluloses in fibers and fines would be expected to give higher non-freezing water values. Coupled with these chemical effects is the physical presence of pores in pulps and handsheets. The cell wall is composed of a porous gel-like system (8). In a paper sheet, pores may also exist between the fiber elements as well. The presence of these small pores can also result in non-freezing water. Hence, the non-freezing water in pulp and paper appears as the result of both the physical presence of pores and strong water-surface interactions.

The increase in the non-freezing water values in Figure 1 could originate from the creation of small pores during the beating of the dried pulp as well as from the fines. Little effect on the non-freezing water value is observed after the paper is dried and rewetted. This is the result of pore closing during drying. In contrast, no new small pores are created on beating a never-dried softwood pulp and thus the non-freezing water shows no change.

The increase in non-freezing water on pressing (Figs. 4, 5, and 6) may be attributed to the consolidation of the pore structure. First, at the lower pressures, the larger pores (>40 Å) where water is frozen are consolidated to pores below the critical pore size (40 Å). This leads to an increase in non-freezing water. However, on further increase in plate pressure, the non-freezing water decreases. This decrease may be the direct result of pore closing as a result of drying between the blotter. Indeed, after pressing at 1000 psi, the moisture content for the kraft pulp decreased from 2.97 g/g to 0.84 g/g of dry pulp. It is therefore very conceivable that some of the smaller pores were closed during moisture removal. Indeed, Caulfield and Weatherwax (16) have shown that 20% of water is held in pores 12 Å or less. It seems likely that these pores are closed on pressing and/or drying. Support for this mechanism is based on the fact that the dried and rewetted sheets show a similar pattern with pressing as the non-dried sheets.

The maxima in the pressing curves in Figures 4, 5, and 6 is thus believed to be the result of two effects, the creation of small pores <40 Å and the closing of the <12 Å pores on pressing. The difference in the maxima between the figures is furthermore a function of the type of pulp and its state of dispersion.

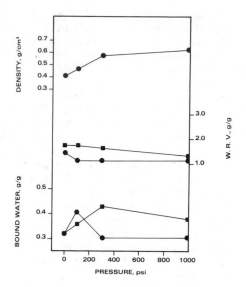

Figure 5.   Effect of pressure on Kraft sheets: (■) sheets pressed and rewetted; (●)
pressed, dried, and rewetted

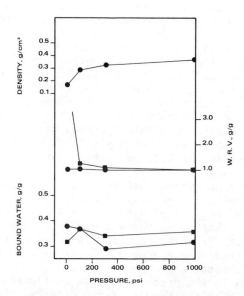

Figure 6.   Effect of pressing on ground wood pulp: (■) sheets pressed and rewetted;
(●) pressed, dried, and rewetted

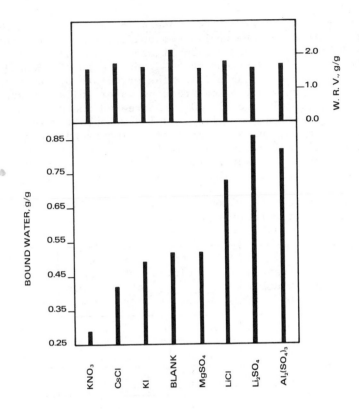

Figure 7.   *Addition of salts to the sulfite pulp*

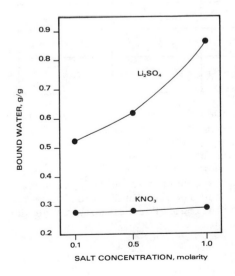

Figure 8.   *Bound water of sulfite pulp as a function of concentration of Li_2SO_4 and KNO_3*

Non-freezing Water in the Presence of Salts. The non-freezing water values for the unbleached sulfite pulp containing 1 molar salt solutions is shown in Figure 7. The viscosities of these aqueous salt solutions at 20°C and the pH of the pulp dispersions are shown in Table III. The structure breakers (KNO_3, CsCl and KI decreased the non-freezing water while the structure makers ($MgSO_4$, LiCl, and $Al_2(SO_4)_3$ increased the non-freezing water. In terms of the capillary model developed for the non-freezing water in pulp, the structure makers increased the critical pore size while a decrease in capillary size occurred with the addition of structure makers. These changes can occur by the adsorption of cations and anions at the fiber surface.

Table III: Properties of salt solutions $Al_2(SO_4)_3$ solution is 0.15M. All other solutions are one molar.

Salt	Viscosity of 1M soln. cp	$\frac{\mu \text{ 1M soln.}}{\mu \text{ water}}$	Surface tension of 1M soln. dynes/cm	pH of salt suspension
KNO_3	0.872	0.976	73.78	6.3
CsCl	0.872	0.975		6.3
KI	0.837	0.936	73.6	7.4
Water	0.894	1.000	72.75	7.0
$MgSO_4$	1.725	1.93	74.85	6.25
LiCl	1.021	1.142	74.68	6.0
Li_2SO_4	1.484	1.66	75.4	8.5
$Al_2(SO_4)_3$	1.207	1.35	79.73	2.6

All the non-freezing water values in Figure 7 were determined in 1 M salt solutions. It was therefore of interest to explore the effect of salt concentration. Figure 8 shows the effect of salt concentration for one structure breaker (KNO_3) and one structure maker (Li_2SO_4). The results show that Li_2SO_4 is strongly concentration dependent while KNO_3 is concentration independent. These results cast some doubt that the results in Figure 7 are due solely to the cation of the salt solution. Further work is needed for clarifying these results.

Abstract

A critical pore size is defined as the largest pore that can carry 100% non-freezing water. Pores of size larger than critical pore size will have freezing and non-freezing water. The critical pore size for pulp fibers is estimated as 40 Å. Since more than 98% of the surface area may be in pores smaller than 40 Å, most of the non-freezing or the bound water will be in pores smaller than critical pore size.

Drying of fibers results in irreversible closing of very small pores. It appears that pores smaller than 12 Å will close irreversibly, which do not open up on reslushing but only if the pulp is beaten. Pressing of pulp sheets alters pore size distribution. Large pores are reduced in size which increases number of pores smaller than critical pore size. The very small pores close on pressing, thereby decreasing the number of pores. These two opposing effects give rise to a maximum in the bound water-pressure curve.

The structure breaker salts decrease the bound water whereas the structure maker salts increase the bound water. To explain the observed bound water trend, it is hypothesized that structure breakers decrease the critical pore size and structure makers increase this value. The viscosities of aqueous salt solutions are used as criteria to separate structure breaker from structure maker salts.

Literature Cited

1. Swanson, T.; Stejshal, E. O.; Tarkow, H., Tappi, 1962, 45(12), 929.
2. Dehl, R. E., J. Chem. Phys., 1968, 48, 831.
3. Ogiwara, Y.; Kubuta, H.; Hayaski, S.; Mitomo, N., J. App. Poly. Sci., 1969, 13, 1689.
4. Frommer, M. A. and Lancet, D., J. App. Poly. Sci., 1972, 16, 1295.
5. Stamm, A. J., Wood Science, 1971, 4(2), 114.
6. Magne, F. C. and Shau, E. L., Textile Research J., 1952, 22, 748.
7. Feist, W. C. and Tarkow, H., Forest Products Journal, 1967, 17 (10), 65.
8. Stone, J. C. and Scallan, A. M., Tappi, 1967, 50(10), 496.
9. Ayer, J. E., Tappi, 1958, 41(5), 237.
10.Law, K. N.; Garceau, J. J.; Kokta, B. V., J. Text. Res., 1975, 45(2), 127.
11.Jayme, G., Tappi, 1958, 14, 180A.
12.Scallan, A. M. and Carles, J. E., Svensk. Paperstidn., 1972, 75 (17), 699.
13. Robertson, A. A., Tappi, 1959, 42, 969.

14. Nordman, L. and Aaltonen, P., *Papier*, 1960, 14, 565.
15. Perry, R. H., "Chemical Engineer's Handbook", 5th ed., McGraw-Hill Book Company, New York, 1973, PP. 3-45.
16. Caulfield, D. F. and Weatherwax, R. C., *Tappi*, 1976, 59(7), 114.

RECEIVED January 4, 1980.

Water in Mucopolysaccharides

Y. IKADA, M. SUZUKI, and H. IWATA

Institute for Chemical Research, Kyoto University, Uji, Kyoto 611, Japan

Many investigations have been devoted to exploration of the structure of water in aqueous solutions of polymers as well as in water-adsorbed or water-swollen polymeric substances. Although there is some controversy among researchers regarding the actual structure, it is generally accepted that water molecules in the vicinity of the polymer segments behave somewhat differently from the normal "bulk" water because of their interaction with the polymer (1,2,3,4). This anomalous water is often called "bound", "non-freezing", "hydrated", "ordered", and so on. Moreover, some workers have pointed out that there may be present another type of water which is neither identical to the bulk nor to the bound water (5,6,7). The amount of these anomalous waters is apparently dependent on the experimental methods employed. Most of the works on the structure of water have been carried out using gravimetric, calorimetric, infrared, dielectric, NMR, or ultrasonic velocity measurements.

Recently we have studied the water structure in water-swollen gel membranes prepared from water-soluble, non-ionizable polymers by means of differential scanning calorimetry (DSC). This work has revealed that all of the gels studied have the non-freezing water and, in addition, that some of them give a endothermic peak at lower temperatures than 0°C as the water content of the gels increases beyond the content of the non-freezing water (8).

This work is a continuation of the preceding study on the synthetic polymer-water system and is concerned with the organization of water in aqueous solutions of mucopolysaccharides. Extensive studies have been conducted on the hydration of proteins and polypeptides (9), but very few exist on the hydration of mucopolysaccharides (10-15), though their biological activity is always exerted in aqueous environments. For instance, chondroitin homologs are typical mucopolysaccharides distributing in connective tissues in a highly hydrated state. Hyaluronic acid, consisting of a repeating disaccharide unit from D-glucuronic acid and 2-acetamide-2-deoxy-D-glucose, is found to exist in biological fluids such as vitreous humour, umbilical cord and synovial fluid.

0-8412-0559-0/80/47-127-287$05.00/0
© 1980 American Chemical Society

Heparin, a natural mannalian mucopolysaccharide, has attracted the attention of biologists because of its specific interaction with various physiological entities including antithrombin III, platelets, and lipoprotein lipase (16).

In order to shed more light on the interaction of water with these mucopolysaccharides, we employ the DSC method to follow the melting behavior of aqueous solutions of the polysaccharides cooled to -50°C. The polymers chosen here are chondroitin sulfate A (Chn S-A), chondroitin sulfate C (Chn S-C), chondroitin (Chn), heparin (Hpn), and hyaluronic acid (HyA). Their chemical structures are shown in Figure 1. The DSC curves allow us to determine the amount of the non-freezing water in highly concentrated solutions, since any endothermic peak is not observed for such solutions over a wide temperature range (17,18). This paper will also describe the presence of more than one endothermic peak in the DSC curves for solutions of relatively low polymer concentrations. To the best of our knowledge observation of such multiple peaks has not yet been reported anywhere.

Experimental

Materials. Na salts of the mucopolysaccharides were purchased from Seikagaku Kogyo Co., Tokyo, Japan and used as received. Ca salts of these polymers were prepared by addition of a saturated solution of $CaCl_2$ to the aqueous solutions of their Na salts, followed by dialysis with deionized water. Intrinsic viscosities of the Na salts in buffered aqueous solutions (19) are noted in Table I, together with the degree of conversion of Na to Ca for the Ca salts, determined by atomic adsorption spectroscopy. A partially sulfated poly(vinyl alcohol) (PVA-S) was synthesized by esterification of PVA with sulfuric acid to a 20.7 mol % conversion and then neutralized with $NaHCO_3$.

TABLE I.

CHARACTERISTICS OF MUCOPOLYSACCHARIDES

Mucopolysaccharide	Intrinsic viscosity[1] (dl/g)	Degree of Conversion of Na to Ca for the Ca salt (mol. %)
Chn S-A	0.450	98
Chn S-C	1.09	98
Hpn	0.177	96
Chn	0.243	98
HyA	—	—

1) 25°C, in 0.15M sodium phosphate buffer +0.2M NaCl, pH 7.0.

DSC Measurements. The Perkin-Elmer Model DSC-1B differential scanning calorimeter was used to determine the heats of fusion of

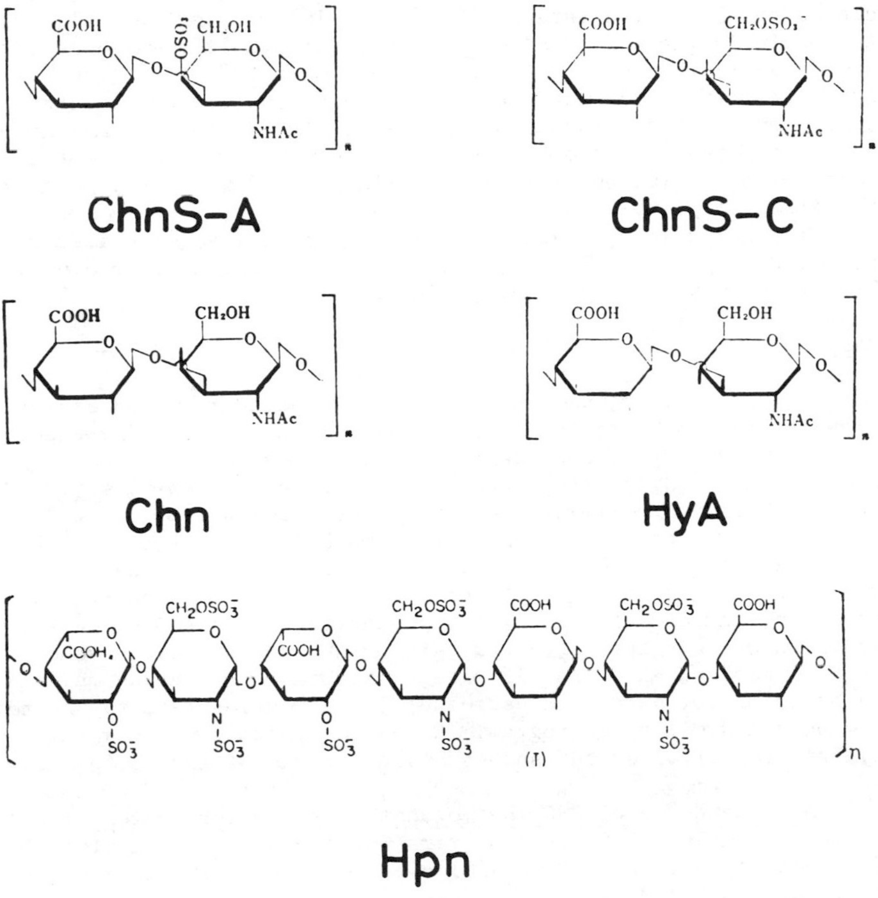

Figure 1. Structures of repeating units of mucopolysaccharides

water associated with these mucopolysaccharides. 1-5 mg of the
aqueous solutions with different polymer concentrations were
placed in aluminium pans to be used for volatile samples and
hermetically sealed. The sealed capsules were put on the sample
holder, on which the Dewar flask supplied by the Perkin-Elmer
Corporation was placed; then dry N_2 gas was purged for 5 mins at
room temperature to remove the moisture in the holder. After that,
the Dewar flask was filled with liquid N_2 and the samples were
cooled to -50°C at a rate of 10°C/min. The samples were allowed
to remain in the cavity at -50°C for 15 mins, after which scanning
was conducted by heating them, from -50°C to 30°C at a rate of
10°C/min. The total water content in each capsule was determined
subsequent to the calorimetric procedure by puncturing the capsule
and drying the sample overnight at 110°C. The water content of
each polymer solution will be given in gram of water per gram of
the dried polymer.

The temperature scale for the DSC curves was calibrated using
the melting point of bulk ice and the peak areas above the base-
line were measured with a planimeter.

Results

In general, thermal changes such as endothermic and exothermic
transitions can be followed either by raising or lowering the
sample temperature. Andrade and his coworkers obtained DSC
thermograms of synthetic hydrogels by cooling the samples (20).
In preliminary experiments we always observed an exothermic
transition to occur at a temperature much lower than the expected
freezing point, when aqueous solutions of mucopolysaccharides
were cooled. As it was difficult to avoid the supercooling even
at a cooling rate of 1.25°C/min, we carried out the DSC measurements throughout this study by raising the temperature of the
samples kept at -50°C for 15 mins. The heating rate was maintained at 10°C/min, because a decrease of the heating rate to 5°C/min
produced a similar thermogram. Repeated freezing and thawing
cycles did not alter the shapes and positions of the DSC peaks.

A. Dependence of DSC Thermograms on the Water Content. As
typical examples of DSC thermograms for the aqueous solutions of
the mucopolysaccharides, we will cite representative results for
two, Chn S-A and Hpn.

Chn S-A: Figure 2 shows the DSC curves for aqueous solutions
of Na salt of Chn S-A with different water contents. The samples
contain in every case 3.10 mg of water. It is clear that the
solution with a water content of 0.22 exhibits no endothermic
transition. This provides a strong evidence that the water present
in such highly concentrated solutions is not freezable even when
cooled to -50°C. This kind of water is called "non-freezing" or
"bound". As can be seen from Figure 2, peaks appear in the DSC
thermograms as the water content becomes higher. For instance,

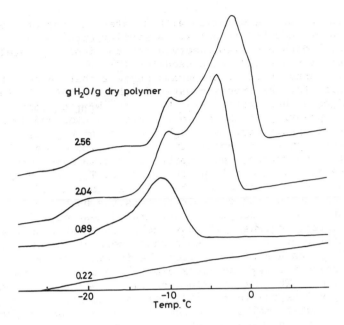

Figure 2. *DSC thermograms for the Na salt of Chn S-A*

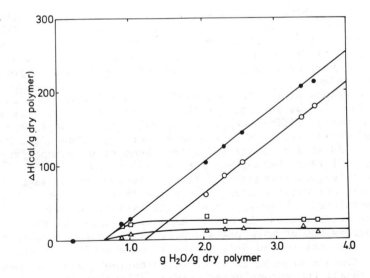

Figure 3. *Integrated heats of fusion (ΔH) vs. sample water content for the Na salt of Chn S-A: (○) W_{fI}; (□) W_{fII}; (△) W_{fIII}; (●) total*

the DSC curve for the solution with a water content of 2.04 has clearly three endothermic peaks. A similar appearance of multiple endothermic peaks was also observed for non-ionic hydrogels, but the number of peaks did not exceed two (8).

Three peaks in the thermograms suggest that three different states exist for the freezing water in the Chn S-A solutions. We will designate these waters W_{fI}, W_{fII}, and W_{fIII} corresponding to the peaks appearing at the highest, intermediate, and lowest temperatures, respectively. Although the peak of W_{fIII} is very weak, it is beyond the experimental error. To estimate the calorific values of these waters, the endothermic thermograms were resolved into three areas. The melting temperature, T_m, of each peak was defined as the point at which the steepest tangent of the left side of the endothermic curve intersects the baseline.

Figure 3 shows the plot of integrated heat of fusion for each freezing water, together with the sum of the heats, against the sample water content. It is seen that the heats of fusion for W_{fII} and W_{fIII} reach a plateau after the initial increase, while the heat of fusion for W_{fI} increases almost linearly with increasing water content. The initial slope of this linear curve gives 75.5 cal/g H_2O as the incremental heat of fusion, ΔH_f, for W_{fI}, which is somewhat smaller than that for the bulk water (79.7 cal/g). The content of non-freezing water, designated hereafter W_{nf}, can be readily obtained from the point of intersection of the total heat of fusion extrapolated to zero. The observed W_{nf} content is 0.64.

Figure 4 shows T_m,s for W_{fI}, W_{fII}, and W_{fIII} as a function of the water content. As can be seen, the variation of T_m with the water content is approximately similar to that of the heat of fusion. T_m of W_{fI} monotonously increases as the water content becomes higher, whereas T_m,s for W_{fII} and W_{fIII} tend to approach a limiting value.

The shapes of the DSC peaks of Ca salt of Chn S-A were substantially similar to those of the Na salt (see Figures 11 and 12). Also, no marked differences were observed for the dependencies of heats of fusion and T_m,s on the water content between the Na and Ca salt. As will be described later, some other mucopolysaccharides give different thermograms for the Na and Ca salts.

Hpn: The DSC thermograms of the solutions of Na salt of Hpn with different water contents are given in Figure 5 as a function of temperature. Obviously, the curves exhibit no endothermic peak when the water content is as low as 0.48, similar to those of Chn S-A, whereas one or more peaks appear as the water content increases. In contrast to Chn S-A, the peak corresponding to W_{fII} is not clear. Absence of peaks at lower water contents was the result common to all the mucopolysaccharides. The plot of heats of fusion against the water content is given in Figure 6. Also in this case, the relation between the sum of the integrated heats of fusion and the water content gives a linear plot, from which the content of W_{nf} is found to be 0.48 and ΔH_f for W_{fI} to be 75.6 cal/g.

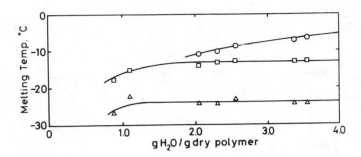

Figure 4.   Melting temperature vs. sample water content for the Na salt of Chn
S-A: (○) W_{fI}; (□) W_{fII}; (△) W_{fIII}

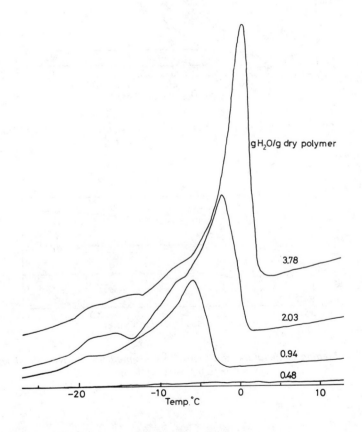

Figure 5.   DSC therograms for the Na salt of Hpn

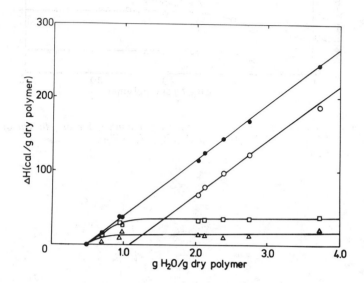

*Figure 6.   Integrated heats of fusion (ΔH) vs. sample water content for the Na salt
of Hpn: (○) W_{fI}; (□ W_{fII}; (△) W_{fIII}; (●) total*

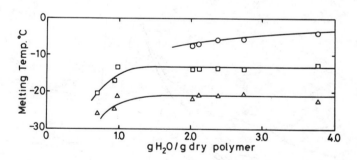

*Figure 7.   Melting temperature vs. sample water content for the Na salt of Hpn:
(○) W_{fI}; (□) W_{fII}; (△) W_{fIII}*

T_m of W_{fI}, as shown in Figure 7, is below 0°C, but varies with the increase of water content, probably approaching 0°C, while for W_{fII} and W_{fIII} both the T_m values and the heats of fusion remain almost constant after certain values. This is also the trend characteristic to all the mucopolysaccharides.

The results for the Ca salt are given in Figures 8, 9, and 10. In this case, we observe no distinct multiple peaks even at higher water contents. The W_{nf} content is 0.54 and ΔH_f is 77.0 cal/g.

B. Comparison of DSC Thermograms for Mucopolysaccharide Solutions. Sets of thermograms for the Na and Ca salts of all the mucopolysaccharides are shown in Figures 11 and 12, respectively. All of the samples have water contents of approximately two. The result for PVA-S is also shown in Figure 11. Analysis of thermograms in Figures 11 and 12 reveals that the number of endothermic peaks depends not only on the nature of mucopolysaccharides but also on the type of gegen-ion. The most striking example of the latter is Hpn, which has three peaks in the thermograms for the Na salt, but one peak for the Ca salt. The Na salt of Chn also gives three endothermic peaks, while the Ca salt has two. On the other hand, both of the Na and Ca salts of Chn S-C seem to have two peaks in their DSC thermograms, in contrast to Chn S-A with three peaks for both the Na and Ca salts. The PVA-S gives the thermogram with one peak, similar to the Ca salt of Hpn.

Discussion.

As exemplified in Figures 2,5, and 8, scanning aqueous solutions with low water contents shows neither endothermic nor exothermic transitions. This strongly indicates that the water is not to be frozen nor melted at least in the temperature range -50°C to +30°C. This non-freezing behavior is explicable in terms of strong binding of the water molecules to the hydrophilic sites of the polymer chains. As illustrated in Figures 3, 6, and 9, this capacity of polymer chains to bind water can be easily determined from the plot of the total heat of fusion against the water content. The W_{nf} contents of the mucopolysaccharides are given in Table II .

The DSC measurements provide useful information on the weak interaction of water with the substrate. It is because peaks appear in the DSC endotherms of the polymer solutions containing some amounts of water in addition to the strongly bound water (W_{nf}), as shown in Figures 11 and 12. The additional water involves W_{fI}, W_{fII}, or W_{fIII}. With the increase of the water content the heat of fusion of W_{fI} monotonously increases and T_m of W_{fI} approaches the melting point of bulk ice. These facts, together with ΔH_f for W_{fI} being very close to that of bulk ice, suggest that W_{fI} is nearly identical to bulk water. There is no reason to suspect that the nature of W_{fII} and W_{fIII} is intermediate between that of W_{nf} and W_{fI}. Interestingly, the integrated heats of

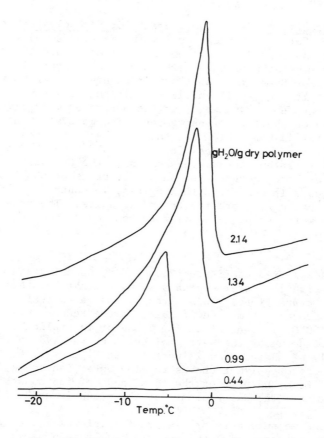

Figure 8.   DSC thermograms for the Ca salt of Hpn

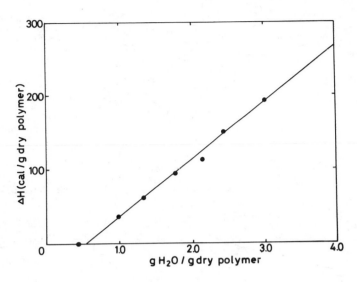

Figure 9.   Integrated heat of fusion (ΔH) vs. sample water content for the Ca salt of Hpn

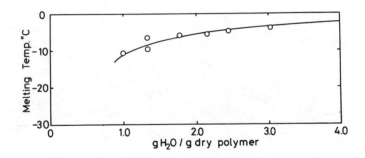

Figure 10.   Melting temperature of W_{fI} vs. sample water content for the Ca salt of Hpn

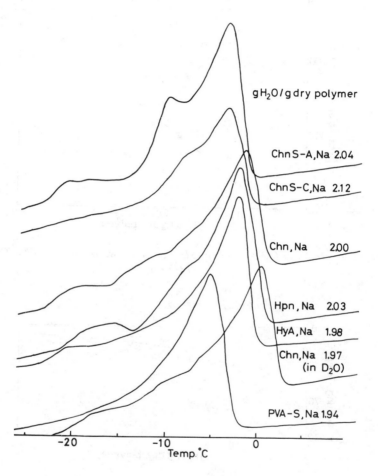

Figure 11.   DSC thermograms for the Na salts of mucopolysaccharides having
approximately the same water content

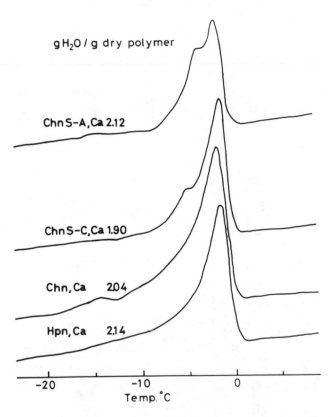

g H$_2$O / g dry polymer

Chn S–A, Ca 2.12

Chn S–C, Ca 1.90

Chn, Ca   2.04

Hpn, Ca   2.14

−20   −10   0

Temp. °C

Figure 12. DSC thermograms for the Ca salts of mucopolysaccharides having approximately the same water content

TABLE II.

CONTENTS OF ANOMALOUS WATERS, ΔH_f, AND LEVELED-OFF T_m

Mucopolysaccharide	g H$_2$O/g dry polymer			ΔH_f (cal/g)	T_m (°C)	
	W_{nf}	W_{fII}	W_{fIII}		W_{fII}	W_{fIII}
Chn S-A, Na	0.64	0.30	0.17	75.5	-12.8	-23.8
" , Ca	0.45	0.44	0.10	78.7	- 6.4	-17.0
Chn S-C, Na	0.71	0.57	0	77.0	-12.0	—
" , Ca	0.52	0.59	0	79.5	- 8.3	—
Hpn , Na	0.48	0.45	0.16	75.6	-13.8	-21.3
" , Ca	0.54	0	0	77.0	—	—
Chn , Na	0.59	0.28	0.10	74.9	-17.6	-23.4
" , Na1)	0.65	0.22	0.11	72.32)	-12.6	-18.8
" , Ca	0.66	0.41	0	79.0	-15.4	—
HyA , Na	0.51	0.59	0	75.9	-25.4	—
PVA-S , Na	0.73	0	0	71.0	—	—

1) In D$_2$O

2) ΔH_f for bulk D$_2$O is 75.4 cal/g.

fusion for W_{fII} and W_{fIII} as well as their T_m,s are in all cases practically independent of the water content except in the very low range, implying that a saturation will take place for both W_{fII} and W_{fIII}, similar to W_{nf}.

The saturated contents of W_{fII} and W_{fIII} are given in Table II, which also includes their leveled-off T_m,s and ΔH_f for W_{fI}. The contents of W_{fII} and W_{fIII} were evaluated under the assumption that the ΔH_f values for W_{fII} and W_{fIII} are both equal to ΔH_f for W_{fI}. If the sample does not contain W_{fIII}, the amount of W_{fII} would be directly estimated in the similar fashion as that of W_{nf}. As can be seen from Table II , the content of W_{nf}, as well as ΔH_f of W_{fI}, does not differ largely from polymer to polymer, ranging between 0.4 and 0.7. These values are in agreement with those reported for mucopolysaccharides ([13],[14],[15]). It may be more convenient to express the unit of the water content in the molar basis instead of the weight base in order to discuss the content of the anomalous waters in terms of the chemical structure of polymers. Table III summarizes moles of the anomalous waters per repeating unit. For the sake of comparison, the contents for Hpn are also based on an averaged disaccharide unit, though its repeating unit is reported to be a heptasaccharide ([21]).

TABLE III .

MOLES OF ANOMALOUS WATERS PER MOLE OF DISACCHARIDE UNIT

Mucopolysaccharide	W_{nf}	W_{fII}	W_{fIII}	$W_{nf} + W_{fII} + W_{fIII}$
Chn S-A, Na	17.9	8.4	2.8	29.1
" , Ca	12.4	12.1	2.7	27.2
Chn S-C, Na	19.9	15.9	0	35.8
" , Ca	14.4	16.2	0	30.6
Hpn , Na	16.4	15.4	5.5	37.3
" , Ca	18.2	0	0	18.2
Chn , Na	13.2	6.3	2.2	21.7
" , Na[1]	13.1	4.4	2.2	19.7
" , Ca	14.7	9.1	0	23.8
HyA , Na	11.5	13.3	0	24.8

1) in D_2O

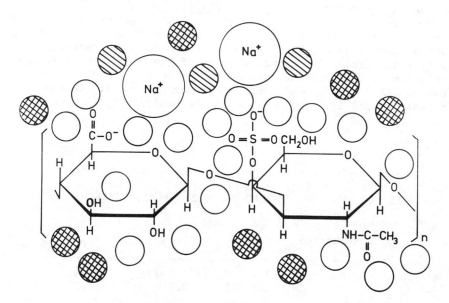

Figure 13.   Hydration model of mucopolysaccharide

It is obvious from Table III that the number of molecules of non-freezing water (W_{nf}) is mostly larger than that of W_{fII}, while little or very small amounts of W_{fIII} are associated with the disaccharide unit. The anomalities of water in solutions are supposed to result from the interactions of water molecules with ionizable and polar groups of the substrate. As Figure 1 demonstrates, the mucopolysaccharide molecules possess a variety of ionizable and polar groups. For instance, Hpn has ionic groups such as carboxyl, O-sulfate and N-sulfate and polar groups such as hydroxyl and hemiacetal. Instead of N-sulfate, Chn S-A and Chn S-C contain N-acetate. On the other hand, carboxyl is the only ionizable group in the molecules of Chn and HyA. When the W_{nf} content or the sum of the contents of the three anomalous waters is taken as a measure representing the extent of the interaction of water with the Na salt of mucopolysaccharide, it appears that Chn and HyA, devoid of sulfate groups, show low contents of anomalous waters compared with the other three mucopolysaccharides having the sulfate groups. Importance of the ionizable groups in the hydration is also supported by the fact that the Na salts mostly lead to higher hydration than the Ca salts. It is noteworthy that T_m,s for W_{fII} and W_{fIII} are higher for the Ca salt than for the Na salt (see Table II).

Comparison between two polymers having the same chemical moiety may be helpful for a better understanding of the hydration of mucopolysaccharides. Chn S-A is different from Chn S-C only in the position of O-sulfate, which is linked to C_4 in Chn S-A and C_6 in Chn S-C. On the other hand, HyA is a epimer of Chn with regard to the hydroxyl group linked to C_4 in the glucosamine unit. In spite of such a slight difference in the chemical structure, each of the two polymers has significantly different DSC thermograms. This is a good example to show that interference of the ionizable and polar groups affects hydration (22).

The results for Hpn provide important information about the influence of ionizable groups on hydration. As is obvious from Figures 5 and 8, conversion of the Na to the Ca salt of Hpn decreases the three peaks in the DSC thermograms to one broad peak, accompanied by reduction in hydration. This may be explained in terms of difference in degree of dissociation between the Na and Ca salts (23). It is interesting to point out that, except for the Ca salt of Hpn, only PVA-S was devoid of both W_{fII} and W_{fIII}.

Finally, we attempt to propose a hydration model of the mucopolysaccharides, at least in the deeply cooled solutions. The model is shown in Figure 13. The data in Table III clearly indicate that, similar to non-ionizable hydrophilic polymers (8), the mucopolysaccharides chains are associated with a rather small number of water molecules, so that it may not be adequate to use the word "shell" to show their sequence of association. Not only the ionic but also hydroxyl groups strongly bind a few (approximately two) molecules of water (W_{nf}) per hydrophilic group and are further

surrounded by smaller numbers of water molecules(W_{fII}) charactariz-
ed by weaker interactions. It is at present not clear where the
W_{fIII} molecules exist. Knowledge of the organization of the water
molecules could be obtained from $\Delta H_f/T_m$ if the exact values of
ΔH_f for W_{nf}, W_{fII}, and W_{fIII} become available.

Abstract

DSC thermograms are reported for aqueous solutions of the Na
and Ca salts of five kinds of mucopolysaccharides. Among all the
thermograms, those for the aqueous solutions of low water content
do not show endothermic or exothermic peaks over a temperature
range -50°C to 30°C, while one, two, or three peaks appear as the
water content gradually increases. The number of peaks depends on
the type of mucopolysaccharides as well as the nature of the salt.
The content of the non-freezing water, determined from the dependence of the heat of fusion on the sample water content, ranges
between 0.4 and 0.7 for the mucopolysaccharides studied. A
hydration model is proposed, taking into account the existence
of both the non-freezing water and more weakly associated waters.

Literature Cited

1. Franks, F.,Ed. "Water, a Comprehensive Treatise", volume 4;
    Plenum Press: New York-London, 1975.
2. Drost-Hansen, W.; Ind. Eng. Chem., 1969, 61, 10.
3. Conway, B.E.; Rev. Macromol. Chem., 1972, 7, 113.
4. Jellinek, H.H.G.,Ed. "Water Structure"; Plenum Press: New York
    -London, 1972.
5. Haly, A.R.; Snaith, J.W.; Biopolymers, 1971, 10, 1681.
6. Hazlewood, C.F.; Chang, D.C.; Nichols, F.L.; Woessner, D.E.;
    Biochem.J., 1974, 14, 583.
7. Taniguchi, Y.; Horigome, S.; J. Appl. Polym. Sci., 1975. 19,
    2743.
8. Takami, S.; Horii, F.; Kitamaru, R.; Ikada, Y.; Polymer
    Preprints, Japan, 1976, 25(2), 426.
9. Kuntz, J_R., I.D.; Kauzmann, W.; Adv. Protein Chem., 1974, 28,
    239.
10. Ehrlich, S.H.; Bettelheim, F.A.; J. Phys. Chem.,1963, 67,
    1954.
11. Bettelheim, F.A.; Ehrlich, S.H.; J. Phys. Chem., 1963, 67,
    1948.
12. Lubezky, I.; Bettelheim, F.A.; Folman, M.; Trans. Faraday Soc.,
    1967, 63, 1794.
13. Suzuki, Y; Uedaira, H.; Bull. Chem. Soc. Japan, 1970, 43, 1892.
14. Suzuki, Y; Uedaira, H.; Nippon Kagaku Kaishi, 1974, 830.
15. Atkins, E.D.T.; Isaac, D.H.; Nieduszynski, I.A.; Phelps, C.F.;
    Sheehan, J.K.; Polymer, 1974, 15, 263.
16. Kakkar, V.V.; Thomas, D.P., Eds. "Heparin"; Academic Press,
    London - New York - San Francisco, 1976.

17. Yasuda, H; Olf, H.G.; Crist, B; Lamaze, C.E.; Peterlin, A.; in "Water Structure"; Jellinek, H.H.G., Ed., Plenum Press, New York - London, 1972; p.39.
18. Nelson, R.A.; J. Appl. Polym. Sci., 1977, 21, 645.
19. Mathews, M.B.; Arch, Biochem. Biophys., 1956, 61, 367.
20. Lee, H.B.; Jhon, M.S.; Andrade, J.D.; J. Colloid Interface Sci., 1975, 51, 225.

RECEIVED January 4, 1980.

PERMEATION, TRANSPORT, AND ION SELECTIVITY

Volume Changes During Water Binding to Hair Fibers

M. BREUER, EDMUND M. BURAS, JR., and A. FOOKSON

Gillette Research Institute, 1413 Research Boulevard, Rockville, MD 20850

Unraveling the molecular mechanism of water binding by keratins (e.g., wool, hair, nails, etc.) has interested chemists for half a century (1). Essentially, two types of models have been suggested for explaining water absorption isotherms of keratins: one that postulates the binding of water molecules on well-defined discrete sites (e.g., polar side chains, peptide bonds)(2), and the other that maintains that swelling of the polypeptide network is the primary mechanism responsible for the absorption of water (3).

The validity of neither of these models has been established beyond doubt, owing mainly to the lack of reliable data on the magnitudes of the changes in the thermodynamic quantities that accompany the binding of water molecules to keratin fibers. In particular, none of the treatments have given adequate considerations to the swelling of the keratin structure and to the contribution that this process makes to the overall free energy changes accompanying the water absorption. No doubt, this omission has been due to the lack of precise data on the volume changes occuring in hair fibers during the binding of water molecules.

Recently, we have developed an optical method capable of measuring hair fiber diameters as a function of ambient humidities with high reproducibility (4). Therefore, we feel that we are now in a position to carry out a rigorous thermodynamic analysis of the water-keratin interaction process, and to examine critically the various water binding theories by comparing the experimentally determined thermodynamic changes at constant volume with those predicted by the various theories.

0-8412-0559-0/80/47-127-309$05.00/0
© 1980 American Chemical Society

The Method for Measuring Fiber Diameters at Ambient Humidities

The arrangement of apparatus is shown schematically in Figure 1. A helium-neon laser (Spectra-Physics, Inc., Mountain View, California, Model 145-01) emitting 2mW at 632.8 nm continuously, was mounted on an optical bench so as to impinge upon hair specimens and cast diffraction patterns upon a screen approximately one meter from the specimens.

The hairs, mounted horizontally in an enclosure made of acrylic sheet with two "windows" of 0.15 mm thick glass coverslips for transmission of the beam, is translatable in two directions perpendicular to the beam, and rotatable within the chamber (a feature not used in this study). Air is circulated through the enclosure at about 120 ml/min by a peristaltic pump (Masterflex Cole-Palmer Instrument Co., Chicago, Illinois). The relative humidity (RH) of the circulated air is controlled by passage through Drierite (W.A. Hammond Drierite Co., Xenia, Ohio) for complete dryness, or through one of the saturated salt solutions in equilibrium with excess solid to give the required humidity. Hair-atmosphere equilibria were verified in 24 to 36 hours, being indicated by no further change in diameter over an additional 5 to 6 hours, using Student's "t" test and requiring a p-null of 0.05 or less. All apparatus and materials are maintained in a room closely regulated at 21 ± 1°C. Successive measurements were made on each type of hair at increasing humidities.

The pattern, oriented vertically on the target plane, was sharp and bright so that at least 10 orders of diffraction could be seen with minimal darkening of the room. The size of the pattern was such that visual examination permitted 4 to 6 orders of minima to be pricked into index-card stock held against the target plane and conveniently measured with 10 cm dial calipers readable to 0.001 cm.

Hair diameters were calculated using the equation:

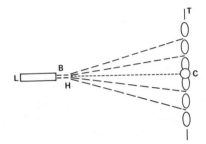

Figure 1. Geometry of apparatus for precise measurements of hair diameter by optical diffraction: H, hair fiber; L, laser; B, laser beam; T, plane target with diffraction pattern; C, central beam

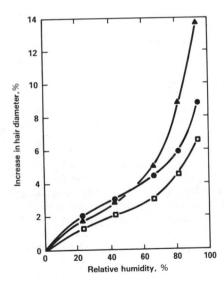

Figure 2. Increase of hair diameter as a function of relative humidity: (●) intact hair; (▲) bleached hair; (□) descaled hair

$$d = \left(n \lambda \sqrt{4D^2 + 1^2} \right)/1 \qquad (1)$$

where d is the hair diameter, n is the diffraction order, λ is the laser light wavelength, D is the distance hair to target, and 1 is the distance between the n-order diffraction minima on either side of the central beam.

Experimental Determination of Hair Fiber Diameter Changes at Various Humidities

First, we checked the reproducibility of the method under constant experimental conditions; i.e., carrying out the measurement at a given, fixed point of the fiber at constant humidity and temperature. The reproducibility of the method in this mode of operation proved extremely good, i.e., the measurements fell within ±0.1% standard error (4).

When we proceeded, however, to measure the changes of the hair diameter as a function of relative humidity, it became obvious that our first hope, i.e., to measure the change of the fiber diameter at the same point along the fiber axis, was unrealistic. Alteration of humidity affected the length of fiber, making the measurements of the diameter at the same point along the fiber axis at different humidities a virtual impossibility. The problem was compounded by the fact that hair fibers neither had uniform diameters along the fiber axes, nor did they possess circular cross sections. Since changes of humidity caused both axial elongations and radial twists of the fibers, our attempts to determine the changes in the fiber diameters at the same points were frustrated. Our methods essentially gauge the diameter of the fiber perpendicular to the direction of the light beam. Consequently, instead of attempting to determine changes of the fiber diameter at given points, we decided to measure fibers at various, randomly chosen points along their lengths.

Figure 2 shows the results obtained with different hair types in terms of their mean increases in diameters as a function of the relative humidity (RH). It can be seen that all three hair types increased in diameter by about 8-9% as the relative humidity was increased from 0 to 93%. These data may be compared to those of Meredith who reported a

diameter increase of 16% for wet wool (5). The
curves, which are sigmoid in shape, resemble those
of the moisture regain (1). Using the available
data from the literature on the axial elongation (5)
and our own data on the radial swelling of hair as a
function of water uptake, we calculated V_{sp} the specific volume of the hair fibers as a function of
their water contents. (We assumed that the radial
swellings of the hair fibers are isotropic).
The values of V_{sp} for the various type of hair
fibers could be expressed by polynomials

$$V_{sp} = V_0 + Bn + Cn^2 \qquad (2)$$

where n is the water content of hair in moles g^{-1}
units and the values of the coefficients for the
various hair types are given in Table I.

TABLE I

Coefficients of Equation (2)

	V_0	B	C
Intact Hair	0.757	10.00	-85.68
Descaled Hair	0.758	3.84	932.4
Bleached Hair	0.757	5.17	156.0

Differentiation of Equation 2 yields $\overline{V}_W$, the partial
molal volume of water in hair as a function of n,
the water content of hair.

$$\overline{V}_W = B + 2Cn \qquad (3)$$

The values of $\overline{V}_W$ for the various hair types are
plotted in Figure 3. A number of interesting points
emerge from these data: First, in all cases the
absolute values of $\overline{V}_W$ are lower than the molal
volume of liquid water (i.e., 18 cm^3 $mole^{-1}$);
second, the limiting values (i.e., when n → 0) of $\overline{V}_W$
for bleached and descaled hair are smaller than that
of virgin hair; and third, whereas the $\overline{V}_W$ for intact
and descaled intact hairs are only slightly
dependent on n, in the case of bleached hair $\overline{V}_W$
increases fast with n.

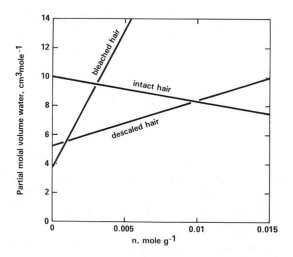

Figure 3. V_W, partial molal volume of water in hair as a function of n, *the water content of hair*

A plausible explanation for these results is that inside the keratin structure water is molecularly dispersed and forms monomolecular layers around the various protein structural units, i.e., the micro-or protofibrils of the keratin. The low values obtained for $\bar{V}_W$ can be explained by assuming that when water molecules penetrate the hair structure, they fill, at least in part, pre-existing voids. The results also suggest that the cortex of the hair structure is more porous than the cuticle, since removal of the cuticle (i.e., descaling of the hair) reduces the value of $\bar{V}_W$. Bleaching seems to have an even larger effect--in addition to reducing the value of $\bar{V}_W$, it also affects strongly its rate of change with increasing water uptake, probably by altering the distribution of the pore sizes in hair. Whereas the pore size distribution in intact hair appears to be fairly uniform, i.e., $\bar{V}_W$ is only weakly dependent on n, bleached hair appears to have a wider pore distribution with pore sizes rapidly surpassing the magnitudes of those of intact hair as n, the water uptake, reaches higher values.

Free Energy Changes Accompanying The Binding of Water to Hair

When hair absorbs water, two processes occur simultaneously: a, water molecules interact with the polypeptide backbones or their side chains, and b, the hair fibers expand due to the incorporation of water molecules into their structures. Thus, the total free energy change can be expressed as:

$$\Delta G_T = \Delta G_B + \Delta G_E \qquad (4)$$

where ΔG_B and ΔG_E are the free energies of binding and of expansion of the hair structure, respectively.

Since statistical mechanical models for water binding are generally derived for constant volume condition, a comparison of experimental data with the calculated models is only possible provided the value of ΔG_E can be calculated or estimated. So far, this has not been the case for the water-keratin interaction.

To obtain ΔG_E we undertook the following steps: The value of ΔG_T was obtained by integrating the water adsorption isotherm from $n = 0$ to $n = n$, thus

$$\Delta G_T = RT \int_0^n \ln \frac{p}{p_0} \, dn \qquad (5)$$

where p and p_0 are the vapor pressures of the absorbed water at a given value of n and of liquid water, respectively. To obtain ΔG_E we made use of the thermodynamic relationship:

$$\left(\frac{\partial G}{\partial V}\right)_{T,n} = V\left(\frac{\partial p}{\partial V}\right)_{T,n} = \frac{1}{\beta} \qquad (6)$$

where β denotes the isothermic volume compressibility and V denotes the volume of the fiber. Assuming that at a selected p/p_0 the compressibility is independent of the fiber compression, i.e.,

$$\beta = \text{constant} \qquad (7)$$

Equation 6 can be integrated to give

$$\Delta G_E = G_{(n = m)} - G_{(n = o)} \qquad (8)$$

$$\Delta G_E = \int_{V_o}^{V_n} 1/\beta \, dV = 1/\beta \, (V - V_o) \qquad (9)$$

where V_o and V_n denote the specific volumes of the dry hair fiber and of a fiber containing n moles per gram of water, respectively. Thus, provided the water absorption isotherm and the compressibility of keratin are known, the value of ΔG_B can be computed from Equation 4.

Computation of ΔG_B From Experimental Data

To obtain ΔG_B, we first computed the value of ΔG_T from Watt and D'Arcy's data (1) by means of graphical integration of Equation 5. The results are plotted in Figure 4.

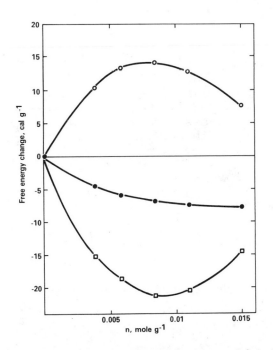

Figure 4. Free energies of binding of water per gram of hair (experimental): (○) ΔG_E; (●) ΔG_T; (□) ΔG_B

To compute ΔG_T we calculated the value of β from the values of ε_1, the axial, and ε_2, the radial, compression moduli obtained by Bendit and Kelly ([6]), by means of the approximation:

$$\beta = \varepsilon_1 + 2\varepsilon_2 \tag{10}$$

The respective values of the bulk modulus (i.e. $1/\beta$) as a function of n are given in Figure 5. The value of ΔG_B could then be computed by means of Equation 4. The results are given in Figure 4.

Comparison of Measured Values of ΔG_B With Those Calculated From Various Models

As mentioned before, essentially two molecular models have been put forward for explaining the water binding processes in hair: a, water molecules bind to discrete, independent sites attached to the polypeptide chains or b, water is absorbed by a swelling process of the polymeric network as described by Flory's polymer theories ([8]).

For the site binding model the free energy change is given by Steinhardt ([7]):

$$\Delta G_B = nRT \left\{ \ln K + \ln\left(\frac{an/m}{1 - n/m}\right) \right\} \tag{11}$$

whereas for the polymer-swelling process

$$\Delta G_B = RT(n \ln v + n'\ln v' + \mathsf{X}nv') \tag{12}$$

where

n = water bound to hair, mole g^{-1}

m = water binding sites hair, mole g^{-1}

a = p/p_0 = water activity

K = binding constant of water to a binding site

v = volume fraction of water in hair

v'= 1 - v, volume fraction of peptide residues in hair

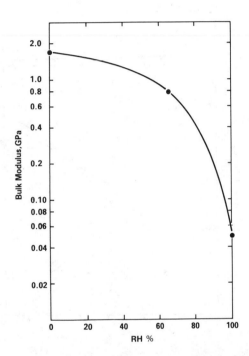

Figure 5.   Bulk modulus as a function of relative humidity

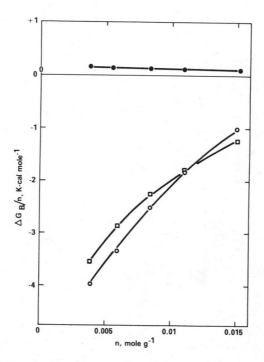

Figure 6.   Comparison of experimental $\Delta G_B/n$, *integral free energy changes per mole of water absorbed in hair, with theoretical values: (●) from Equation 12; (□) from Equation 11; (○) experimental*

χ = Flory-Huggins interaction parameter

n'= peptide residues in hair, mole g^{-1}

To compare the experimental results with the theoretical models, we postulated in equation (11) that at a = 1 and all the available binding sites are occupied, i.e. m = 2.00 x 10^{-2} mole g^{-1}, and that the theoretical and experimental ΔG_B values are equal when n/m = 0.5, i.e., when a = 0.75. In this way the calculated and the experimentally measured ΔG_B vs n curves intersect at this point.

In calculating ΔG_B by equation (12), we assumed that n' = 8.92 x 10^{-2} mole g^{-1} and χ = 1.00 (3).

The integral free energy changes, calculated on the basis of equations (11) and (12) together with the experimental values, per mole of bound water, are given in Figure 6.

Discussion and Conclusions

The results of this investigation suggest that neglecting the free energy changes occuring consequent to the expansion of hair fiber during water absorption introduces a considerable error in the assessment of the total free energy change of water binding. Rosenbaum's conclusions (3), that the Flory's polymer swelling theory accounts better for the water binding data than does a model based on the assumption that water binds to discrete sites, does not seem to be borne out when the thermodynamic work required for expanding the hair fiber is also taken in account. The very low values of the partial molal volume of water in hair, which we found, also suggest that the mechanism is essentially different from the one that was postulated by Flory for the swelling of polymeric gels (8).

Wheras the apparent agreement obtained between the experimentally measured integral free energy changes and those calculated on the basis of Equation 11 is most interesting, it should not be taken as a proof for the validity of the Site Binding Model. The need for this cautionary statement becomes evident after a closer examination of the data presented in Figure 4. The curve ΔG_B vs. n has a minimum suggesting that the differential binding free energy, i.e. $(\partial \Delta G_B / \partial n)$ becomes positive at values n > 0.008 mole g^{-1}; Equation 11 cannot predict this type of behavior. We feel that the mechanism which explains the volume change data most satisfactorily is the following: dry hair is a fairly rigid semi-crystalline porous solid. Water penetrates into the pores between various fibrils of the hair structure and pries them apart, thus bringing about a gradual increase of the hair volume. The thermodynamic driving force for the water absorption is a combination of three processes: interaction with discrete polar side chains (acidic and basic groups) and peptide bonds, capillary condensation, and entropic gains owing to the mixing of water with the polypeptide chains, with the site binding being the dominant process.

It is interesting that descaling of hair brings about an increase in the value of $\bar{V}_W$. This result suggests that the cuticle is probably less porous than the cortex and that the value of the partial molal volume of water in the cuticle is near to the molal volume of liquid water.

Finally it seems that chemical treatment of hair (i.e., bleaching) changes the pore size distribution of hair, bringing about a wider distribution of pore sizes.

Literature Cited

1. Watt, I.C.; D'Arcy, R.L. Polymer Sci. Symposium 1976, 55, 155.

2. Watt, I.C.; Leeder, J.D. J. Text. Inst. 1968, 55, 353.

3. Rosenbaum, S., J. Polymer Sci.: Part C 1970, 31, 45.

4. Buras, E.; Fookson, A.; Breuer, M.M. Proceedings the First International Congress on Human Hair, Hamburg, 1979, in press.

5. Meredith, R., In "Fiber Science", Preston, J.M., Ed.; The Textile Institute: Manchester, 1958.

6. Bendit, E.G.; Kelly, M. Textile Res. J. 1978, 48, 674.

7. Steinhardt, J.; Reynolds, J.A. "Multiple Equilibria in Proteins"; Academic Press: New York, 1969.

8. Flory, P. "Principles of Polymer Chemistry"; Cornell University Press: Ithaca, N.Y., 1953.

RECEIVED January 4, 1980.

Relaxation Studies of Adsorbed Water on Porous Glass

Varying Temperature and Pore Size at Constant Coverages

GEORGES BELFORT[1]

Rensselaer Polytechnic Institute, Troy, NY 12181

NAOMI SINAI

University of Utah, Salt Lake City, UT 84112

Recent studies using Infrared Spectrocopy IR to characterize the state of water in desalination membranes have concluded that the water sorbed in these membranes has a low degree of association and that bonds between water and the membrane are considerably weaker than those in liquid water (1,2). These conclusions have been made for widely differing membrane materials such as cellulose acetate (1,2),polyimide, and porous glass (1) and appear to contradict the conclusions obtained from pNMR (3-8), differential scanning calorimetry (9-11) and transport (12,13) studies. These latter studies suggest that water molecules in the vicinity of the membrane are motionally restricted with respect to free bulk water. Several investigators have thus proposed multi-state models to characterize the occluded water in the porous media (3-8,10-13). The terms "phase" and "exchanging fractions" are sometimes used. To minimize confusion, since a thermodynamic phase is not what is being considered, "environmental state" should and will be used here. An explanation of the mechanism of solute selectivity and water transport in desalination membranes clearly depends on a resolution of the above apparent contradiction. A qualitative model of the state of water inside desalination membranes considering both the IR and pNMR results (including those presented here) will be proposed.

This study is a direct continuation of our previous work in which pNMR was used to measure the proton relaxation times of water adsorbed on four powdered porous glasses ranging in pore size from 29 to 189 Å as a function of coverage at room temperature (8).

Interpretation of pNMR data for adsorbed systems is

[1] To whom correspondence should be sent.

The research reported here was conducted at the School of Applied Science and Technology, Hebrew University of Jerusalem, Israel.

0-8412-0559-0/80/47-127-323$05.75/0
© 1980 American Chemical Society

difficult at best. Thus, independent variables such as temperature (6), degree of loading (or coverage) of the adsorbed species (14,15), operating frequency (16,17) and isotopic substitution (16) are usually used to assist and clarify the interpretation of the researcher. According to Resing (18) "variation of the temperature over as broad a range as possible for a given system offers the greatest probability of successful interpretation of NMR relaxation data." For this reason, the research reported here was conducted as a complimentary study to that reported earlier (8).

Here we report results of the pNMR relaxation times of water adsorbed on the smallest (29 Å) and the largest (189 Å) pore size porous glass desalination membranes studied earlier (8) as a function of three coverages for the temperature range -80 to +90°C. Although the approach and the relaxation model used here is similar to that used by Belfort et al (6) to study water adsorption at two coverages on three different pore size porous glass desalination membranes, the following differences should be emphasized.

(i)   The Vycor-type (96% silica) porous glass used here was prepared by the same procedure as the capillary porous glass desalination membranes used by Schnabel (19). Care was taken to prevent the introduction of paramagnetic centers into the glass melt during production (Schnabel, private communication). With the two large pore size porous glasses (designated CPG-10-125 and CPG-10-240) used by Belfort et al (6), this was not the case and the relaxation model could not be fitted to the data.

(ii) Coverages at 100% RH and 50% RH were arbitrarily chosen for the earlier study (6), while detailed adsorption isotherms and associated BET results reported earlier (8) are used here to rationally choose the desired coverages.

(iii) Porous glass from the same production batch was divided into two lots by the manufacturer. One lot was used in this study while the other lot was used by Luck (Marburg, West Germany) for the IR investigations (1). A proposed model would therefore have to be consistent with the results from both studies.

(iv) The computer fitting procedure was improved over that previously used (6) resulting in significant time savings.

Methods

Samples and Water Sorption Details of the preparation and cleaning procedures of the porous glass obtained from Jenaer-Glaswerk Schott, Mainz, Germany, have already been described (8).

Constant temperature adsorption isotherms and BET results for the two porous glasses used here are reproduced in Table I from Ref. 8.

Nuclear Magnetic Resonance-Apparatus and Procedure The nuclear magnetic relaxation times T_1 and T_2 and the mobile fraction f_m were measured with a Spin-lock pulse-spectrometer (model CPS-2, Port Credit, Ontario, Canada) at 33 Mhz. Temperature was controlled with a Brucker Temperature Controller (Brucker, Germany). The magnet was 15 inches in diameter (Varian V3800, Palo Alto, CA, USA). The logic for the control of the pulse spectrometer was fully automated and the data was accumulated, stored and averaged by a Control Data Mini-Computer. Hard copies (paper-tapes) for each run was supplied automatically together with x-t plot of the averaged relaxation data.

The transverse relaxation times T_2 were measured by "spin-echo" method (2 pulses: 90-t-180°). The "echo" signal s(t), appearing at t = 2t, satisfied the equation:

$$S(t) = s(0) \exp(-t/T_2) \qquad (1)$$

The longitudinal relaxation times T_1 of the protons were measured by saturating the line with a "comb" of 180° pulses, followed by a sequence of 2 pulses 90-t-180°. The intensity M_z (t) of the "echo" appearing after the 180° pulse is a function of the time t between the "comb" and the sequence (90-t-180°-t-echo where t is constant), and satisfied the equation:

$$M_0 - M_z(t) = M \exp(-t/T_1) \qquad (2)$$

where $M_0 = M_z(\infty)$ and $M_z(0) = 0$. T_2 and T_1 are obtained according to Equations (1) and (2) from a computer non-linear least-squares-fit program.

The mobile fraction f_m [$= (I_{m,T} \cdot T)/(I_{m,298} \cdot 298)$] was obtained for each sample as follows. The intercepts in amplitudes on the vertical axis (at unit gain), $I_{m,T}$ and $I_{m,298}$ were measured by extending the free induction decay curve back to the zero time at T°K and 298°K, respectively.

Experimental Results

Magnetic Relaxation Measurements The longitudinal T_1 and transverse T_2 relaxation data for the two porous glasses (samples 2 and 5) described in Table I at 88 and 100% are presented in Figure 1 as a function of reciprocal temperature. The 88% RH was chosen so as to compare the effect of removing the long component from the porous for glass sample 5. The amounts of water adsorbed in g H_2O/g glass at 100 and 88% RH for samples 2 and 5 were 0.184 and 0.182, and 0.33 and 0.05, respectively. Based on their respective monolayer volumes as determined by the BET analysis of water sorption (see column 6 in Table I), the

TABLE I

MEMBRANE CHARACTERISTICS DETERMINED BY ADSORPTION ISOTHERMS[a] (8)

Sample No.	Average pore diameter[b], d $\overset{\circ}{A}$	Average pore volume[c] at 100% RH, V $cm^3 g^{-1}$	Surface area, S_{BET} $m^2 g^{-1}$	Porosity[d], ε	Monolayer Water Volume, V_m $cm^3 g^{-1}$
2	74(29)	0.186(0.149)	100(210)	0.207	0.024
5	197(189)	0.330(0.331)	67(70)	0.316	0.016

[a] All values shown in brackets were obtained from the porous glass supplier, Dr. Roland Schnabel, Jenaer Glaswerk Schott, Mainz, Germany, who used nitrogen adsorption isotherms to obtain d from the BET method.

[b] $d = (4V/S_{BET}) \times 10^4$, $\overset{\circ}{A}$ for the cylindrical model. Since $S_{BET} = V_m/t \times 10^4$, where t = statistical layer thickness for adsorbed water $\approx$ 2.39 $\overset{\circ}{A}$. The BET method is best suited to adsorbed inert gases such as N_2 since problems arise as to a suitable choice of t for the hydrogen-bonding H_2O. In this study a value of $t = 2.39$ $\overset{\circ}{A}$ was derived from the surface area of a water molecule adsorbed on amorphous silica, where 12.5 $\overset{\circ}{A}^2$ was used.

[c] Assuming $\rho_w = 1$ g/cc.

[d] ε = void fractional volume = $1/[1 + \rho_w/V\rho_g]$, $\rho_g = 1.4$ g/cc, $\rho_w = 1.0$ g/cc.

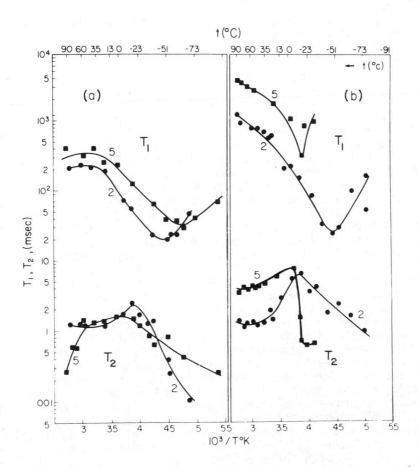

Figure 1.   Proton relaxation times vs. $10^3/T$ of adsorbed water in porous glass samples 2 and 5 at (a) $P/P_0 = 0.88$ and (b) $P/P_0 = 1.00$; (—) indicates the experimental trends.

relaxation data was also measured at a coverage of two layers for each glass sample. This corresponded to a relative humidity of 62 and 40% for the samples 5 and 2 respectively. The two layer coverage results are presented in Figure 2 as a function of reciprocal temperature. Results for the two sample glasses have been extracted from Figures 1 and 2 and are summarized in Table II.

A typical example of the primary relaxation data was given in Figure 2 of Reference 8 where one component exponential plots for T_1 and T_2 were observed at room temperature at coverages above 30% RH.

The longitudinal and transverse relaxation times for both samples 2 and 5 show in Figures 1 and 2 inverse temperature profiles typical of adsorbed systems for all the coverages except the 100% RH for sample 5. This implies a shallow T_1 minimum, a drastic spreading out of the T_1 curve, and may include a shoulder effect or maximum as T_2 increases with temperature (20,21). Logarithmic Gaussian temperature-independent (B = constant) distributions have been used to model these systems and is discussed below. Because of the similarity of these profiles, the motional characteristics of the adsorbed water is probably similar for the different conditions listed in Figures 1 and 2. The data in Table II support this view since (except for sample 5 at 100% RH) (i) T_{1min} varies between 20 and 30 msec for all the cases listed (ii) the temperature at which T_{min} occurs, Θ_{min} is also in a very narrow range and varies only from 210 to 230°K.

For sample 5 at 100% RH, the T_1 minimum is deep and the T_2 drops precipitously between about -8 and -20°C indicating freezing (23). This implies liquid-like behavior of a dominating water fraction with a liquid-like single correlation time as predicted by the theory of Bloembergen, Purcell, and Pound (BPP) (22).

Because of the symmetry of the T_1 versus $10^3/T°K$ curves even at low temperatures where paramagnetic impurities could dominate the relaxation process causing the right arm of the T_1 profile to drop (see Figure 1e and 1f in Reference 6), intramolecular proton-proton interaction is assumed to dominate.

The BPP theory predicts a T_1/T_2 at Θ_{min} of 1.6 and is applicable for one environmental state with a single correlation time. From Table II, the T_1/T_2 ratio at Θ_{min} for sample 2 is lowest at 100% RH, when the most bulk-like water is probably present. It increases at intermediate coverage when more than one state of water is present and decreases again at low coverage when another state dominates. Another interesting result is that the T_1/T_2 ratio at Θ_{min} for both samples 2 and 5 at 2-layer coverage is indentical. In fact both the T_1 and T_2 curves shown in Figure 2 are superimposable implying adsorbed water with the same motional characteristics. Thus, at low coverages, the average pore size diameter has little effect on the motional behavior of the water in the porous glass studied here.

The large values for T_1/T_2 at Θ_{min} together with the

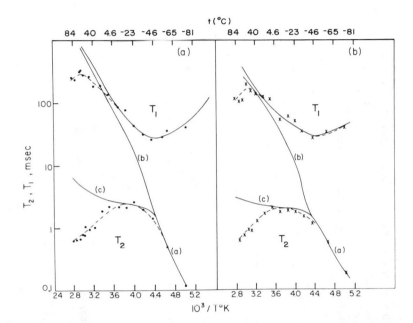

Figure 2.   NMR relaxation times for water adsorbed on porous glass: (a) sample 2 at $P/P_o = 0.4$; *(b) sample 5 at* $P/P_o = 0.62$; *(···) experiment; (—) theory; curves a and b; least-squares fit to the Resing model (21); curve c obtained by using the parameters derived from the least-squares fit to adjust the* T_2 *behavior at high temperatures to produce a shoulder effect caused by the presence of high energy sites (21)*

TABLE II

PULSED NMR EXPERIMENTAL PARAMETERS FOR WATER ADSORBED IN POROUS GLASS

Porous Glass	Mean Pore Diameter d_p Å	Relative Humidity P/P_o -	θ_{min} °K	T_1min msec	T_1/T_2 at θ_m -
5	197(189)	1.00	254	320	–
		0.88	210	30	70
		0.62[a]	230	28	22
2	74(29)	1.00	226	26	5
		0.88	227	20	40
		0.40[a]	230	28	22

[a] Assumed to be two-layer coverage.

shallowness of the T_1 minimum and the fall-off of the mobile fraction at lower temperatures (discussed later), can all be accounted for by assuming a broad distribution of correlation (in molecular jump) times (24).

In terms of the simple model of adsorbed water (8), environmental state B (bulk-like water) is first removed from the pore during dehydration leaving behind the two bound states A_1 and A_2 (or the combined state A). The combined state A is assumed to occupy two layers of water and is independent of porous glass pore diameter. See Table II in Reference (8).

Initial Amplitude (of Free Induction Decay) Measurements

In the above model the mobile or slow decay fraction, f_m is state B, and the less mobile fraction ($1-f_m$) is the combined bound state A. The intercept of a free induction decay with the y- axis is proportional to the paramagnetic susceptibility of the specimen, and hence the number of protons in the specimen is proportional to inverse temperature. Since the less mobile fraction has decayed before the instrument has recovered from its 90° pulse (in < 20 μ sec) the magnitude of the intercept I_m at zero time for the free induction decay reflects the number of protons in the mobile fraction only. Thus, as the temperature, T is varied, the product I_mT, is proportional only to the mobile water molecules.

The plot of $f_m = I_mT/(I_m298°K)$ versus reciprocal temperature for 100% RH and for 2-layer coverage for porous glass samples numbers 2 and 5 are shown in Figures 3 and 4. For sample number 2 (mean pore diameter 29 Å) at 100% RH f_m is constant at temperatures above about -16°C. Below this temperature the mobile water fraction steadily diminishes without a sudden drop expected from a freezing phenomena. This slow transition from a mobile to less mobile fraction is either due to super-cooling and slow freezing or the apparent phase transition effect (2,5). For sample 5 at 100% RH on the other hand f_m drops precipitously around 0°C. Together with the results presented in Figure 1 and Table II, this drop in f_m appears to be due to freezing of the slow decay fraction or state B.

Both curves appear to be similar for both porous glass samples at 2-layer coverage as shown in Figure 4. This, once again, confirms that the adsorbed water on both glass samples is motionally similar at 2-layer coverage.

Interpretation of Results

The Nuclear Magnetic Resonance Model   In this section a nuclear magnetic resonance multistate model is fitted to the experimental relaxation data shown in Figures 1 and 2. A fit is not attempted for sample 5 at 100% RH because of the anomolous T_1, T_2 versus $10^3/T°K$ profiles produced by freezing (see Figure 1).

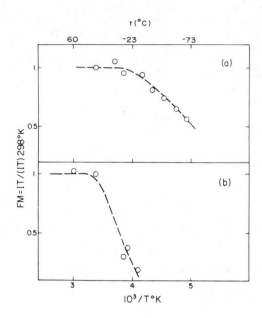

Figure 3.   Plot of mobile fraction f_m $(= \dfrac{I(T) \cdot T}{I(298) \cdot 298\ K})$ vs. $10^3/T$ of adsorbed water at $P/P_o = 1.00$ for porous glass: (a) sample 2; and (b) sample 5

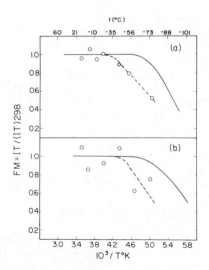

Figure 4.   Plot of the mobile fraction (f_m) vs. reciprocal temperature of adsorbed water on: (a) sample 2 at $P/P_o = 0.40$ and (b) sample 5 at $P/P_o = 0.62;$ $(\cdots)$ experiment; (—) theory

For details of the log-normal distribution model and the appropriate activation law, the reader is referred to References 6 and 21. In these references details are provided as to the method of calculation. Three computer programs were used sequentially to fit the model to the experimental data in Figures 1 and 2. The total number of parameters in the model are five: three (τ_o, H, To) to fit the activation law for τ^*, one to characterize the distribution width B, and σ_0^2 the second moment.

The first program (obtained from Dr. Leo J. Lynch, Division of Textile Physics Wool Research Labs., 338 Blaxland Rd., Rydel Sydney, NSW Australia.) uses the model to predict σ_0^2 and B at Θ_{min} as estimates for the second program. The second program (obtained from Dr. Henry A. Resing, Dept. of Chemistry, Code 6173, Naval Research Labs., Washington, DC 20390) uses numerical integration techniques to calculate the parameters for the best least squares fit [see Figure 2 solid lines for T_1 and T_2 curves (a) and (b)]. These parameters are summarized in Table III. The third program (also obtained from Dr. Henry A. Resing) uses the parameters derived from the least squares fit to adjust T_2 behavior at high temperatures as to produce a shoulder effect (see Figure 2 solid line for T_2 curve (c)).

The following comments regarding the parameters can be made:

(i) The criteria for an acceptable fit was based first on the value obtained for the second moment σ_0^2 and then on the reasonableness of the other parameters. Also see footnote d in Table III. Three of the five cases produced acceptable fits without fixing σ_0^2 at its physical minimum of 1.57×10^{10} rad^2 sec^{-2}. The proton-proton distance r_{pp} calculated from σ_0^2 (see footnote c in Table III) for samples 5 and 2 at 2-layer coverages (p/p_0 = 0.62 and 0.40 respectively) are very close to r_{pp} = 1.54 Å for a free water molecule, implying an open rather than dense packing. On adding water a denser packing (r_{pp} = 1.47 Å) is observed for sample 2 at 88% RH, while relatively large r_{pp} are calculated for sample 5 at 88% RH and sample 2 at 100%.

The second moment values σ_0^2 for sample 5 at 62% RH and sample 2 at 40% RH are close to the values obtained for water adsorbed on bacterial cell walls (26) and zeolite 13-X. (27).

(ii) At lowest coverage (2 layers) the activation enthalpy H is less for sample 5 than for sample 2. For sample 2 at 88% RH where r_{pp} = 1.47 Å was the smallest, the H-value was the highest (=1836 cal $mole^{-1}$).

(iii) For sample 2, all the preexponents τ_0 lie within the range $10^{-12} > \tau_0 > 10^{-15}$ sec in accordance with the activated state theory. The τ_0 values for sample 5 appear to be out of this range maybe due to the presence of residual free water.

(iv) The B-values are about the same for the two-layer coverage and increase as expected to the highest spread at the

TABLE III

PARAMETER SUMMARY

| Porous glass sample no. | Mean pore diameter[b] d_p Å | Relative humidity P/Po | Parameter[a] | | | | |
			Second[c] moment $10^{-10}\sigma_o^2$ rad² sec⁻² [d]	Pre-exponent $10^{12}\tau_o$ sec	Activation enthalpy H cal mole⁻¹	Spread parameter B	Proton-Proton distance r_{pp} Å
5	189	0.88	1.570 (1.268)[d]	2.144 (2.034)	784 (892)	0.2419 (0.2880)	1.59 (1.65)
		(2 layer) 0.62	2.093	1.618	968	0.2248	1.52
2	29	1.00	1.570 (1.327)	1.089 (0.966)	1290 (1336)	0.6600 (0.6783)	1.59 (1.64)
		0.88	2.572	0.029	1836	0.1955	1.47
		(2 layer) 0.40	1.882	0.256	1406	0.2730	1.55
AG-39 (from Ref. 6)	24	1.00	2.301	0.163	1645	0.174	1.50

[a] The glass temperature for the activation law was $T_g = 140°K$.

[b] Pore diameters from the nitrogen adsorption isotherm and BET analysis.

[c] Second moment σ_o^2 for water adsorbed in other surfaces aid the proton–proton distance $r_{pp}^b = \frac{9}{20}\frac{\gamma^2\hbar^2}{\sigma_o^2}$, where σ_o^2 is in Gauss², $\gamma = 2.675 \times 10^4$ Gauss⁻¹ sec⁻¹, $\hbar = 1.054 \times 10^{-27}$ erg sec and $|rad^2 sec^{-2}| \equiv 7.137 \times 10^{-2} |Gauss^2| \times 10^{10}$:

Other surface	$\sigma_o^2 \times 10^{-10}$ rad² sec⁻²	r_{pp} Å
Isolated molecule	1.57	1.59
Free water molecule	1.93	1.54
Bacterial cells walls	1.83	1.55
Zeolite 13X	2.00 - 2.29	1.53 - 1.49
Ice	2.63	1.46
Wool 98.5% RH	3.20	1.42

[d] Numbers in the brackets were the result of the best fit. Where $\sigma_o^2 < 1.57 \times 10^{10}$ rad² sec⁻², σ_o^2 was fixed at 1.57×10^{10} and the best fit for the other parameters was obtained.

highest % RH.

For the cases where the model fits the data best, i.e. without forcing σ_0^2 to its minimum value of $1.570 \times 10^{10} rad^2 sec^{-2}$, and providing $\tau_c \ll T_2$, the proton-proton distances, r_{pp} at 2 layer coverages gave values (1.52 and 1.55Å) essentially equal to that of free water molecules (1.54Å). For porous glass sample 2, increasing the coverage to 99% of complete filling at 88% RH, without introducing significant bulk water, the r_{pp} value drops significantly to 1.47Å very near that of ice and the activation enthalpy H also increases to a high at 1836 calmole^{-1}

Relaxation Analysis

Assuming that proton interaction with both paramagnetic impurities and surface hydroxyl groups are negligible, the above experimental results could in principal be explained by each or both possibilities (28):

(i) A distribution of correlation times for the adsorbed waters. Thus, a two-fraction fast exchange model (TFE) with rapid exchange between the molecules of water in region B'(=A₂+B) and region A₁ could be postulated. This model gives the following equation for special case of a sample with a very small $(P_{A_1} \ll P_{B'})$ but efficient $(T_{1A_1}^{-1} \gg T_{1B'}^{-1})$ region (adsorbent with a small fraction of high energy sites):

$$T_2^{-1} = T_{2,B'}^{-1} + \frac{P_{A_1}/P_{B'}}{T_{2A_1} + \tau_{A_1}} \tag{3}$$

or for $T_{2A_1} \ll \tau_{A_1}$ at high temperatures where the slope of log T_2 decreases with increasing temperature,

$$\Delta \frac{1}{T_2} = \frac{P_{A_1}}{P_{B'} \cdot \tau_{A_1}}. \tag{4}$$

$T_{2,B'}$ and T_2 (actual observed value) are obtained from Figures 1a and 1b at the highest temperature ($\approx 90°C$). By extrapolating the linear low temperature negative slope log T_2 vs $10^3/T°K$ line backward to the highest temperature, an estimate of $T_{2,B'}$ is obtained. The results are summarized in Table IV.

Thus, the mean life-time of a proton τ_{A_1} in the first bound layer A₁ adjacent to the bulk water B', decreases when the average pore diameter increases at both 88% RH and 2 layer coverage. This can be understood in light of the fact that the relative amount of bulk water (state B') increases with increasing pore diameter and thus the opportunity of this high energy proton (in state A₁) to interact with state B is increased. For both samples of porous glass, an increase in % RH decreases τ_{A_1}, the mean

TABLE IV

MEAN LIFE-TIME OF PROTONS IN THE "SHORT" REGION A_1 AT 90°C

Porous glass	Relative humidity %	$\Delta(\frac{1}{T_2})$ [a] sec^{-1}	P_{A_1} [b] –	$P_{B'}$ [c] –	τ_{A_1} sec	Interaction
5	88	3.519	0.3	0.7	0.122	$A_1 + [A_2+B(less)]$
	62	1.648	0.5	0.5	0.607	$A_1 + A_2$
2	100	0.741	0.136	0.864	0.212	$A_1 + (A_2+B)$
	88	0.823	0.137	0.863	0.193	$A_1 + [A_2+B(less)]$
	40	1.554	0.5	0.5	0.644	$A_1 + A_2$

[a] $\Delta(\frac{1}{T_2}) = T_2^{-1}$ (observed) $- T_{2B}^{-1}$, where T_{2B}^{-1}, is obtained from Figures 1 and 2 by extrapolating the cold T_2 curve linearly to 90°C.

[b] Measured V_{A_1} = 0.015 ml/g and volumes adsorbed were 0.029 and 0.05 at 88 and 62% RH respectively for sample 5.
V_{A_1} = 0.025 ml/g and volumes adsorbed were 0.184, 0.182, and 0.05 at 100, 88 and 40% RH respectively for sample 2.

[c] $B' = A_2 + B$

life-time of the protons in the short A_1 region at 90°C. The
difference between τ_{A_1} at 100 and 88% RH for sample 2 is probably
not significant. The same explanation given above applies here.
Finally, the value τ_A for both samples is similar supporting
earlier findings with [1] T_1/T_2 at Θ_m, and f_m behavior at two layer
coverage.
    (ii) An anisotropic reorientation of the water molecules in
region A. The IR results showed a weakening of the OH-bond at
low water coverages suggesting an <u>increase</u> in rotational diffusion
([1]). The pNMR results, on the other hand, suggest a <u>decrease</u> in
translation diffusion for low water coverages. A combined picture
may be constructed as follows: at low coverages the bulk water
hydrated proton complex $H_9O_4{}^+$ is reduced by the wall effects to
$H_5O_2{}^+$ and H_3O^+ complexes ([29,30]). These smaller water clusters
are held tightly to the solid surface impeding translation diffusion but allowing rotation about their axis. This could
possibly explain both the pNMR and IR results. Additional
verification should be sought via the deuterium NMR studies where
rotational diffusion from <u>intra</u>-relaxation effects are separated
from <u>inter</u>-relaxation effects.

Discussion

Proposed Model for Adsorbed Water in Desalination Membranes
These results suggest that some minimum coverage of water is necessary with a particular pore size of porous glass before the first
layers of adsorbed water can structure themselves into a motionally restricted state. At coverages below this critical value, the
adsorbed water molecules or monomers are relatively isolated with
large r_{pp} values, (Table III), and characterized by low T_1/T_2
values (i.e., 22 in Table II). As adsorbed water is added, T_1/T_2
values increase (i.e., to 70 and 40 in Table I for samples 5 and 2,
respectively) and the r_{pp} values shorten (to 1.47Å for 88% RH on
sample 2 in Table III) with an increase in H value (to 1836 Cal
mole[-1] for 88% RH in Sample 2, See Table III). Above this minimum
coverage for restricted water to establish itself, less restricted
and probably dimers and higher clusters of water enter the pores
during adsorption. However, like the restricted water, a minimum
amount (volume) of water must be adsorbed before bulk-like properties can establish. Luck ([1]) has termed the two types of waters
hydrate and transition stage water.
    Both the lack of freezing and the inability of ions to enter
the pores containing motionally restricted water can be explained
by the existence of fragmented clusters such as monomers, dimers
etc. Thus, the presence of these fragmented clusters prevent the
necessary aggregation and co-operative expansion needed for an
ice-like structure to exist, while at the same time they are less
able to hydrate ions resulting in low solubilities and consequently
low rejections in the desalination sense ([1,2]). This could be the
microscopic mechanistic basis for the solution-diffusion model so

successfully used in desalination.

Clearly, the absolute amount and ratio of restricted water to bulk-like water will depend inversely on the pore-size of the porous glass desalination membrane (see column 6, Table I). By normalizing the samples, an average thickness of an annulus of the first adsorbed layer, $\bar{t}$ (in angstroms can be estimated by $\bar{t} = V_m \times 10^4/S_{BET\_water}$. From Table I $\bar{t}$ (sample 2) = $0.024 \times 10^4/100 = 2.40$ Å, and $\bar{t}$ (sample 5) $=0.016 \times 10^4/67=2.39$ Å, ie results in the same $\bar{t}$ for both porous glasses. Without the establishment of bulk-like water, the pNMR results presented here and previously ($\underline{6},\underline{8}$), show typical adsorbed behavior with T_1 and T_2 relaxation times well below that of bulk water. A precipitous drop in T_2 is also not observed for these cases indicating the absence of freezing. Clearly, the presence and amount of bulk-like water with motionally restricted underlayers of water presents a complicated situation and any measurement of both waters represents a superposition of each contribution.

For cases with hardly any or no bulk-like water such as with small pores (<50Å diameter) at full coverage, or larger pores (sample 5 - 189 Å) at lower coverages (<88% RH), typical and predictable multi-state adsorbed behavior is possible. See Table III. As soon as a significant (with respect to the method of measurement) amount of bulk-like water is established a freezing phenomenon is noticed using pNMR (with T_2 for sample 5 in Figure 1b, and with f_m for sample 5 in Fugure 3b) and the available multi-state models are inapplicable.

The *Fragmented Cluster Model* proposed above is also consistant with the IR results of Luck, et. al. for porous glass ($\underline{1}$), and Toprak, $\underline{et}$. $\underline{al}$., for cellulose acetate ($\underline{2}$). Indeed, the IR results confirm the original model of Belfort $\underline{et\ al}$., for explaining the desalting mechanism of porous glass desalination membranes ($\underline{31}$). Thus, at low coverages, below the initial minimum for the establishment of motionally restricted water, a mixture of isolated monomers (H_3O^+) and small clusters ($H_5O_2^+$) are adsorbed onto the surface with restricted translational but increased rotational diffusion. Additional adsorption increases hydrogen bonding and decreases rotational diffusion. After the establishment of the motionally restricted adsorbed layers, additional monomers, dimers and other small clusters enter adsorbed in the pores. These species should also show a lower degree of association between each other than fully clustered bulk-like water.

The "barrier-effect" also reduces the average translational self-diffusion of protons and is most acute in this case with low coverages. The presence of high concentrations of solute, the physical obstruction of the pore surfaces, and the presence of almost impermeable barriers (such as the liquid-air surface) will result in a pure steric effect reducing the self-diffusion coefficient of water molecules. This is termed the barrier effect ($\underline{32}$). Belfort and Segvers (unpublished results) have seen for adsorbed water in 24 Å diameter porous glass, reductions in the measured (by pulsed gradient spin echo NMR) water self-diffusion

coefficient of 80%. This occurred when the experiment duration
was lengthened from several milliseconds to hundreds of milliseconds. Even for a measurement interval of 2 msec, the self-
diffusion coefficient of water was reduced to about 0.63 for that
of bulk water at 20°C. This reduction in D_{water} has been attributed to both a barrier-effect and the so-called "ordered" water
effect. The contribution of each effect is very difficult to
resolve, although at very short times, D_{water} appeared to approach
its bulk value indicating that the former effect was probably
dominant.

   If this were true, why then are the values of the spin-latice
T_1 and spin-spin T_2 relaxation times for protons in water greatly
reduced for adsorbed water with respect to their values in bulk
water at the same temperature (6,21). In fact, water in some
adsorbed systems is thought to supercool without formally freezing
to temperatures as low as 183°K (-90°C) (16). Increasing the
volume of adsorbed water in a particular porous system usually
results in an increase in T_1 and T_2 for protons. At 100% RH,
or when the pores are full, the T_1 and T_2 values for protons are
still well below that of bulk water (see Figure 1). Resing, et al.
(33,34) have explained this for muscle-bound water by involing
the "intermediate exchange rate" model in which one molecule per
thousand, "irrotationally bound" with very short transverse relaxation time $T_2 + 3.5 \pm 1.5$ μsec, exchanges with the great majority
of the rest of the water. Chang and Woessner (35) have seriously
questioned the basic assumption of the irrotationally bound water
(IBW) model in terms of the two-fraction fast exchange (TFE) model,
and have shown that the non-freezing bound water is sufficient to
explain the reduction in T_2 from its bulk value. Resing, et al
(36) have questioned this criticism since it is based on what they
considered an incorrect model (i.e. the TFE model) for water
associated with (rigid) muscle tissue.

   This controversy underscores the difficulty of interpretation
of pNMR results for adsorbed water systems and supports the need
for several experimental approaches for additional clarification.
For this reason, in addition to comprehensive pNMR studies, infrared spectroscopy was also used [by Prof. Dr. W.A.P. Luck (1)] in a
combined approach to better understand the motional characteristics
of occluded water in exactly the same batch of porous glass used
here.

   If in addition to the motionally restricted layers described
above, the pores are totally filled with water, the amount of
water added in excess of the restricted two layers will depend on
the free equivalent diameter, t_3, shown in Figure 5. If d_4 is the
diameter of the bulk water cluster or aggregate (nH_2O) at room
temperature, then two ranges can be defined for the size of t_3.
For $2t_3 < d_4$ and $2t_3 \geq d_4$. Only for the second case can bulk
water structure establish itself along the centerline of the pore.
Referring to Figure 5, with estimates of $t_1 \approx t_2 \approx 3Å$ and $2t_3 \approx$
$20-40_\circ Å$ [for 10-20 H_2O (37)], the pore diameter $d=2$ $(t_1+t_2+t_3) \approx$
$32-52Å$.

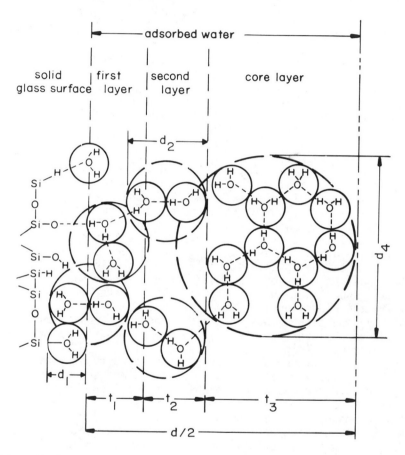

Figure 5.   Fragmented cluster model of adsorbed water showing hydrogen-bonded proton clusters in a porous glass pore

Unmodified porous glass desalination membranes that have shown moderate to high salt rejections have had measured pore diameters of 24 ± 2 Å, 41 ± 3 Å, and 50 ± 5 Å with approximate salt rejections of 65, 35 and 48%, respectively (6, 38, 39). Schultz and Asunmaa (40) have suggested that there exists a salt-free ~ 22 Å thick liquid-crystalline hydration sheath lining the pores of both cellulose acetate and porous glass membranes, while Ohya, et. al have recently shown that salt rejection improves significantly with a decrease in membrane pore radius below 20-22 Å (41).

The present *Fragmented Cluster Model* predicts a priori the size range of the pore diameter below which bulk water is not able to establish itself. Above this size range, porous glass and dense cellulose acetate desalination membranes have not shown significant salt rejection.

Conclusions

A variable temperature at constant adsorbed water coverage study was conducted for two different pore size glass desalination membranes at three different relative humidities. This compliments the earlier constant temperature and variable water coverage study (8). The following can be stated:

(i) Experimental observations of the relaxation rates T_1^{-1} and T_2^{-1} with inverse temperature indicate that there apparently exist a distribution of correlation times and hence of environmental states for the adsorbed water. A freezing phenomenon was only observed at 100% RH for the larger pore size (189 Å). Otherwise a slow transition from a mobile to a non-mobile state occurred below 0°C. This can be explained by super-cooling and slow-freezing or an apparent phase transition effect. (Depending on the validity and applicability of the model, this difference in freezing behavior could be used as a quick diagnostic tool to determine whether a membrane is suitable for desalination or not). At low coverages (2 layers), the water behaved very similarly on both pore-size glasses. For example T_1/T_2, f_m profiles and τ_{A_1} were all similar.

(ii) Parameters obtained by fitting the log-normal multi-state pNMR relaxation model show a reduction of the proton-proton distance and an increase in activation enthalpy in absorbing water from 2 layers to almost saturated coverage for the small pore-size glass.

(iii) Using relaxation analysis of the experimental results, both a distribution of correlation times and anisotropic reorientation could possibly explain the results. The latter effect is supported by IR studies, while the former is strongly indicated from the pNMR observations. Both effects could concurrently be responsible for the apparent contradiction between IR and pNMR conclusions.

(iv) Using a "Two Fraction Fact Exchange" Model, the estimated mean life-time of a proton in the bound state A_1, τ_{A_1}, deceases both with an increase in mean glass pore diameter and an increase in percent relative humidity. Based on the pNMR studies presented here, a hydration model *The Fragmented Cluster Model* for adsorbed water in porous glass (and possibly cellulose acetate) membranes is proposed. It predicts a priori, that for pore sizes above about 32 to 52 Å, bulk water structure can be established. Observations from the literature indicate that above this pore size range significant desalting is not expected.

Abstract

The proton magnetic resonance relaxation times of water adsorbed on two different pore-size porous glasses (29 and 189 Å) has been measured as a function of three coverages for the temperature range - 80 to 90°C. Both a distribution of correlation times and anisotropic reorientation are used to reconcile differing conclusions from pNMR and IR spectroscopy. A least squares fit of Resing's log-normal multi-state model for magnetic relaxation resulted in parameters that indicated a reduction in proton-proton distance with an increase in activation enthalpy on adsorbing water from about 2 layers to almost saturation.

Based on these results, a hydration model - *The Fragmented Cluster Model* for adsorbed water in porous glass (and possibly cellulose acetate) membranes is proposed. It predicts a priori, that for glass pore sizes above 32 to 52 Å, bulk water structure can be established. Observations from the literature indicate that above this pore size range significant desalting is not expected.

Acknowledgment

We gratefully acknowledge the assistance of Isser Goldberg and Ofer Eidelman in adapting and improving the computer programs for fitting the multi-state model. We thank our scientific partner, Professor Dr. W.A.P. Luck, Philipps Universitat, Marburg, Germany, for stimulating discussions and encouragement. Appreciation and thanks are also due to Dr. Roland Schnabel, Jenner GLaswerk Schott, Mainz, Germany for supplying the porous glass, to Uri Schmitt for helping us with the vacuum system. GB would also like to thank Professor David Shaltiel for managing the project during his absence on Sabbatical Leave. This research was supported by a grant from the National Council for Research and Development, Israel, and the GKSS Geesthacht-Tesperhude, Germany.

Literature Cited

1. Luck, W.A.P.; Schioberg, D.; Sieman, U., "Infrared Investigations of Water Structure in Desalination Membranes" presented at the Faraday Society Meeting, "Transport Across Synthetic Membranes", Cranfield Institute, UK (March 1, 1979),

and W.A.P. Luck, Topics in Current Chem., 1976, 64, 113.

2. Toprak, C.; Aga, J.N.; Falk, M.Trans. Faraday Soc. I. 1979, 803-815.

3. Belfort, G. Nature Phys. Sci., 1972, 237, p. 60-61.

4. Frommer, M.A.; Shporer, M.; Messalem, R.M. J. Appl. Polymer Sci., 1973, 17, 2263.

5. Frommer, M.A.; Murday, J.S.; Messalem, R.M. European Polym. Journal, 1973, 9, 367.

6. Belfort, G.; Scherfig, J.; Seevers, D.O. J. Coll. & Intf. Sci., 1974, 47, 1, 106.

7. Shporer, M.; Frommer, M.A. J. Macromol. Sci - Phys. BIO, 1974, (3), 529-542.

8. Almagor, E.; Belfort, G. J. Colloid Interface Sci., 1978, 66, (1), 146.

9. Kawaguchi, M.; Taniguchi, T.; Tochigi, K.; Takizawa, A. J. Appl. Polymer Sci., 1975, 19, 2515.

10. Frommer, M.A.; Lancet, J. Appl. Polym. Sci., 1972, 16, 1295.

11. Burghoff, H.G.; Pusch, W. J. Appl. Polym. Sci., 1979, 23, 473.

12. Krishnamoorthy, B.; Saraf, D.N. Indian J of Techn., 1972, 10, 59.

13. Chang, Y.J.; Chen, C.T.; Tobolsky, A.V. J of Polym. Sci., Polym-Physics Ed., 1974, 12, 1.

14. Zimmerman, J.R.; Brittin, W.E. J. Phys. Chem. 1957, 61, 1328.

15. Woessner, D.E. J. Chem. Phys, 1963, 39, 2783.

16. Resing, H.A. J. Phys. Chem., 1976, 80, 2, 186.

17. Resing, H.A.; Garroway, A.N.; Foster, K.R. Chapt. 42 in "Magn. Resonance in Colloid & Interface Science," ACS Symp. Ser., 1976, 34.

18. Resing, H.A. In "Advances in Molecular Relaxation Processes", 1972, 3, 199, Elsevier Publ. Co.

19. Schnabel, R., "Proceedings of the Fifth Int. Symp. on Fresh Water from the Sea", Sardinia, 16-20 May, 1976, 4, 409-413.

20. Woessner, D.E. J. Chem Phys. 1963, 39, 2783.

21. Resing, H.A., in "Advances in Molecular Relaxation Studies", 1967, Elsevier Publ. Co., p. 1.

22. Bloembergen, N.; Purcell, E.M.; Pound, R.V. Phys. Rev., 1948, 73, 679.

23. Antoniou, A.A. J. Phys. Chem., 1964, 68, 2755.

24. Resing, H.A. J. Chem. Phys., 1965, 43, 669.

25. Resing, H.A. J. Chem. Phys., 1965, 43, 669.

26. Resing, H.A.; Neihof, R.A. J. Coll. Interf. Sci., 1970, 34, 480.

27. Kvlividze, V.I.; Kiseler, V.F; Serpinski, V.V. Dold. Akad. Nauk SSSR, 1965, 165, 1111.

28. Pfeifer, H., "Nuclear Magnetic Resonance and Relaxation of Molecules Adsorbed in Solids", in NMR, Basic Principles and Progress, 1972, 5, 53 (ed. P. Dahl, E. Fleich, and R. Kosfeld) Springer Verlay.

29. Roberts, N.K. J. Phys. Chem. 1976, 80, 10, 1117.

30. Eigen, N.; De Maeyer, L. Proc. R. Soc. London, Ser A, 1958, 247, 505-533.

31. Belfort, G.; Sinai, N.; Sterling, D., "The State of Water in Synthetic Membranes and Aqueous Solutions." Second Annual Report, Jan., 1977 - Sept. 1977, project no. 848092, prepared for the NCRD, Prime Ministers Office, Jerusalem, Israel and GKSS, Baudesministerium fur Forshung und Technologie, West Germany, (March 1978).

32. Neuman, C.H. J. Chem. Phys., 1978, 60, 4508.

33. Resing, H.A.; Garroway, A.N.; Foster, K.R. Chapt. 42 in "Magn. Resonance in Colloid & Interface Science," ACS Symp. Ser., 1976, 34.

34. Foster, K.R.; Resing, H.A.; Garroway, A.N. Science, 1976, 194, 4262, 324.

35. Chang, D.C.; Woessner, D.E. Science, 1977, 198, 1180-1181.

36. Resing, H.A.; Foster, K.R.; Garroway, A.N. Science, 1977, 198, 1181-1182.

37. Luck, W.A.P. Fortschr. Chem. Forschung, 1964, $\underline{4}$, 693.

38. Littman, F.E.; Kleist, F.D.; Croopnick, G.A. "Research on Porous Glass Desalination Membranes", Office of Saline Water Res. Dev. Prog., 1971, Rep. No. 720.

39. Ballou, E.V.; Wydeven, T.; Leban, M.I. Env. Science Techn., 1971, $\underline{5}$, 1032.

40. Schultz, R.; Asunmaa, S. Rec. Progr. Surface Sci., 1970, $\underline{3}$, 291.

41. Ohya, H.; Konuema, H.; Negishi, Y. J. Applied Polym. Sci., 1977, $\underline{21}$, 2515-2527.

RECEIVED January 4, 1980.

Solute Permeation Through Hydrogel Membranes

Hydrophilic vs. Hydrophobic Solutes

S. W. KIM, J. R. CARDINAL, S. WISNIEWSKI, and G. M. ZENTNER

Department of Pharmaceutics, University of Utah, Salt Lake City, UT 84112

Poly(2-hydroxyethyl methacrylate) (p-HEMA) is a hydrophilic methacrylate polymer which was first prepared by Wichterle and Lim (1). This polymer, and many other synthetic hydrogels, has been extensively examined for potential biomedical applications (2). Although many studies have focused on the physicochemical nature of these hydrogels, many questions remain unanswered. Among these are the nature, organization, and role of water in determining such properties as interfacial and transport phenomena.

Problems which deal with the presence of water and the structure of water at the molecular level are often complex. For hydrogels, it has been proposed (3) that water can be treated in terms of a three state model. These include: bound, interfacial, and "bulk-like" water. Bound water is strongly associated with the polymer, probably as water hydrating the hydrophilic groups of the polymer. Interfacial water is probably associated with hydrophobic interactions between the polymer segments. Finally, "bulk-like" water is that with properties which are similar to that of bulk water in aqueous solution. Several studies have been designed in an effort to verify this model. The total gel water content was estimated semiquantitatively using NMR (4,5). Similar approaches were made to investigate the state of water in p-HEMA gels using the techniques of dilatometry, specific conductivity and differential scanning calorimetry (6).

Recently, we have examined solute permeation through hydrogel membranes in an effort to develop models which describe in detail the transport phenomena with particular emphasis on the role of water in this process. These studies have utilized p-HEMA and its copolymers, and both hydrophobic and hydrophilic solutes (7,8,9). It was determined that p-HEMA and its copolymers are permeable to both hydrophobic and hydrophilic solutes. The factors which influence the permeabilities include the nature and per cent of crosslinkers and the water content of the hydrogel.

In this manuscript, the permeabilities of water soluble non-

0-8412-0559-0/80/47-127-347$05.00/0
© 1980 American Chemical Society

electrolytes and hydrophobic solutes in p-HEMA and crosslinked p-HEMA are examined from a mechanistic point of view. For hydrophilic solutes, it was found that permeation probably occurs via the "bulk-like" water regions of the hydrogels. For hydrophobic solutes, analyses of permeation data indicate that solutes diffuse predominantly via a "pore" type mechanism in p-HEMA and via a "partition" mechanism in p-HEMA highly crosslinked with ethylene glycol dimethacrylate (EGDMA). The analyses of permeation data was based on the assumption that the porous flux of a solute is associated with the "bulk-like" water regions of the hydrogels and the partition flux with the polymer matrix, "interfacial" and "bound" water regions of the hydrogels.

Materials and Methods

Materials. HEMA was a highly purified sample (gift of Hydron Laboratories, New Brunswick, N.J.) containing the following levels of impurities: methacrylic acid 0.06%, ethylene glycol dimethacrylate 0.024%, and diethylene glycol methacrylate 0.24%. EGDMA (Monomer Polymer Laboratories, Philadelphia, PA) was purified by base extraction and distillation. The initiator, azobis(methylisobutyrate) was prepared by the method of Mortimer (10). Poly-HEMA films and films containing 1 mole % EGDMA were synthesized in the presence of their equilibrium water contents. Films with 5.25 mole % EGDMA were synthesized in 40% (v/v) ethanol as the solvent. All films were equilibrated in water (changed repeatedly) for three to four weeks prior to use.

All solutes were used as received. All steroids produced a single spot from TLC. Radiolabeled steroids had the same R_f values as the unlabeled materials with >95% of the detectable activity associated with the primary spot.

Methods. The diffusion experiments were performed at room temperature (23°C) utilizing a glass diffusion cell consisting of two compartments each with a volume of 175 ml. Each chamber was stirred at a constant rate to reduce boundary layer effects. Solute concentrations were monitored by ^3H or ^{14}C tracers, refractive index, or U.V. spectroscopy. Partition coefficients, defined as the ratio of the concentrations in the membrane and in the bulk aqueous phase were determined by solution depletion technique. The thickness of water swollen membranes were measured using a lightwave micrometer (Van Kueren Co., Watertown, MA).

Permeation coefficients for hydrophilic solutes were obtained through the use of the following equation (11):

$$\ln(1-2\ C_t/C_0) = -(1/V_1 + 1/V_2)\ AUt \qquad \text{Eq. 1}$$

where C_t = concentration at time t; C_0 = initial concentration; $V_1 = V_2$ = compartment volume (175 ml); A = membrane area (14.9 cm^2); U = permeability (cm/sec); and t = time (seconds). Diffusion coefficients are given by $D_m = Ud/K_D$ where d = wet membrane thickness and K_D is the partition coefficient.

Under the conditions of the experiments used in our laboratories, Eq. 1 is not valid for hydrophobic solutes due to high partitioning of these species into the membrane. In a previous publication (8), it was shown that the following equation is valid for the hydrophobic solutes:

$$\ln \frac{(C_iV - (2V + K_D/V_m) \; C_t)}{(C_iV - (2V + K_D/V_m) \; C_o)} = - \; \frac{2 \; UA}{V} (t - t_o) \qquad \text{Eq. 2}$$

where C_i = initial concentration in compartment I; C_o = concentration in compartment II at the onset of steady state (t_o), C_t = concentration in compartment II at any time t which is greater than t_o; V_m = membrane volume, and V = compartment volume (175 ml). In the limit that K_D is small, Eq. 2 reduces to Eq. 1.

Results and Discussion

1) Hydrophilic Solutes. The mechanisms of permeation of hydrophilic solutes in hydrogel films has been considered previously by Yasuda et al. (12). These authors utilized the "free volume" model for solute permeation in hydrogel films in which it was assumed that: i) the effective free volume for solute diffusion corresponds to the free volume of the aqueous phase; ii) the solute diffuses through "fluctuating pores" by successive jumps through "holes" which are larger than the solute; iii) the solute permeates only through aqueous regions and solute-polymer interactions are minimal. Based on this model, the diffusion coefficient, D_m, in the hydrated membrane is given by:

$$\frac{D_m}{D_o} \; \alpha \; - \; \frac{Bq_2}{V_f} \left(\frac{1}{H} - 1 \right) \qquad \text{Eq. 3}$$

where Bq_2 is proportional to solute cross sectional area (πr^2), D_o is the diffusion coefficient for the solute in water, V_f is the free volume, and H is the volume fraction of water in the hydrated membrane. From Eq. 3 it is apparent that D_m should be dependent upon both the cross sectional radius of the diffusing solutes and the membrane hydration.

Values of D_m for the hydrophilic solutes in p-HEMA and p-HEMA crosslinked with 1 mole % EGDMA are shown in Tables I and II. It is evident that the D_m values in the crosslinked membrane are smaller than in p-HEMA. Plots of these values according to Eq. 3 are shown in Fig. 1. A semiemperical equation developed by Wilke and Chang (13) was utilized to calculate D_o. The molar volume of the solute was estimated from atomic contributions according to LeBas (14). The molecular radii, r_o, given in Table I were calculated assuming that the solutes were spherical (15).

TABLE I

Transport Parameters of Hydrophilic Solutes in p-HEMA

Solute	r_0^2 $(\text{\AA}^2)$	$D_0 \times 10^5$ (cm^2/sec)	K_D	$D_m \times 10^{-8}$ (cm^2/sec)	$\ln \frac{D_m}{D_0}$
Na Methotrexate	----	0.38	5.57	1.95	-5.27
Sucrose	27.7	0.41	0.23	4.96	-4.41
Lactose	27.4	0.41	0.22	3.28	-4.75
Inositol	18.1	0.64	0.20	17.4	-3.61
Glucose	18.1	0.64	0.23	26.0	-3.47
Thiourea	8.8	1.30	1.27	97.0	-2.59
Raffinose	35.7	0.31	0.14	1.81	-5.14
Urea	8.1	1.39	0.53	149.	-2.23

TABLE II

Transport Parameters of Hydrophilic Solutes in
p-HEMA with 1 Mole % EGDMA

Solute	K_D	$D_m \times 10^8$ (cm^2/sec)	$\ln \frac{D_m}{D_0}$
Na Methotrexate	5.84	1.09	-5.85
Sucrose	0.25	3.53	-4.75
Lactose	0.21	2.45	-5.12
Inositol	0.18	13.1	-3.89
Glucose	0.24	15.6	-3.71
Thiourea	1.14	88.0	-2.70
Raffinose	0.12	1.31	-5.47
Urea	0.48	128	-2.39

From the plots shown in Fig. 1, it is evident that Eq. 3 is valid for the hydrophilic solutes examined in the present study. The dependence of D_m on cross sectional radius is evident from the linearity of the plots. The water contents of p-HEMA and p-HEMA crosslinked with 1 mole % EGDMA are 42% (w) and 37% (w) respectively. This effect of membrane hydration is contained in the slope of the plots given in Fig. 1. It is apparent that as the membrane hydration is increased, D_m is less sensitive to changes in the size of the permeating solute.

From these results, it may be concluded that hydrophilic solutes permeate p-HEMA and p-HEMA crosslinked with 1 mole % EGDMA primarily via the water filled channels or "pores" within the hydrogel films. This conclusion does not appear to be valid,

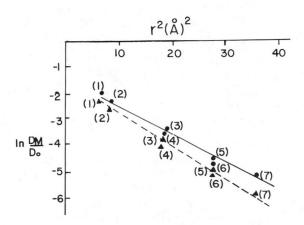

Figure 1. Dependence of diffusion coefficients of hydrophilic solutes on molecular size in (—) p-HEMA and (---) p-HEMA crosslinked with 1 mol % EGDMA: (1) urea; (2) thiourea; (3) glucose; (4) inositol; (5) sucrose; (6) lactose; (7) raffinose.

however, for p-HEMA containing higher concentrations of the cross-linker EGDMA. For example, in films containing 2.5 mole % EGDMA deviations from linearity in plots of the type shown in Fig. 1 were noted. For films prepared from 5 mole % EGDMA, no detectable diffusion was found after two weeks for the solutes raffinose and inositol.

From the data shown in Tables I and II, and that given in a thesis by Sung (5), it is possible to further define the model for hydrophilic solute permeation in hydrogel films via an analysis of the K_D values.

The partition coefficients for water in p-HEMA and p-HEMA crosslinked with 1 mole % EGDMA are 0.52 and 0.51 respectively. However, from Sung's data (5), it is possible to define partition coefficients for water into the various subclasses of water in the hydrogel membranes. For p-HEMA, these values of K_D are: bulk water 0.30, bulk + intermediate water 0.41, and bulk + intermediate + bound water 0.52. For p-HEMA with 1 mole % EGDMA, the values are: bulk water 0.21, bulk + intermediate water 0.36, and bulk + intermediate + bound water 0.50. A comparison of these values with the experimental values found in Tables I and II indicates that the sugars partition primarily into bulk water of both membranes and, therefore, that the diffusion of these solutes occurs primarily in the bulk water of the membranes. This result is consistent with the observed very low permeability of inositol and raffinose in p-HEMA with 5 mole % EGDMA. These membranes have little or no bulk water (5).

Thiourea and Na methotrexate show large deviations for K_D from values expected based on partition solely into the water fraction of the membrane. This phenomena may be due to specific interactions of the solutes with the macromolecular chains. The increase in polarizability in going from urea to thiourea and the presence of polarizable aromatic groups in Na methotrexate indicates that this interaction may be dispersive in nature. From these results, it may be inferred also that some portion of the total transport of these solutes may occur in regions other than the bulk water regions of the hydrogel membranes. It is interesting to note that the total volume fraction of H_2O in p-HEMA and p-HEMA with 1 mole % EGDMA may be available for the transport of urea.

2) Hydrophobic Solutes. The values of D_0, K_D, D_m and ln D_m/D_0 for several hydrophobic solutes in p-HEMA and p-HEMA cross-linked with 5.25 mole % EGDMA are given in Tables III and IV.

TABLE III

Transport Parameters of Hydrophobic Solutes in p-HEMA

Solute	$D_0 \times 10^6$ (cm^2/sec)	K_D	$D_m \times 10^9$ (cm^2/sec)	$\ln \frac{D_m}{D_0}$
Testosterone	6.04	48	18.8	-5.78
Norethindrone	5.83	70	13.5	-6.06
Progesterone	5.59	129	7.04	-6.68
17-Hydroxy progesterone	5.56	83	10.6	-6.27
Estradiol	6.41	177	5.48	-7.07
Cortisol	5.55	27	8.86	-6.44

TABLE IV

Transport Parameters of Hydrophobic Solutes in
p-HEMA Crosslinked with 5.25 Mole % EGDMA

Solute	K_D	$D_m \times 10^9$ (cm^2/sec)	$\ln \frac{D_m}{D_0}$
Testosterone	89	2.70	-7.71
Norethindrone	131	1.16	-7.91
Progesterone	232	1.12	-8.52
17-Hydroxy progesterone	132	1.53	-8.20
Estradiol	235	1.21	-8.57
Cortisol	20	1.78	-8.04

By comparison of D_m values given in Tables III and IV with those found in Tables I and II, it may be noted that D_m values for hydrophobic solutes are approximately two orders of magnitude less than for the hydrophilic solutes. Conversely, K_D values are about two orders of magnitude greater for the hydrophobic solutes indicating that very strong interactions occur between these hydrophobic solutes and the macromolecular segments of the hydrogel membranes.

In a previous publication (7) it was concluded that hydrophobic solutes, such as progesterone, permeate p-HEMA primarily via the "pore" mechanism. However, for progesterone in p-HEMA with 5.25 mole % EGDMA, it was found that the "partition" mechanism dominates permeation. In this mechanism, it is presumed that the solutes permeate by dissolution and diffusion within the macromolecular segments of the polymer backbone. In the "pore" mechanism K_D values are expected to be less than one and reflect

the distribution of solute between membrane solvent and bulk solvent. In the "partition" mechanism polymer segment/solute interaction is high which leads to wide range of K_D values which are
much greater than one.

From the data given in Tables III and IV, it is apparent that
K_D values for both membranes are much greater than one. These
high K_D values for p-HEMA appear to be inconsistent with the
"pore" mechanism for solute transport. This contradiction can be
solved assuming that K_D is dominated by the high solubility of the
hydrophobic solute within the hydrophobic region of the hydrogel,
whereas, permeation is carried by diffusion within "fluctuating
pores" as described in hydrophilic solute permeation. It is,
therefore, important to separate "pore" and "partition" contributions to the total permeation in order to determine the dominant
mechanisms for hydrophobic solutes in p-HEMA.

As noted in previous sections of this manuscript, it has been
proposed that three types of water exist within hydrogel films,
namely, bound, intermediate, and bulk water. From this model, it
is proposed that hydrogel membranes prepared without crosslinker
are composed of two domains designated A and B (8). Domain A consists of polymer segments, bound water and interfacial water.
Domain B is considered to be bulk water which forms the "fluctuating pores." P-HEMA with 5.25 mole % EGDMA, having no "bulk-like"
water, is assumed to be composed entirely of A type domains.
Transport in the A domains of the hydrogel occurs through the
bound and interfacial water and/or through the hydrophobic polymeric regions. Therefore, permeation in the A domains will occur
via the "partition" mechanism as previously described. K_D values
for transport in A domains vary widely depending on the solute
solubilities. Transport in the B domains occurs by diffusion of
the solute in "bulk-like" water. K_D values are ideally one in
this case since the solute is simply partitioning from bulk water
into hydrogel domains of "bulk-like" water. Diffusion composed
exclusively of A-type domains, as postulated in p-HEMA with 5.25
mole % EGDMA, occurs only by the partition mechanism which has no
"bulk-like" water. This model was successfully applied to estimate the contribution of each domain to the total permeabilities
of steroids in hydrogel membranes. It can be shown that p-HEMA
contains 22.8% "bulk-like" water (8). As noted, K_D for partition
into this water must be 1.0. With this information, the ratio of
solute diffusion due to B domains to total solute permeability,
D_B/P_T, which represents the fraction of "pore-type" permeation in
p-HEMA membranes, can be determined (8). These values are given
in Table V. It may be noted that the values of D_B/P_T are all
approximately 0.80 except cortisol which is 0.88. This indicates
that the "pore" contribution to transport in p-HEMA is similar for
all hydrophobic steroids except the relatively water soluble steroid, cortisol, which permeates by "pores" to a greater extent.

TABLE V

Transport Parameters of Steroids in
p-HEMA by Modelistic Analysis

Solute	Diffusion Coefficient (D_B) in "Bulk-like" Water (cm^2/sec)	$\ln \dfrac{D_B}{D_0}$	$\dfrac{D_B}{P_T}$	$\dfrac{K_D (I)}{K_D (II)}$
Testosterone	7.20×10^{-7}	-5.78	0.80	0.54
Norethindrone	7.29×10^{-7}	-6.06	0.77	0.54
Progesterone	5.59×10^{-7}	-6.68	0.76	0.56
17-Hydroxy progesterone	5.56×10^{-7}	-6.27	0.82	0.63
Estradiol	6.41×10^{-7}	-7.07	0.77	0.75
Cortisol	5.95×10^{-7}	-6.44	0.88	1.35

The relatively high fractions of D_B/P_T for all steroids suggest that permeation through p-HEMA membrane is dominated by the
"pore" mechanism. The high K_D values are consistent with the proposed model. According to the model and data obtained in the
p-HEMA membrane, partitioning of hydrophobic solutes is governed
predominantly by A type domains. Solute within these domains
makes a small contribution to permeability. Solute permeation is
dominated by the "pore" mechanism.

The assumption made in the model is that A domains are of the
same nature in both p-HEMA and crosslinked p-HEMA. If this is
strictly true, the partition coefficients which are dominated by
A type domains in p-HEMA should be related to partition coefficients in p-HEMA with 5.25 mole % EGDMA according to the volume
fraction of A type domains present in p-HEMA. (This volume fraction is 0.772 since "bulk-like" water is 22.8% in p-HEMA.) These
values are given in Table V as the ratio $K_D(I)/K_D(II)$. $K_D(I)$ and
$K_D(II)$ are partition coefficients of steroids in p-HEMA and p-
HEMA with 5.25 mole % EGDMA respectively. The estradiol value of
0.75 is in close agreement with the predicted value of 0.772.
However, this ratio for the other steroids except cortisol is
approximately 0.54 to 0.63. Though the quantitative agreement is
not good, qualitative agreement with the predicted value is obtained. This indicates that differences exist between the A
domains in p-HEMA and p-HEMA crosslinked with EGDMA.

It was discussed previously that diffusion coefficients of
hydrophilic solutes in p-HEMA according to Eq. 3 showed a straight
line correlation provided the free volumes accessible to the various solutes are equal. Diffusion coefficients of these steroids
(from Table III) are plotted in Fig. 2 using the r^2 value of 11.45
$\AA^2$ (16). Experimental values of $\ln D_m/D_0$ for these steroids
deviate substantially from the linear line obtained from hydro-

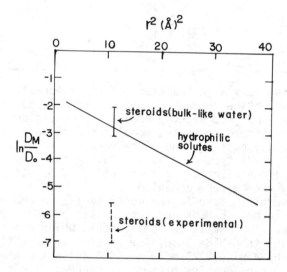

Figure 2. Dependence of diffusion coefficients on solute molecular size in p-HEMA: (—) correlation of steroid diffusion in B-type domains with water-soluble solutes; (---) experimental values of steroid diffusion.

philic solutes. However, if the calculated values of ln D_B/D_O are fit with the linear correlation obtained from the hydrophilic solutes, excellent agreement is obtained (Fig. 2). These results provide further substantiation of the model presented above, i.e., that hydrophobic solutes permeate p-HEMA membranes primarily via the "bulk-like" water regions.

In conclusion, 1) Hydrophilic solutes permeate p-HEMA and p-HEMA crosslinked with lower mole % EGDMA via the "pore" mechanism. The diffusion coefficients of the solutes depend on the molecular size and may utilize the "bulk-like" water in the hydrogels. As the water content of hydrogel increases, the solute permeability increases. 2) Hydrophobic solutes permeate p-HEMA and p-HEMA crosslinked with EGDMA via either the "pore" or "partition" mechanisms. Diffusion coefficients are lower than those of hydrophilic solutes; however, steroids can permeate even in p-HEMA with 5.25 mole % EGDMA due to the predominant "partition" mechanism for hydrophobic solute permeation in this membrane. Hydrophilic solutes fail to permeate the high crosslinked hydrogels. 3) Based on partition coefficient data, the hydrophilic solutes examined appear to permeate p-HEMA and p-HEMA with 1 mole % EGDMA via "bulk-like" water regions. Partition coefficients of steroids in p-HEMA are dominated by their high solubility of the steroids in the hydrophobic regions of the hydrogels or A domains, whereas, permeation is dominated by diffusion within "fluctuating pores."

Acknowledgements

Stimulating discussions with Drs. J. D. Andrade, D. G. Gregonis, and J. Feijen made this work possible. We greatfully acknowledge Dr. S. Ronel, Hydron Med. Sci., Inc., for his generous donations of pure HEMA.

Supported by NIH grants HD 09791 and HL 13738.

Abstract

The permeabilities of water soluble nonelectrolytes and several hydrophobic steroids in poly(hydroxyethyl methacrylate) hydrogel films were determined. The effects of crosslinking and variations in equilibrium water content of the films, on the observed permeabilities, were investigated. For hydrophilic solutes the permeation and partition coefficients are consistent with transport via the "bulk-like" water regions of the hydrogel films. These "bulk-like" water regions probably exist within the porous regions of the film. Decreases in the "bulk-like" water via copolymerization or crosslinking reduce both the partition and permeation coefficients, indicating exclusion of hydrophilic solutes from non "bulk-like" water regions. For hydrophobic solutes, permeability coefficients are smaller and partition coefficients are much larger relative to the hydrophilic solutes. For the hydrophobic solutes modelistic analysis of the permeation and partition

data indicate permeation occurs predominantly by a pore-type
mechanism in poly(hydroxyethyl methacrylate) and by a partition
mechanism in highly crosslinked poly(hydroxyethyl methacrylate)
films. The porous flux was associated with the "bulk-like" water
regions of the hydrogel films and the partition flux with the
collective polymer matrix, "interfacial" and "bound" water region
of the films.

Literature Cited

1. Wichterle, O.; Lim, D. Nature, 1960, 185, 117.

2. Ratner, B. D.; Hoffman, A. S., "Hydrogel for Medical and
    Related Applications"; ACS Symposium Series, Ed. by J. D.
    Andrade, 1976, 31, (1-36).

3. Jhon, M. S.; Andrade, J. D. J. Biomed. Mater. Res., 1973,
    7, 509.

4. Lee, H. B.; Andrade, J. D.; Jhon, M. S. Polymer Preprints,
    1974, 15, 706.

5. Sung, Y. K., Ph.D. Dissertation, University of Utah, Salt
    Lake City, Utah, 1978.

6. Lee, H. B.; Jhon, M. S.; Andrade, J. D. J. Colloid. Inter-
    face Sci., 1975, 51, 225.

7. Zentner, G. M.; Cardinal, J. R.; Kim, S. W. J. Pharm. Sci.,
    1978, 67, 1352.

8. Zentner, G. M.; Cardinal, J. R.; Feijen, J.; Song, S. J.
    Pharm. Sci., 1979, 68, 970.

9. Wisniewski, S.; Kim, S. W. J. Membrane Sci. (submitted).

10. Mortimer, G. A., J. Org. Chem., 1964, 30, 1632.

11. Lyman, D. J.; Kim, S. W. J. Polymer Sci., Symposium, 1973,
    41, 139.

12. Yasuda, H.; Peterlin, A.; Colton, C. K.; Smith, K. A.;
    Merrill, E. W. Die. Makromolecular Chemie, 1969, 126, 177.

13. Wilke, C. R.; Chang, P. A.I.Ch.E. Journal, 1955, 1, 264.

14. LeBas, G., "The Molecular Volume of Liquid Chemical Com-
    pounds"; Logmans, Green and Co.: London and New York, 1951.

15. Wisniewski, S., Ph.D. Dissertation, University of Utah, Salt Lake City, Utah, 1978.

16. Lacey, R. E.; Cowsar, D. R., "Controlled Release of Biologically Active Agents"; Ed. by A. C. Tanquary and R. E. Lacey, Plenum Press: New York and London, 1974, (117-144).

RECEIVED January 4, 1980.

Water Binding in Regular Copolyoxamide Membranes

S. GROSSMAN, D. TIRRELL, and O. VOGL

Polymer Science and Engineering, University of Massachusetts, Amherst, MA 01003

The development of a new class of regular copolyoxamides has been reported recently from this laboratory (1,2,3). The characteristic structural feature of these materials is the regular alternation of oxamide and normal amide units in the chain backbone:

$$\text{+HN-R-NHCCNH-R-NHC-R'-C+}_n$$
$$\overset{\text{OO}}{\underset{}{}} \qquad \overset{\text{O}}{\underset{}{}} \quad \overset{\text{O}}{\underset{}{}}$$

I

We have prepared a rather large number of regular copolyoxamides, many of which have shown useful membrane (4) or complexation (5) properties. The present paper deals with membrane applications of the regular copolyoxamides, with particular emphasis on the significance of water-polymer interactions.

Synthesis of Regular Copolyoxamides. The synthetic route to the regular copolyoxamides is given in Scheme I (1,2,3). The important intermediate N,N'-bis(ω-aminoalkyl)oxamides are obtained through the condensation of the appropriate diamines (in large excess) with diethyl oxalate in petroleum ether. The product is extracted with THF from a small amount of linear oligomer, and crystallizes from the extract in yields of approximately 80%. The polymerization is then carried out using either interfacial or solution techniques in which the diamine-oxamide is condensed with the acid chloride of choice. The resulting polymers are soluble in H_2SO_4, trifluoroacetic acid, or N,N-dimethylacetamide(DMAc)/LiCl mixtures. The development of good mechanical properties seems to require relatively high molecular weights, as judged by the inherent viscosity of the polymer; inherent viscosities of 0.8-1 dℓ/g are desirable and are often quite readily obtained.

0-8412-0559-0/80/47-127-361$05.00/0
© 1980 American Chemical Society

Scheme I

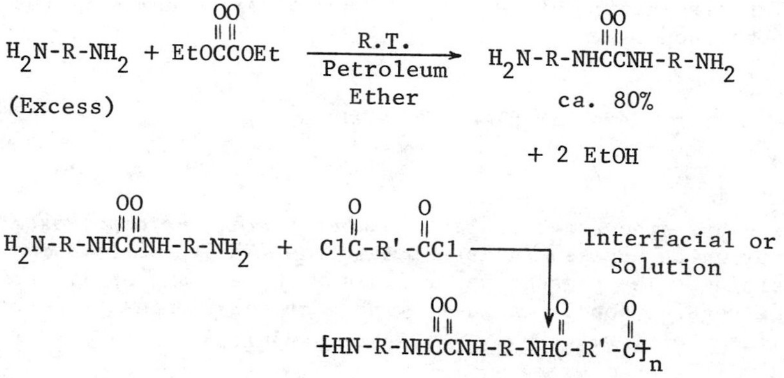

$$H_2N\text{-}R\text{-}NH_2 + EtO\overset{\overset{O}{\|}}{C}\overset{\overset{O}{\|}}{C}OEt \xrightarrow[\text{Petroleum Ether}]{\text{R.T.}} H_2N\text{-}R\text{-}NH\overset{\overset{O}{\|}}{C}\overset{\overset{O}{\|}}{C}NH\text{-}R\text{-}NH_2$$

(Excess)   ca. 80%

                                                         + 2 EtOH

$$H_2N\text{-}R\text{-}NH\overset{\overset{O}{\|}}{C}\overset{\overset{O}{\|}}{C}NH\text{-}R\text{-}NH_2 + Cl\overset{\overset{O}{\|}}{C}\text{-}R'\text{-}\overset{\overset{O}{\|}}{C}Cl \longrightarrow \begin{array}{c}\text{Interfacial or}\\ \text{Solution}\end{array}$$

$$\xrightarrow{} \left[\!HN\text{-}R\text{-}NH\overset{\overset{O}{\|}}{C}\overset{\overset{O}{\|}}{C}NH\text{-}R\text{-}NH\overset{\overset{O}{\|}}{C}\text{-}R'\text{-}\overset{\overset{O}{\|}}{C}\!\right]_n$$

Nomenclature of Regular Copolyoxamides. Because the use of systematic polymer nomenclature becomes rather cumbersome in describing the regular copolyoxamides, it is more convenient to use names similar to those used for aliphatic nylons. Thus, for Structure I, we use the number of carbon atoms in R, followed by 2 for -COCO-, followed again by the number of carbon atoms in R, and finally by a code for the remaining diacid unit. The diacid codes are given in Table I. The polymer designation uses a

Table I
Common Diacid Units in Regular Copolyoxamides

Structure	Code	Structure	Code
$-C(CH_2)_n C-$ (with two C=O)	$n+2$	pyridine diacyl structure	P
m-phenylene diacyl structure	I		

lower-case p- as prefix, in order to distinguish it from similar monomer structures (given the prefix m-). Structure II, the most thoroughly studied of the regular copolyoxamides, is thus given the designation p-222I:

$$\left[\!\!\left[HNCH_2CH_2NHCCNHCH_2CH_2NHC \text{—} C\right]\!\!\right]$$

II

Membrane Properties of p-222I. Copolyoxamide p-222I was selected for thorough study on the basis of encouraging results of measurements of transport parameters for this and related polymers. These results, obtained on ultrathin films prepared from trifluoroacetic acid solutions, are shown in Table II.

Table II
Transport Parameters of Regular Copolyoxamides ($\underline{6}$)

Polymer	D_1C_1 g/cm-sec $\times 10^8$	D_2K g/cm-sec $\times 10^8$
p-222I	18.6	1.04
p-323I	5.6	0.64
p-424I	0.4	0.02
p-424P	1.7	0.09

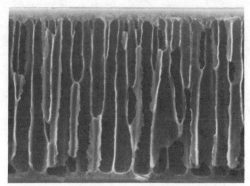

Figure 1a.   Cross section of p-222I film cast from DMAc–LiCl, evaporated 20 min at 80°C, cooled 2 min, and gelled at 2°C in 1.5% aqueous NaCl (2)

Journal of Polymer Science

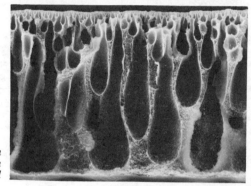

Figure 1b.   Cross section of p-222I film cast from TFA, evaporated 2 min at room temperature, and gelled at 3°C in distilled water (2)

Journal of Polymer Science

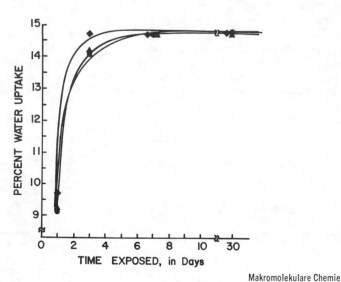

Makromolekulare Chemie

Figure 2.   Water uptake of p-222I powder (7): (●) first run; (▲) 2nd run; (♦) third run

The high value of the water flux parameter D_1C_1, along with adequately low NaCl transport, motivated the fabrication of p-222I into asymmetric membranes. The preparation of asymmetric membranes is a multistep process involving solution casting, partial solvent evaporation, and gelation as the most critical operations. The resulting asymmetric structure (a thin dense "skin" supported by a porous substructure 50-100 μm in thickness) combines the high flux of a thin membrane with the mechanical properties of a much thicker film.

Asymmetric films of p-222I have been cast from solutions in either DMAc/LiCl mixtures or trifluoroacetic acid. These two casting solvents produce very different membrane structures, as shown in Figures 1a and 1b (4). Under a rather wide variety of casting conditions, the DMAc/LiCl system affords a very regular capillary pore structure with capillary diameter of approximately 5 μm. Small-angle light scattering by water-swollen films suggests that this structure exists in the wet state as well; it is not an artifact of the SEM sample preparation (4). The structure shown in Figure 1b is an example of the kinds of morphologies obtained for membranes cast from trifluoroacetic acid. This casting solvent allows much greater control of the membrane structure; Figures 7a-d illustrate the effect of solvent evaporation time.

Water Sorption by Regular Copolyoxamides. High rates of water transport in polymeric membranes require a high degree of hydrophilicity, while the maintenance of good mechanical properties in the swollen state is favored by a high glass transition temperature or by the formation of hydrophobic domains. The synthetic route to the regular copolyoxamides allows a good deal of control over these properties, so it was of interest to determine the dependence of hydrophilicity on polymer chemical structure. The equilibrium water uptake at 93% relative humidity at room temperature was thus measured for a series of 13 regular copolyoxamides, with the results shown in Table III. It should

Table III
Equilibrium Water Uptake of Regular Copolyoxamides (7)

Polymer	Water Uptake (%)	Polymer	Water Uptake (%)	Polymer	Water Uptake (%)
p-020I	32.1 + 2.3				
p-222I	14.7 + 0.4	p-222P	17.3 + 0.4	p-2226	13.5 + 0.5
p-323I	8.5 + 0.0	p-323P	14.7 + 0.3		
p-424I	14.2 + 0.6	p-424P	10.6 + 1.0	p-4246	15.2 + 1.2
p-626I	7.7 + 0.4	p-626P	7.9 + 0.9	p-6266	7.0 + 0.3
				p-22210	11.8 + 1.4

be pointed out that these equilibrium water uptake values were quite reproducible over several sorption-desorption cycles; typical data for p-222I are shown in Figure 2.

In the first two columns of Table III are given the equilibrium values for water uptake for the copolyoxamides prepared from isophthaloyl chloride and N,N'-bis(ω-aminoalkyl)-oxamides of varying methylene chain length. The amount of absorbed water decreases with increasing aliphatic chain length, as expected, but not strictly monotonically; the percent water uptake is plotted as a function of the number of methylene carbon atoms in the alkyl group in Figure 3.

The uptake by p-323I appears to be anomalously low. This suggests differences in the degrees of crystallinity of the samples, and the higher crystallinity of p-323I (vs. p-222I and p-424I) is verified by wide-angle x-ray scattering, as shown in Figures 4a-c. It seems likely that crystallization of p-323I is promoted by forcing successive N-H bonds to point in the same direction, but we have not determined the crystal structure of the polymer. Wide-angle x-ray scattering also suggests that the absorption of water in p-323I occurs only in non-crystalline regions—x-ray scattering patterns are identical for anhydrous polymer and polymer equilibrated at 80% relative humidity (7).

Columns 2 and 3 of Table III give the equilibrium values for water uptake by copolyoxamides incorporating the 2,6-pyridinedicarbonyl group. Again hydrophilicity decreases with increasing methylene content, rather more regularly than in the previous series (Figure 5). Crystallinity undoubtedly is of importance here as well; p-222P, p-323P and p-424P all show at least some crystallinity by the x-ray technique, and p-323P is again the most highly crystalline. The differences in crystallinity are not as pronounced as in the isophthalamide polymer series, and equilibrium water uptake depends primarily on the chemical structures of the polymers. The pyridine ring seems to promote both the crystallization of the polymer, and also water absorption. Water absorption is higher in the P series than in the I series, except for the 424 polymers, in which the higher crystallinity of 424P depresses the equilibrium water uptake.

Columns 5 and 6 give the equilibrium uptake values for several wholly aliphatic copolyoxamides. No x-ray data are available for these polymers.

Calorimetric Studies of Water-Polymer Interactions. Additional information concerning water-polymer interactions was obtained through analysis of wet copolyoxamide samples by differential scanning calorimetry. Calorimetric experiments were performed both on polymer powders and on porous polymer films, with primary attention given to p-222I.

Typical results of DSC experiments on wet copolyoxamide powders are shown in Figure 6. The Figure shows thermograms for p-222I, but the other copolyoxamides behaved similarly. Samples were equilibrated at 93% relative humidity and sealed into aluminum pans to prevent loss of water during the experiment.

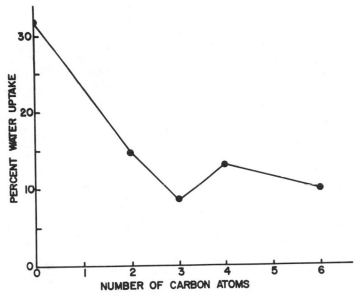

Makromolekulare Chemie

Figure 3. Percent water uptake of copolyoxamides p-A2AI as a function of chain length of diamine unit A (7)

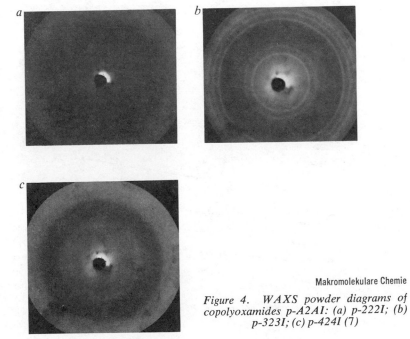

Makromolekulare Chemie

Figure 4. WAXS powder diagrams of copolyoxamides p-A2AI: (a) p-222I; (b) p-323I; (c) p-424I (7)

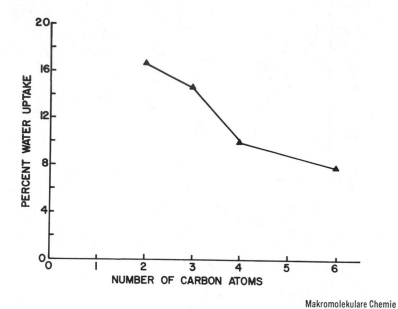

Makromolekulare Chemie

Figure 5.   *Percent water uptake of copolyoxamides p-A2AP as a function of chain length of diamine unit A (7)*

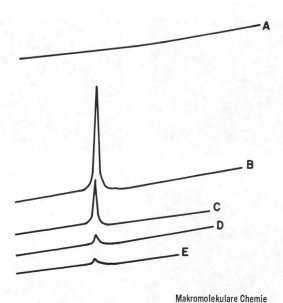

Makromolekulare Chemie

Figure 6.   *DSC scans of water in p-222I powder (7)*

On cooling to -33°C, followed by heating to 123°C at a rate of 10°C/min, we observed no melting transition for the absorbed water (A in Figure 6). This suggests that none of the absorbed water is able to freeze on cooling to -33°C, probably as a result of strong water-polymer interaction. When the sample was then held at 123°C for 10 min and recooled to -33°C, a second heating scan produced thermogram B. A sharp melting endotherm at 0°C was observed. Assuming the normal heat of fusion for water, the endotherm represented the melting of an amount of water equal to 12.6% of the sample weight. Upon successive scanning (to a maximum temperature of about 50°C) the endotherm decreased in size (thermograms C-E), and after one hour had nearly disappeared. We did not run any longer experiments to determine whether or not the endotherm would disappear completely at equilibrium.

These results may be interpreted in the following way. As stated above, it appears that all of the water absorbed by the polymer at room temperature and 93% relative humidity is bound to the polymer in such a way that it cannot freeze on cooling of the sample to -33°C. Heating to above 100°C releases this bound water, and a melting endotherm is observed on the subsequent scan. Under our conditions, we observed melting of 12.6% of the sample weight as free water, as compared to a total water content of the sample of 14.7% (determined gravimetrically). Perhaps faster experiments would have allowed observation of melting of this total water content; we did not investigate this possibility. In subsequent scans, the maximum temperature was limited to 50°C, and at these temperatures, rebinding of water occurs at a measurable rate. The endotherms in thermograms C-E represent free water contents of 6.1%, 1.2% and 0.9%, respectively. It is expected that at long times, complete rebinding would be observed.

These results suggest an interesting experiment dealing with the kinetics of water binding. The calorimetric sample can be prepared under the desired conditions (partial pressure of water vapor), sealed, and heated to release any bound water. After rapid cooling to the desired "binding" temperature, the melting endotherm can be used to follow the disappearance of free water as a function of time. Of course, the usual cautions about events occurring during the scan must be observed. This kind of experiment can then be combined with a gravimetric determination of the kinetics of water uptake to provide a more complete picture of the uptake/binding process.

Water-Polymer Interaction in p-222I Membranes. A second set of calorimetric experiments employed asymmetric membranes of p-222I, in an attempt to assess the dependence of water-polymer interaction on membrane structure. The membranes used are shown in Figures 7a-d; the photographs are scanning electron micrographs of membrane cross-sections, and the conditions of membrane preparation are given in the Figure captions. The photographs show a decreasing porosity of the films with increasing solvent evapora-

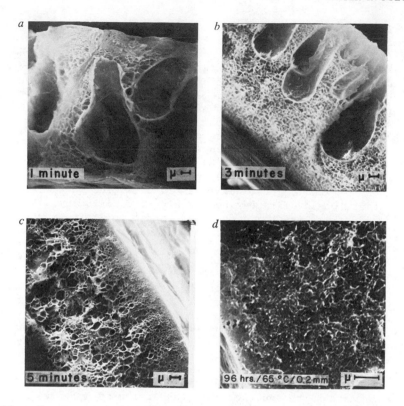

Figure 7.   Cross sections of p-222I films used for DSC and surface area measurements.   Evaporation conditions: (a) 1 min, room temperature; (b) 3 min, room temperature; (c) 5 min, room temperature; (d) 96 hr, 65°C, 0.2 mm.

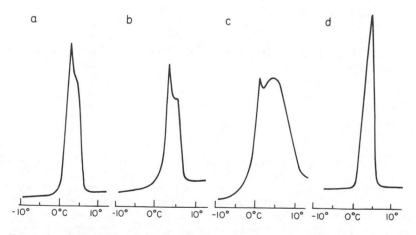

Figure 8.   DSC melting endotherms of water in p-222I membranes and of pure water.   Water content: (a) 68% (b) 61%; (c) 51%; (d) pure water.

tion time, and this is reflected in the water contents of the
films (determined on films immersed in water, from which excess
surface water was removed by blotting with filter paper):   67.7%
for membrame 1 (Figure 7a), 61.1% for membrane 2 (Figure 7b),
50.6% for membrane 3 (Figure 7c), and 41.0% in membrane 4
(Figure 7d).

Figures 8a-c show typical DSC melting endotherms for water
in membranes 1-3. The shapes of the peaks, compared to that for
pure water in Figure 8d, are quite complex, and give the appearance
of two endothermic events near 0°C. This may be due to interaction of water with the polymer film as suggested by earlier
workers who have made similar observations in hydrogels (8,9) and
in cellulose acetate membranes (10), or perhaps to a change in
thermal conductivity of the sample during melting (11). In any
case, the shape of the endotherm does depend on the thermal
history of the sample, as shown in Figure 9, while its total size
remains unchanged within the precision of the experiment.

As in our earlier calorimetric experiments on p-222I powders,
we assumed a normal heat of fusion for water in calculating the
amount of water melting near 0°C. The difference between this
amount and the total water content of the membrane is given as
bound water in column 3 of Table IV.

Table IV
Binding of Water in Copolyoxamide Films

Film	Total Water Content (%)	Bound Water per g of Dry Polymer (in g)	Surface Area (m^2/g)
1	67.7	0.74	14.8
2	61.1	0.67	13.9
3	50.6	0.50	8.07
4	41.0	0.36	< 1.0

The amount of bound water per gram of dry polymer decreases
with increasing solvent evaporation time, as does the total water
content of the film. A similar observation has again been made
for cellulose acetate membranes. Frommer and Lancet (12)
suggested that this was a consequence of densification of the
polymeric phase of the water-swollen membrane with increasing
solvent evaporation time. Rapid gelation was thought to produce
a highly swollen polymeric phase which can accommodate (and bind)
a large amount of water. More complete solvent evaporation may
allow the development of a comparatively dense phase, reducing
the amount of bound water. A similar explanation may suffice for
p-222I as well.

In order to characterize our membrane structures as completely as possible, we determined their total surface areas by
nitrogen adsorption. The surface areas measured in this way are
shown in the last column of Table IV, and include not only the

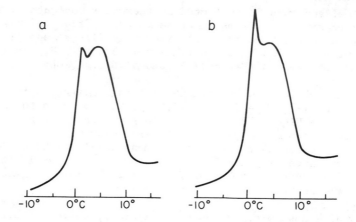

*Figure 9.   DSC melting endotherms of water in membrane of water content 51%:
(a) annealed at —30°C for several minutes; (b) quenched and rescanned*

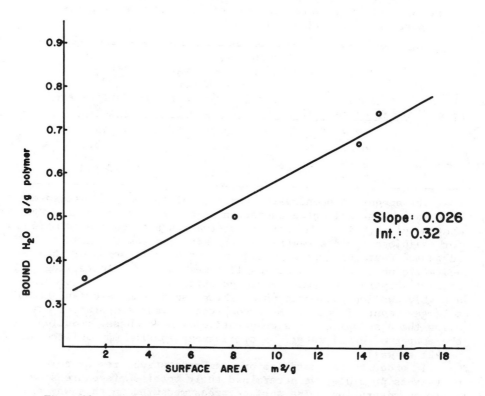

*Figure 10.   "Bound water' in p-2221 membranes vs. membrane surface area by
nitrogen adsorption*

upper and lower surfaces of the film, but the (accessible) pore structure as well. As expected from the electron microscopic examination of the films, surface area decreases with increasing solvent evaporation time.

Thus a second possible explanation for the variation in bound water content is suggested. If water is prevented from freezing not only by interaction with the bulk polymeric phase, but also by surface interactions, a linear dependence of bound water on membrane surface area would be expected. The intercept of the line would be the amount of water bound by bulk polymer, and the amount of "surface-bound" water would be obtained from the slope. Figure 10 is a plot of bound water vs. membrane surface area for the four copolyoxamide films used in this work. The data are very limited, but they do fit the best straight line with a correlation coefficient of 0.99. The best intercept is 0.32 g/g, suggesting that the bulk polymer binds 0.32 g of water per gram of dry polymer, and the best slope is 0.026 g/m^2, corresponding to the binding of 2.6×10^{-8} m^3 of membrane surface.

This highly speculative suggestion requires three assumptions: (i) that the densities of the polymeric phases of the four films examined are essentially equal. Although none of the films show any crystallinity by x-ray diffraction, we have no direct evidence that the bulk densities are the same; (ii) that the surface area measured by nitrogen adsorption is a true measure of the surface exposed to water in the swollen membrane; and (iii) that there is no quenching of water in the membrane, wrongly interpreted as bound water. Quenching does not appear to be likely under the conditions of the experiment; the rates of cooling in the DSC are probably much too low (11).

If surface binding does account for our observations, the thickness of the "bound water layer" at the membrane surface can be calculated, assuming that the density of the layer does not differ greatly from 1 g/cm^3. The estimated "layer thickness" calculated in this way is 200-300Å—very large indeed. We do not attach quantitative significance to this calculated value, but we would like to suggest that water-polymer interaction at the membrane surface may extend over distances greater than a few molecular diameters, and that this interaction may contribute significantly to the separation process.

Conclusions

Water sorption in a series of regular copolyoxamides was investigated by gravimetric determination of equilibrium water uptake at 93% relative humidity at 25°C, and by calorimetric measurement of freezing water in hydrated polymer samples. Equilibrium water uptake was determined as a function of the chemical structures of the polymers, with the following results: i) all of the regular copolyoxamides showed high water sorption

(7-30% by weight); ii) sorption/desorption cycles were repeatable
for powders and for asymmetric membranes, i.e. no irreversible
dehydration was observed; iii) equilibrium water uptake increased
with increasing amide content within a homologous series;
iv) sorption occurred in non-crystalline fractions of semicrystalline polymers; and v) time-dependent binding of water was
observed by DSC. Differential scanning calorimetry was used to
determine relative amounts of freezing and non-freezing water in
membranes of copolyoxamide p-222I. The thermal behavior of water
in these systems was very complex. A possible correlation of
non-freezing water with membrane surface area was suggested.

Literature Cited

1. Chang, H. J.; Vogl, O. J. Polym. Sci. Polym. Chem. Ed.,
    1977, 15, 311.
2. Chang, H. J.; Vogl, O. J. Polym. Sci. Polym. Chem. Ed.,
    1977, 15, 1043.
3. Stevenson, D.; Beeber, A.; Gaudiana, R.; Vogl, O. J.
    Macromol. Sci.-Chem., 1977, A11(4), 779.
4. Tirrell, D.; Vogl, O. J. Polym. Sci. Polym. Chem. Ed.,
    1977, 15, 1889.
5. Siggia, S.; Beeber, A. H.; Vogl, O. Anal. Chim. Acta.,
    1978, 96, 367.
6. Report, Tubular Reverse Osmosis Membrane Research, OSW
    14-30-2884, 1971-72.
7. Tirrell, D.; Grossman, S.; Vogl, O. Makromol. Chem., 1979,
    180, 721.
8. Lee, H. B.; Jhon, M. S.; Andrade, J. D. J. Coll. Interf.
    Sci., 1975, 51, 225.
9. Ahad, E. J. Appl. Polym. Sci., 1978, 22, 1665.
10. Hovigome, S.; Taniguchi, Y. J. Appl. Polym. Sci., 1975, 19,
    2743.
11. The authors are indebted to Professor Felix Franks for very
    helpful discussions on this and other points.
12. Frommer, M. A.; Lancet, D. J. Appl. Polym. Sci., 1972, 16,
    1295.

RECEIVED January 4, 1980.

Diffusion of Water in Rubbers

E. SOUTHERN
National College of Rubber Technology, Holloway, London, England

A. G. THOMAS
Malaysian Rubber Producers Research Association, Brickendonbury, Hertford, England

The amount of water absorbed by rubbers is strongly influenced by hyrophilic impurities present in the rubber (1,2). These impurities are much more prevalent in emulsion polymerised synthetic rubbers and natural rubber than in solution polymerised rubbers. The former group of rubbers may absorb several percent of water if left immersed for a long enough period (the maximum amount observed in these experiments was 14% over a period of 15 months) whereas the latter group absorbs very little water (usually less than 1%). The vulcanising ingredients and fillers may absorb water (2) so the rubbers used in these experiments were cured with dicumyl peroxide and contained no other curing ingredients or fillers (see Appendix) so that the vulcanisates were transparent. It was found that a white bloom appeared on the surface of the rubber after immersion in water for several days unless the samples were acetone extracted before use and so this procedure was used throughout. This bloom should not be confused with the progressive whitening which occurs in the bulk of the rubber as described below.

The theory proposed for equilibrium swelling and diffusion is based on the assumption that the hydrophilic impurities are present in particulate form and are dispersed throughout the rubber. The precise nature of this impurity in natural rubber is not known so it was decided to make a model rubber by adding 0.1% of a hydrophilic impurity (sodium chloride) to a solution polymerised synthetic rubber (cis-polyisoprene) which is chemically the same as natural rubber. Using this model rubber it is possible to check the theory more precisely since both the nature and concentration of the hydrophilic impurity in the model rubber are known. It is proposed that the water diffuses through the rubber and forms droplets of solution inside the rubber where there are particles of the hydrophilic impurity thereby causing a non-uniform distribution of water in the rubber. The

0-8412-0559-0/80/47-127-375$05.00/0
© 1980 American Chemical Society

individual droplets of solution scatter light and cause a
transparent rubber compound to become white and opaque as a
result of immersion in water.

Experimental Techniques

The equilibrium swelling of rubber depends on the vapour
pressure of the water which is in contact with the rubber
surface. The vapour pressure was varied using solutions of
sodium chloride in water and concentration was expressed as
grams of salt in 100 cm^3 of solution. The rubber samples were
flat sheets measuring 100 x 50 x 2 mm and absorption
measurements were carried out by immersing them in about
500 cm^3 of solution. Desorption measurements were carried out
by hanging the samples so that their major surfaces were
parallel to a dry airstream thus ensuring that both surfaces
were desorbed at the same rate. This was achieved by setting
up the samples and a small centrifugal fan in a constant
temperature enclosure ($25 \pm 0.5^{o}C$) containing about 1 kg of
silica gel in gauze containers.

In both cases the samples were weighed periodically and
the mass of water absorbed or desorbed, M_t, against the square
root of the time, t, was plotted. The average diffusion
coefficient, D, was calculated from the slope of this graph
using the solution to the diffusion equation for a
semi-infinite medium (3)

$$M_t = 2 (C_\infty - C_o) D (t/\pi)^{1/2}$$

where C_∞ is the overall concentration of water in the rubber
at equilibrium and C_O is the initial concentration of water in
the rubber.

Model Rubber

A model rubber was made from cis-polyisoprene, a
synthetic rubber which is chemically similar to natural
rubber, to which had been added a small amount of hydrophilic
impurity. This was 0.1% of sodium chloride in most of the
experiments but 1% of an animal protein (bovine albumen) was
also used for one set of experiments and this gave similar
results to those obtained when sodium chloride was used,
demonstrating that the phenomenon is not a feature of one type
of impurity only. Since the cis-polyisoprene used was
solution polymerized it was relatively free from hydrophilic
impurities before mixing. The desired amount of sodium
chloride was dissolved in water to form a concentrated
solution. This solution was added to the rubber on a heated
mill, the water then evaporated producing a fine dispersion of

sodium chloride in the rubber. The absorption curves of the model rubber containing 0.1% sodium chloride (vulcanizate D) and that of pure polyisoprene (vulcanizate C) are shown together with the curve for natural rubber (vulcanizate B) in Figure 1. It can be seen that the addition of the salt profoundly affects the absorption rate and that the curve for the model rubber closely resembles that of natural rubber. The rubbers were all transparent before immersion in water but the model rubber and the natural rubber rapidly became cloudy in appearance and were completely opaque after 24 hours immersion. The pure polyisoprene sample remained transparent although it showed faint cloudiness after 72 hours immersion. It seems possible therefore that the small amount of water absorbed by the polyisoprene was at least in part due to impurities. The behaviour of the model rubber was thought to be sufficiently similar to that of natural rubber for the present purpose.

Equilibrium Water Absorption

Theory. It is assumed that the absorption of water by rubbers is due to the presence of water soluble impurities. The water forms droplets of solution around the impurities inside the rubber thereby enlarging the cavity in the rubber around the impurity and equilibrium is reached when the osmotic pressure of the solution outside the rubber π_o is equal to the difference between the osmotic pressure of the solution in the droplets π_i and the pressure P exerted by the rubber on each droplet thus

$$\pi_o = \pi_i - P \tag{1}$$

The droplets are assumed to be of equal size and the concentration of the solution in each droplet is assumed to be the same. Classically the osmotic pressure π is given by

$$\pi = CRT/M$$

where C is the concentration of the solute in gm cm^{-3}, M is the molecular weight of the solute, R is the gas content and T is the absolute temperature.

$$\pi_o = \frac{C_o RT}{M_o} \quad \text{and} \quad \pi_i = \frac{C_i\, \rho_w\, RT}{C_w\, M_i} \tag{2}$$

where C_o and C_i are the concentrations of the impurity in the external solution and in the rubber respectively, assuming that all the water (density ρ_w) present in the rubber at a concentration C_w is in the droplets, M_o and M_i are the molecular weights of the impurity in the external solution and

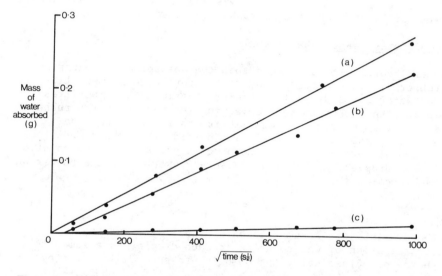

Figure 1.   *Absorption of water by (a) natural rubber, (b) model rubber, and (c) polyisoprene*

in the rubber respectively, R is the gas constant and T is the absolute temperature. The pressure P exerted by the rubber can be calculated from the pressure required to enlarge a spherical hole in an infinite block of rubber (4) and is given by

$$P = \frac{E}{6} \left(5 - \frac{4}{\lambda} - \frac{1}{\lambda^4} \right) \tag{3}$$

where E is the Young's Modulus of the rubber and λ is the extension ratio of the radius of the hole. The extension ratio λ is related to the concentrations of impurity and water by

$$\lambda^3 - 1 = C_w/C_i \tag{4}$$

Substituting equations 2, 3, and 4 in equation 1 gives

$$\frac{C_o RT}{M_o} = C \rho_w RT \quad - \frac{E}{6} \left[5 - 4 \left(\frac{C_w}{C_i} + 1 \right)^{-\frac{1}{3}} - \left(\frac{C_w}{C_i} + 1 \right)^{-\frac{4}{3}} \right] \tag{5}$$

Equations 2 are strictly true only for dilute solutions and equation 3 is subject to the limitation that the strains in the rubber are not too large (ie $\lambda < 3$ or $C_w/C_i < 26$).
If the contribution of the rubber pressure is neglected, equation 5 becomes

$$C_w = \frac{1}{C_o} \left(\frac{C_i \, \rho_w \, M}{M_i} \right)$$

$$= A/C_o \tag{6}$$

where $A = C \rho_w M_o/M_i$ is a constant for a particular rubber and type of external solution.

Experimental Results. Equation 5 cannot be used to calculate C unless the values of the constants are known. An explicit relation for C is given by equation 6 in which the effect of the rubber pressure has been neglected. Equation 6 has been used to calculate the equilibrium amount of water absorbed by natural rubber (vulcanizate B) from a salt solution containing 10% sodium chloride. The value of A cannot be calculated since the nature and concentration of the impurity in rubber is unknown but its value was estimated from the observed equilibrium swelling of rubber in a saturated salt solution. The value of A was found to be 1.14×10^{-3} and the calculated value of the concentration of water in the rubber at equilibrium in the 10% salt solution was 0.0140 gm cm^{-3} which was in excellent agreement with the experimental value of 0.0105 gm cm^{-3}. A more stringent test of the theory has been

made by calculating the equilibrium concentration of water in
the model rubber (vulcanizate D) when immersed in sodium
chloride solutions. All the constants in equation 5 for this
system are known and the experimental and theoretical values of
C_W are given in Table I where it can be seen that agreement is
satisfactory. The contribution of the rubber pressure is about
10% and if it is neglected the agreement between theory and
experiment is worse.

Table I   Equilibrium concentration of water in model rubber
            immersed in sodium chloride solutions

C_o (gm cm^{-3})	C_W (gm cm^{-3})		$\dfrac{C_W \text{ (theory)}}{C_W \text{ (experiment)}}$
	Experiment	Theory	
0.100	0.00695	0.00801	1.15
0.200	0.00376	0.00432	1.15

     It seems likely that the formation of droplets of
solution inside the rubber could lead to internal rupture of
the rubber; this possibility has been investigated. A sample
(vulcanizate B) which had absorbed 12% of water was dried and
the tensile strength measured using dumbells. There was no
significant difference between these samples and those taken
from a sheet which had not been immersed in water. It seems
unlikely therefore that the rubber had suffered any internal
damage as a result of water absorption.

Kinetics of Water Absorption

     Theory. As in the Equilibrium Swelling Theory it is
assumed that the rubber contains a number of hydrophilic
impurities and that the water dissolves these impurities
forming droplets of solution inside the rubber. The water
diffuses through the rubber in which it is only slightly
soluble and dissolves the hydrophilic impurities which are
initially present in the rubber, thus forming droplets of
solution. As the diffusion proceeds these droplets gradually
increase in size and finally reach equilibrium when the
conditions given in the Equilibrium Swelling Theory are
satisfied. Since the droplets increase in size as a result of
transport of water through the rubber phase, it can be
anticipated that until equilibrium is reached droplets near the
surface will be generally larger (more dilute solution) than
those in the bulk. The overall concentration of water in the
sample (ie, in the droplets as well as in true solution) thus
decreases with distance into the sample during an absorption
experiment and the overall distribution is similar therefore

to that of a liquid which is completely soluble in the rubber phase. This model thus gives rise to a gradient in the overall concentration of water in the sample, and if it is assumed that at all times the concentration of water in the rubber phase is in local equilibrium with the water in the droplets there is a concentration gradient in the rubber phase also. This is necessary for diffusion in the rubber phase to occur. Since the solubility of water in pure rubber is very low it is a reasonable approximation to neglect the amount of water in true solution in comparison with the water is present in the droplets. The concentration of impurity in the solution forming the droplet is therefore taken as $C_i/(C_i + C_w)$ where C_i and C_w are the concentrations of impurity and water in the rubber compound respectively. In the absence of a known expression for the concentration of water, s, in the rubber phase it has been assumed that it depends linearly on the concentration of impurity in solution, thus

$$s = s'\left[1 - \alpha \, C_i/(C_i + C_w)\right] \tag{7}$$

where s' is solubility of water in pure rubber containing no impurities and α is a constant. The rate of transport of water is $D' \, ds/dx$ where D' is the diffusion coefficient of water in pure rubber. As the true solubility of water in pure rubber is very low, D' may be taken to be constant. From equation 7

$$D' \frac{ds}{dx} = \frac{D' s' \alpha \, Ci}{(C_i + C_w)^2} \frac{dC_w}{dx}$$

$$= D \frac{d\,C_w}{d\,x} \tag{8}$$

where
$$D = \frac{s' \alpha \, C_i}{(C_i + C_w)^2} D' \tag{9}$$

The rate of transport of water is thus expressed by equation 8 in terms of an apparent diffusion coefficient, D, and the gradient of the overall concentration of water in the rubber, dC_w/dx . The theory predicts that the diffusion of water in rubber containing hydrophilic impurities can be expressed in terms of a concentration dependent diffusion coefficient (equation 9).

Experimental Results. Equation 9 is in accord with the observation that the diffusion coefficient decreases with increasing concentration of water (5 - 8). The equation predicts that the diffusion coefficient should be inversely proportional to the square of the concentration of water neglecting the term C_i which is small. Experiments on natural rubber and the model rubber (vulcanizates B and D respectively) have been carried out at different concentrations of water by immersing the samples in sodium chloride solutions. Absorption measurements were carried out both on samples which were initially dry and on those containing water initially. The latter was obtained by transferring samples which had reached equilibrium in a particular salt solution to a fresh but more dilute salt solution. The concentration dependence of the diffusion coefficient is shown in Figures 2 and 3 where it can be seen that the data is in agreement with the theory. The diffusion coefficients obtained by the above methods were, of course, average values over the concentration range existing in each measurement. This range was of the order of 0.2%, sufficiently small for the value of D obtained to be considered representative of the value at the mean concentration.

If absorption measurements are made on samples, which initially contain no water, by immersing them in salt solutions of different concentrations, the effective absorption diffusion coefficient D_a can be calculated. These measurements are followed by desorption experiments in which the samples are dried by blowing air across both major surfaces and the effective desorption coefficient D_d is calculated. The diffusion coefficients obtained from such absorption and desorption measurements are shown in Table II. It is evident that the ratio of the diffusion coefficient obtained from desorption to that from absorption measurements increases as the concentration of water in the rubber, C, increases. The diffusion coefficient obtained from a desorption experiment can only be expected to be greater than that from an absorption experiment if D decreases with increasing liquid concentration (3).

Table II   Average diffusion coefficients obtained from absorption, D_a, and desorption, D_d, experiments.

Rubber	C_0 (gm cm^{-3})	C_w (gm cm^{-3})	D_a (cm sec^{-1} x 10^{-8})	D_d	$\dfrac{D_d}{D_a}$
natural	0.100	0.01010	0.78	4.0	5.1
natural	0.315	0.00331	7.5	7.5	1.0
natural	0.315	0.00340	9.3	9.3	1.0
model	0.100	0.00695	0.74	5.4	7.2
model	0.200	0.00376	1.35	9.3	6.9
model	0.315	0.00129	17.6	37	2.1

The magnitude of the apparent diffusion coefficient of water in rubber is much lower than would be expected from the viscosity of water assuming that it behaved in the same way as

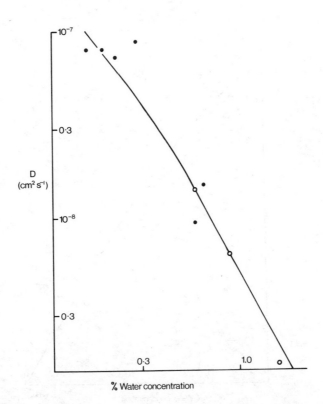

Figure 2.   *Concentration dependence of the diffusion coefficient of water in natural rubber using samples (●) initially dry and (○) initially containing water. Line calculated using Equation 9 with s'α = 6.3 × 10⁻⁸ and Cᵢ = 0.1%.*

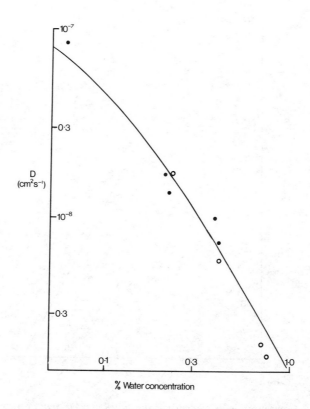

Figure 3.   Concentration dependence of the diffusion coefficient of water in model rubber using samples (●) initially dry and (○) initially containing water. Line calculated using Equation 9 with $s'\alpha = 1.8 \times 10^{-8}$ and $C_i = 0.1\%$.

an organic liquid. Equation 9 shows the relation between the apparent diffusion coefficient, D, and the true diffusion, D', which can be expected to be about 10^{-7} $cm^2 sec^{-1}$ for a liquid having a viscosity of 1 cp (9). It can be seen that the solubility of water in pure rubber, s', is part of the proportionality constant. The value of s' is not known for rubbers but the solubility of water in low molecular weight paraffins is very low (10) about (5×10^{-5}) and it can be anticipated that a value of the same order would be obtained for rubbers. The measured value of the solubility of water in cis-polyisoprene is 1×10^{-3} but the cloudiness of the rubber suggests that impurities may be present thereby increasing the apparent solubility. It is clear therefore that due to the term s' the apparent diffusion coefficient of water in rubber will be considerably lower than the true value. Taking the following estimates for the values of the other constants in equation 9; $s' = 10^{-3}$ gm cm^{-3}, $\alpha = 1$, $C_i = 10^{-3}$ gm cm^{-3} $C_w = 10^{-1}$ gm cm^{-3}; the value of D is found to be 10^{-11} $cm^2 sec^{-1}$ which compares favourably with the experimental value of 5×10^{-11} $cm^2 sec^{-1}$ from measurements on a thin natural rubber sample (vulcanizate A, 0.3 mm thick) immersed in distilled water. The factor of 5 is not regarded as significant in view of the uncertainties in the values of the constants used. It is noteworthy that the apparent diffusion coefficient is four orders of magnitude lower than the estimated true diffusion coefficient in pure rubber.

Conclusion

The diffusion of water in natural rubber is complicated by the presence of water soluble impurities in the rubber. These act as sites for droplets of solution. The impurities in natural rubber have not been identified but experiments with a model rubber have shown that the nature of the impurity, providing that it is hydrophilic, is not critical. No evidence has been found of internal rupturing of the rubber by the formation of water droplets.

Theories have been advanced which account for the equilibrium amount of water absorbed and for the diffusion of water in natural rubber. The equilibrium swelling theory is an improved version of that of Briggs et al. (2) in that a more realisitic calculation of the rubber pressure is used. The diffusion theory accounts for the experimental observations both in predicting the correct order of magnitude of the diffusion coefficient of water in rubber and also its concentration dependence.

The decrease of the diffusion coefficient of water in various hydrophobic polymers with increasing water concentration has been observed by various workers (5-7) and the phenomenon has been attributed to clustering of the water molecules

(6-8,11). While this mechanism may be more important with very pure materials it seems that the impurity mechanism would also account for the observations unless it is certain that all traces of hydrophilic material have been removed.

Literature Cited

1. Lowry, H.H.; Kohman, G.T. J. Phys. Chem., 1927, 31, 23.

2. Briggs, G.T.; Edwards, D.C.; Storey, E.B. Proc. Fourth
    Rubber Technol. Conf.: London, 1962; p.362.

3. Crank, J. "The Mathematics of Diffusion", Oxford
    University Press, 1956.

4. Gent, A.N.; Lindley, P.B. Proc. Roy. Soc. A, 1958, 249,
    195.

5. Crank, J.; Park, G.S. Eds. "Diffusion in Polymers",
    London, Academic Press, 1968.

6. Barrie, J.A.; Platt, B.J. J. Polymer Sci., 1961, 49, 479,
    and 1961, 54, 261.

7. Barrer, R.M.; Barrie, J.A. J. Polymer Sci., 1958, 28, 377.

8. Barrie, J.A.; Machin, D. J. Macromolecular Sci. (Physics),
    1969, 3, 645.

9. Southern, E.; Thomas, A.G. Trans. Far. Soc., 1967, 63,
    1913.

10. Schatzberg, P. J. Phys. Chem., 1963, 67, 776.

11. Rouse, P.E. J. Am. Chem. Soc., 1947, 69, 1068.

Appendix: Rubber Compound Formulations

Compound	A	B	C	D
natural rubber, RSSI	100	100	-	-
cis polyisoprene	-	-	100	100
sodium chloride	-	-	-	0.1
dicumyl peroxide	1	3	1.6	1.6
cure time (mins)	60	50	50	50
cure temperature(^{o}C)	150	140	150	150

Figures are parts by weight and all vulcanizates are heated in the mould for 10 mins at 100°C before the temperature is raised to the value shown in order to minimise anistropy.

RECEIVED January 4, 1980.

Hydration Control of Ion Distribution in Polystyrene Sulfonate Gels and Resins

JACOB A. MARINSKY,[1] M. M. REDDY,[2] and R. S. BALDWIN

Chemistry Department, State University of New York, Buffalo, NY 14214

Numerous attempts to formulate ion-exchange equilibria have been reported ([1-12]) and a rather complete discussion of the spectrum of theoretical approaches and models that have been employed for this purpose has been presented by Helfferich ([13]). One of the most successful of these is the Gibbs-Donnan ([14],[15]) model. In this model, $K_{N}Ex^M$, the selectivity coefficient, i.e.,

the molality product ratio at equilibrium for the Z_M-, Z_N-valent cation exchange reaction, is expressed by

$$K_{N}Ex^M = \frac{\bar{\gamma}_N^{Z_M}}{\bar{\gamma}_M^{Z_N}} \; \frac{\gamma\pm_M^{(Z_N + Z_N Z_M/Z_X)}}{\gamma\pm_N^{(Z_M + Z_N Z_M/Z_X)}} \; \exp\left[-\pi[Z_N\bar{V}_M - Z_M\bar{V}_N]/RT\right] \quad (1)$$

where π is the swelling pressure of the resin, ([1],[2],[6]), $\bar{V}_M$ and $\bar{V}_N$ are the partial molal volumes of the exchanging ions in the resin phase and $\gamma\pm_M$ and $\gamma\pm_N$ are the mean molal activity coefficients of salts $M_{Z_X}X_{Z_M}$ and $N_{Z_X}X_{Z_N}$. Barred symbols denote the resin phase, $\bar{\gamma}_M$ and $\bar{\gamma}_N$ corresponding to the mean molal activity coefficients of the exchanging pairs of ions in the gel. The charge of the coion of salts $M_{Z_X}X_{Z_M}$ and $N_{Z_X}X_{Z_N}$ is designated by Z_X. The validity of this model has been fully demonstrated by us in earlier studies with zeolites ([16]) to support its further application to the examination of the ion-exchange phenomenon in organic resins. Our efforts in this direction have been concerned with the evaluation of the $\pi(\Delta\bar{V})$ and $\bar{\gamma}_M/\bar{\gamma}_N$ terms of Equation 1. With their assessment $K_{M}Ex^N$ is calculable for comparison with experiment, $\gamma\pm_M$ and $\gamma\pm_N$ being accessible in the literature ([17]).

Current addresses:
[1] State of New York, Department of Health, Environmental Health Center, Albany, NY 12201
[2] Hooker Chemicals and Plastics Corporation, Electrochemical Research, Niagara Falls, NY 14302

0-8412-0559-0/80/47-127-387$05.00/0
© 1980 American Chemical Society

The water sorption properties of polystyrene sulfonate (PSS) exchangers with different degrees of cross-linking have been observed to coincide in the lower water activity region where the πV_{H_2O} term is negligibly small ($\underline{6}$). There is coincidence as well between osmotic coefficient data ($\underline{18}$) for the linear PSS and those obtained by Soldano for the $\leq 0.5\%$ cross-linked PSS gel; ($\underline{19}$) only at the highest water activity where a πV_{H_2O} term is operative in the gel as a consequence of solvent restraint is a lower amount of water taken up by the gel. These experimental results showed that the cross-linking agent, divinyl benzene, does not affect the physical chemical properties of the PSS and provided fundamental justification for their use in the computation of resin-phase swelling pressure (π) with Equation 2

$$RTm\emptyset_m W_{H_2O}/1000 = - RT \ln a_{H_2O} = \pi V_{H_2O} \qquad (2)$$

where W_{H_2O} is the molecular weight of water and $\emptyset_m$ is the osmotic coefficient of the polyelectrolyte analogue at m, the counterion molality in the resin. Partial molal volumes of the ions in the resin phase were taken as the partial molal volumes of the ions in aqueous solution at infinite dilution ($\underline{16}$, $\underline{20}$).

There was justification as well for utilization of the osmotic coefficient data for the linear polyelectrolyte analogue in the Gibbs Duhem equation to compute resin-phase activity coefficients for the exchanging ions. However, since the trend, with dilution, of osmotic coefficient data for fully dissociated polyelectrolytes cannot be deduced beyond the lowest measurable concentration as it can with simple electrolytes, where the Debye-Hückel limiting law applies, the computation with this equation of mean molal activity coefficients meaningfully related to a value of unity for the polyelectrolyte at infinite dilution was impossible. It was necessary to use the equation as shown below to compute mean molal activity coefficient values, $\gamma_{\pm m}$, as a function of counterion concentration, m, relative to an indeterminate mean molal activity coefficient, $\gamma_{\pm m_r}$, at the low reference concentration m_r.

$$\ln \gamma_{\pm m}/\gamma_{\pm m_r} = \emptyset_m - \emptyset_{m_r} + \int_{m_r}^{m} (\emptyset_m - 1)d \ln m \qquad (3)$$

By this approach the inherent deficiency of the earlier attempts to evaluate this term by assuming a particular osmotic behavior in the dilute concentration range ($\underline{6},\underline{7},\underline{10}\text{-}\underline{12},\underline{19},\underline{21}$) was avoided.

Only trends in ion-exchange selectivity were predicted by this approach when it was used to examine the interpretive quality of the Gibbs Donnan model for analysis of the ion-exchange phenomenon in flexible, cross-linked ion-exchange resins ($22,23$). To circumvent the intrinsic deficiency of this test of the Gibbs Donnan model for the anticipation of ion-exchange selectivity, the exchange of pairs of ions, one present in macroscopic quantity and the other in trace, between a cross-linked gel and its linear polyelectrolyte analogue was examined (24). In these systems the activity coefficients of the reference state cancel.

In this presumably definitive test of the model the value of $\gamma\pm_m/\gamma_{m_r}$ for the macroion in each phase was obtained from integration of Equation 3 to the experimental concentration of the macro counterion component. Computation of this term for the trace-ion component employed the hybrid function given by Equation 4

$$\ln \gamma\pm_{\overline{m}}/\gamma_{m_r} = \ln (\gamma\pm^{el})_m(\gamma^h)_{\emptyset m}/(\gamma\pm)_{m_r} \tag{4}$$

In Equation 4, $(\gamma\pm^{el})$ and $(\gamma^h)_{\emptyset m}$ refer to deviation from ideality attributed respectively to trace ion-polyion interaction based on the experimental concentration of the macro counter-ion component and to trace ion-solvent interaction presumed to be determined by the molality the trace ion-component would assume at the experimental water activity ($\emptyset m$) of the solvent when present as the macro component. This equation was justified in the following way:

The characteristic shape of the curve obtained for the log $\gamma\pm_m/\gamma_{m_r}$ versus log m plot that is used to characterize these polyelectrolytes is similar to the shape of the curve that is obtained from a plot of log $\gamma\pm_m$ versus $m^{1/2}$ for simple electrolytes. In simple electrolytes this property of log $\gamma\pm_m$ is attributed to the product of competing ion-ion and ion-solvent interactions (17). At low concentrations, $\gamma\pm_m^{el}$ is the dominant factor and $\ln \gamma\pm_m$ is inversely proportional to $m^{1/2}$; as the molality is raised γ_m^h becomes increasingly important and eventually $\ln \gamma\pm_m$ passes through a minimum and then increases exponentially with $m^{1/2}$. By presuming these factors to be similarly operative in the polyelectrolyte as well, $\ln \gamma\pm_m^{el}/\gamma\pm_{m_r}$ is assumed, a priori, to be represented by the initial straight line portion of the curve at low values of m (24). Extension of this line is believed to provide the value of $\ln \gamma\pm_m^{el}/\gamma\pm_{m_r}$ over the complete concentration range. Division of $\gamma\pm_m/\gamma\pm_{m_r}$ by the extrapolated values of $\gamma\pm_m^{el}/\gamma\pm_{m_r}$ over this concentration range yields the corresponding value of γ_m^h which in turn can be analyzed as a function of the water activity associated with the polyelectrolyte at these concentrations. Thus an appropriate estimate of the $\ln\gamma_{\pm m}/\gamma^{\pm}_{m_r}$ term for the trace ion in Equation 1 was thought to be provided by the product of $\gamma^{el}_{\pm m}/\gamma_{\pm m_r}$ obtained by extrapolation of the linear portion of the $\ln \gamma_{\pm m}/\gamma_{\pm m_r}$ versus $\ln m$ curve for the trace-ion form of the polyelectrolyte to the stoichiometrically equivalent concentration of the macro-ion component in the distribution study and $\gamma^{h}_{\emptyset m}$, obtained at the water activity of the equilibrium mixture from the γ^{h} versus $\emptyset m$ curve analyzed as described above, for the trace ion system (24).

There was satisfactory agreement between observation and prediction when computation was based on this presumption that the stoichiometry determines $\gamma^{el}_{\pm m}/\gamma_{\pm m_r}$ and the water activity defines $(\gamma^{h})_{\emptyset m}$. The accurate prediction of ion-exchange selectivity by this application of the Gibbs-Donnan model would appear to provide strong evidence for its validity.

Recently these studies were extended to the examination of the exchange of two counterions over their complete composition range (26). The distribution of Na^+ and Zn^{++}, between polystyrene sulfonate resin crosslinked with divinylbenznene (1%, 2%, 4%, 8%, 12% and 16% by weight) and the linear polystyrene sulfonate analogue of the resin at three different concentration levels (0.02, 0.06 and 0.12 normal) was measured. The equivalent fraction of Zn and Na was varied from 0 to 1 and 1 to 0 to examine the distribution pattern of these exchanging counterions over the complete composition range. The polyelectrolyte analogue was used in these studies to permit direct assessment of selectivity with Equation 1, the reference state mean molal activity coefficient ratio of the sodium and zinc being expected to cancel as before.

Over most of the composition range examined there was good agreement between the computed and measured selectivity values. Only when the fraction of Zn^{2+} approached unity was there sizeable discrepancy between the two values, the predicted value deviating by as much as a factor of five or six. Explanation of the discrepancy between prediction and experiment in the Zn-rich range was sought in the theoretical treatment of Manning (26). These results were rationalized (24) with the "condensation" of counterions concept developed by Manning.

The complication introduced by counter-ion "condensation" in the prediction of ion distribution patterns is, however, sizeably diminished at the concentration levels encountered with gels (24,25), even at their lowest degree of crosslinking; for example, the distribution ratio of ions in the gel in contact with a simple mixed electrolyte is rather accurately predicted with the hybrid activity coefficient function once the reference activity coefficient ratios have been evaluated with one selectivity measurement (24).

An alternative explanation for the discrepancy between the predicted and observed distribution of Na and Zn, when the concentration ratio of Zn to Na is high was suggested by the interpretation given by Boyd and Bunzl (27) to volume changes observed to accompany the selective binding of ions by polystyrene sulfonate gels. They concluded on the basis of the volume changes observed, that complexation of multiply-charged cations was extensive in this ion exchanger. Examination of the complexation of Zn(II), Ca(II) and Co(II) by the linear analogue of the polystyrene sulfonate gels showed, however, that such complexation by the polystyrene sulfonate polyion is very small indeed. The formation constant of 0.1 measured is much too small to account for the discrepancy between prediction and observation (28).

The validity of the approach that has been used to compute the activity coefficient ratio of two exchanging counterions in DVB-crosslinked PSS resins has, we believe, been adequately demonstrated. It is our next objective to obtain directly the absolute value of such ratios without resort (1) to a separate calibration step or (2) to limiting studies to exchange between the resin and its polyelectrolyte analogue.

In order to achieve this objective the following simplification of the model employed has been introduced: Recently we have shown that deviation from ideal behavior of metal ions that is due exclusively to ion-polymer interaction in the highly charged regions of the polymer surface is directly proportional to the charge of each ion as shown: $(\gamma_{\pm M}^{el})_m^{Z_N} = (\gamma_{\pm N}^{el})_m^{Z_M}$. The effect of this source of nonideality in the ion distribution pattern is thereby cancelled (29). Differences in the excess free energy originating from ion-solvent interaction thus become the one important source of difference in the affinity of pairs of ions for the resin; absolute ion-distribution patterns should be directly accessible without calibration of a particular exchange system on this basis. This aspect is fully discussed and tested in the text that follows.

Theoretical

It has been well documented by the research performed in this laboratory that use of the Gibbs-Donnan model for the interpretation of ion-exchange equilibria provides a most useful avenue for the accurate anticipation of counter ion distribution in charged polymeric systems. The activity coefficient ratio of competing ions in the polymeric phase which is the essential parameter to be assessed (see Equation 1) for successful use of this model has, as we have pointed out, so far been unaccessible by a straight-forward computation. This deficiency of our

approach to the anticipation of ion-exchange equilibria is, we
believe, removed in the following way:

i.   The Effective Concentration of Mobile Counterions in Charged
     Polymers.

The first fundamental insight to the analysis of non-ideality
in charged polymeric systems such as proteins was due to Linderstrøm-Lang (30). His complicated treatment was restated much
later in a more convenient form by Scatchard (31) to facilitate
the study of ion-binding in complicated protein systems. The
more rigorous derivations of Scatchard have been essentially
duplicated by Tanford (32) with a simplified version of the
Linderstrøm-Lang treatment. With this treatment the pK of ionizable groups of a protein, assumed to be completely independent of
one another, is a function of the charge, Z, as shown

$$pK = pK_o - \frac{2Zw}{2.303 \, (kT)} \qquad (5)$$

where w is a measure of the electrostatic free energy required to
increase the charge of the polyion from 0 to Z. From the well-
known Hasselbach-Henderson equation

$$pK = pH - \log \frac{\alpha}{1-\alpha} \qquad (6)$$

where, α, represents the degree of dissociation; thus

$$pH - \log \frac{\alpha}{1-\alpha} = pK_o - \frac{2Zw}{2.303 \, (kT)} \qquad (7)$$

The potential difference between the surface of a polyion
and the region in which the potential is measured during pH
measurements in the course of neutralization of a weak polyacid
with standard base provides an experimental evaluation of this
deviation term (0.8686 Zw) in such polymeric systems. The
deviation from ideality of mobile H^+ ions at the site of the
neutralization reaction is obtained from the well-known equation
given below:

$$pH - \log \frac{[A^-]}{[HA]} - pK_{HA} = pK_{HA_{app}} - pK_{HA} = -0.4343 \frac{\varepsilon \Psi_{(a)}}{kT} = -0.886 \frac{Zw}{kT} \qquad (8)$$

In this equation ε is the unit charge of the proton, $\Psi_{(a)}$ corresponds to the potential at the surface of the polyion, K_{HA} is the
intrinsic dissociation constant of the acid, K_{HA} is its
apparent dissociation constant and $[A^-]$ and $[HA]^{app}$ are molar
concentrations of the dissociated and undissociated polyacid
expressed on a monomer basis.

Arnold and Overbeek (33) in their pioneer demonstration of
this plotted potentiometeric titration data obtained with polymethacrylic acid as pH- log $\alpha/1-\alpha$ versus ·α. Ideally pH - log
$\alpha/1-\alpha = pK_{HA}$ and any deviation from a straight line of zero slope

in such a plot is presumed to be a quantitative measure of the deviation from ideal behavior of the system as the polyacid is progressively dissociated. The K_{HA} of 1.48×10^{-5} that is determined from extrapolation of the potentiometric data to intercept the ordinate axis at $\alpha = 0$ is in good agreement with the dissociation constant reported for isobutyric acid, the repeating monomer unit in this polymer. Since there is no ionization of the polymer at $\alpha = 0$ this number should correspond to the negative logarithm of the intrinsic dissociation constant of the repeating acid group of PMA, as it does, if the source of deviation is exclusively electrostatic in nature. Thus the change in the value of pK (ΔpK) with α, is attributed to the change in the electrostatic free energy of the molecule as a consequence of group-group interactions accompanying the ionization process.

ii. Equivalence in the Deviation from Ideality of Counter-Ions:

We have shown that the fundamental deviation term, $\exp-\varepsilon\Psi_{(a)}/kT$, derived from the study of H^+ ion in weak polyacid systems, also provides an accurate basis for estimate of the deviation from ideality of other mobile M^{+z} counterions present concurrently in these complicated systems: The deviation term then is $\exp-Z\varepsilon\Psi_{(a)}/kT$ where Z is the charge of the metal, M^{+Z}, counter-ion. In a study of complexation of Ca^{+2}, Co^{+2} and Zn^{+2} by polymethacrylic and polyacrylic acid as a function of $\underline{A}$ made in our laboratory ($\underline{29}$) the distribution of trace-level concentrations of these respective metal ions between a cation-exchange resin (Na-ion form) and solution (M $NaClO_4$) in the absence and presence of various concentrations of the respective polyacids was measured at different degrees of neutralization to demonstrate this. The partition coefficient D_0 (absence of polyacid) and D (presence of ligand) bear the following relationship to β and $\underline{A}$ for M^{+2}-polyligand systems:

$$\frac{D_0-D}{DA} = \beta_1(\exp-2\varepsilon\Psi_{(a)}/kT) + \beta_2A(\exp-2\varepsilon\Psi_{(a)}/kT) \qquad (9)$$

Analysis of the distribution results, with $\exp-\varepsilon\Psi_{(a)}/kT$ and $\underline{A}$ values directly available from pH measurements (binding of trace metal ion does not affect the stoichiometry of the system) made concurrently with the distribution measurements provided unambiguous verification of the general applicability of the non-ideality term so obtained in these systems. The experimental value of D_0-D/DA were plotted versus A. The data were extrapolated and the intercept of the ordinate defined the value of β_1. A plot of log D_0-D/DAβ_1 versus pH-log $\alpha/1-\alpha$ -pK$_{HA}$ then yielded a line with a slope of 2 which intersected the origin. The ordinate,

$\log D_0 - D/DA\beta_1$ is equal to $-2\varepsilon\psi_{(a)}/2.3kT + \log(1 + \beta_2 A/\beta_1)$ while the abscissa $pH - \log \alpha/1-\alpha - pK_{Ha}$ is equal to $\varepsilon\psi_{(a)}/2.3kT$: the observed result demonstrates that (1) essentially only the MA^+ species exists in these systems and (2) that the non-ideality of the divalent ions is defined by $Z\varepsilon\psi_{(a)}/kT$).

iii. Evaluation of the Molal Activity Coefficient Ratio of Pairs of Mobile Counter Ions in the Charged Polymer Phase.

As has been pointed out earlier the linear portion of a plot of $\log \gamma_{\bar{m}}^{\pm}/\gamma_{\bar{m}}^{\pm}$ versus $\log m$, defined by $(\emptyset_{\ell}-1) \log m/m_r$, is presumed to identify the electrostatic contribution to the mean molal activity coefficient $(\gamma_{\pm_m}^{el}/\gamma_{\pm_{m_r}}^{})$ over the complete concentration range. Subtraction of $(\emptyset_{\ell}^*-1) \log m/m_r$ from $\log (\gamma_{\pm_m}^{}/\gamma_{\pm_{m_r}}^{})$ starting at values of $\log m$ near zero then lead to evaluation of $\log \gamma_{\emptyset m}^{\pm h}/\gamma_{\pm_{m_r}}^{h}$ as a function of the charged polymer concentration at the water activity of the equilibrated system. The $\log \overline{\gamma}_M/\overline{\gamma}_N$ term in Equation 1 is composed of electrostatic and hydration terms as shown below:

$$\log \frac{\overline{\gamma}_M}{\overline{\gamma}_N} = \log \frac{(\gamma_{\pm}^{el}{}_m)_M}{(\gamma_{\pm}^{el}{}_m)_N} \frac{(\gamma_{\emptyset m}^{\pm h})_M}{(\gamma_{\emptyset m}^{\pm h})_N} \frac{(\gamma_{\pm_{m_r}}^{el})_N (\gamma_{\pm_{m_r}}^{h})_N}{(\gamma_{\pm_{m_r}}^{el})_M (\gamma_{\pm_{m_r}}^{h})_M} \tag{10}$$

In the polymer the electrostatic terms cancel: also

$$\gamma_{\pm} = (\gamma^+)^{\nu/\nu+1} (\gamma^-)^{1/\nu+1} \tag{11}$$

where ν is the degree of polymerization. If $\nu > 200$

$$\gamma = (\gamma^+)^1 (\gamma^-)^0 = \gamma_+ \tag{12}$$

and

$$\log \frac{\overline{\gamma}_M}{\overline{\gamma}_N} = \log \frac{(\gamma_{+\emptyset m}^h)_M}{(\gamma_{+\emptyset m}^h)_N} \frac{(\gamma_{+m_r}^h)_N}{(\gamma_{+m_r}^h)_M} \tag{14}$$

At the reference concentration (0.01 for divalent and 0.02 for univalent ions), a_w, the activity of water, is very nearly unity and $\gamma_{+m_r}^h$ has to be unity as well so that $\gamma_{\emptyset m}^h/\gamma_m^h = \gamma_{\emptyset m}^h$ and

$$\log \overline{\gamma}M/\overline{\gamma}N = \log (\gamma_{\emptyset m}^h)/(\gamma_{\emptyset m}^h)_N \tag{15}$$

at the experimental water activity; the value of $\overline{\gamma}_M/\overline{\gamma}_N$ is thus

---

$\emptyset_{\ell}^*$ is the limiting value of the osmotic coefficient at $m < 0.02$

directly calculable in the ion-exchange resin at every experimental condition with the $(\gamma_{+\emptyset m}^{h})$ values so resolved from such analysis of osmotic coefficient data.

Marinsky and Högfeldt (34) have shown that at the higher concentrations of polystyrene sulfonate the polyanion is essentially unhydrated so that $\gamma_{\pm}^{h} \simeq \gamma_{+}^{h}$. There may be a finite but small contribution to γ^{h} from the polyanion, however, at the lower concentrations and the prediction of ion distribution patterns in the lower polymer concentrations with less cross-linked resins may be affected by neglect of such a contribution.

iv. <u>Examinations of the Predictive Quality of the Method Proposed For The Anticipation of the Distribution of Pairs of Ions Between an Ion-Exchange Resin and Simple Neutral Electrolyte.</u>

Recently Boyd et al (35) conducted an extensive study program designed to provide a thermodynamic basis for the prediction of the selectivity pattern of Zn^{2+} and Na^{+} during equilibration in $Zn(NO_3)_2$, $NaNO_3$ mixtures (0.1 N) of polystryene sulfonate exchanger crosslinked to various degrees (0.5% to 24%) with divinylbenzene. The complete composition range of the resin (X_{Zn} from zero to 1) was covered in these studies. In order to facilitate the prediction of ion-exchange selectivity in this dilute electrolyte mixture they measured the water content of the pure ion forms and the mixed ion forms of the resins as a function of degree of crosslinking. These data were used in the Gibbs Duhem equation for ternary mixtures through application of the cross differential identities which apply for exact differentials. In this way, they computed the activity coefficient ratio and the partial molar volume difference in the resin of the exchanging ions to predict ion-exchange selectivity coefficients for comparison with the experimental coefficients. In order to employ these data, however, it was necessary to calibrate first the activity estimates through experimental observations made over the complete composition range of the exchanger at one fixed crosslinking value (0.5% DVB).

We have compared in Table I the selectivity measurements made by Boyd and coworkers of the NaPSS, $Zn(PSS)_2$, $Zn(NO_3)_2$, $NaNO_3$ system with predictions based upon (1) their rigorous thermodynamic analysis of the system and (2) our direct assessment of the ratio of $(\gamma_{Na}^{h})^2/(\gamma_{Zn}^{h})$ as described in the preceeding section where $\log_{Na}K_{EX}^{Zn} = (\log (\gamma_{Na}^{h})^2 (\gamma_{\pm Zn(NO_3)_2})^3/(\gamma_{Zn}^{h})(\gamma_{\pm NaNO_3})^4$, neglecting the $\pi\Delta V/2.3$ RT term which is relatively unimportant.

We see at once from Table I that the predictive quality of the procedure that we have developed here is fully as good as that provided by the Boyd (35) approach. Their predictions are based upon calibration of the system through measurements of

TABLE I

A Comparison of log K_{EX} Predictions with Experimental Measurements

$$\log{}_{Na}K^{Zn}_{EX}$$

| | $X_{Na} = 1$ | | | | $X_{Zn} = 1$ | | |
DVB	Exp.	Boyd (35)	Pred. This Lab.	Expt.	Boyd(35)	Pred. This Lab.
2%	0.096	0.153	0.04	0.10	0.19	-0.01
4%	0.00	0.057	0.00	0.09	0.19	0.00
8%	-0.22	-0.18	-0.13	-0.11	0.01	-0.20
12%	-0.30	-0.33	-0.23	-0.21	-0.11	-0.36
16%	-0.35	-0.47	-0.34	-0.42	-0.18	-0.52

selectivity made with the 0.5% crosslinked resin, whereas our predictions depend only upon resolution of γ_h from the log $\gamma_{\pm m}/\gamma_{\pm m_r}$ versus log m curves derived from our analysis of osmotic coefficient data obtained for the polymer analogues of the cross-linked resins. The indication is that the PSS resins are essentially fully dissociated in their divalent forms as we have shown (28).

In Table II we have presented selectivity predictions as well for several pairs of ions made by the method we have introduced here. In this table we also include the experimentally measured values. Upon comparison of prediction with experiment one observes that when the macroion component is univalent (H^+) and the trace ion component is divalent (Sr^{+2}, Cd^{+2}, Co^{+2}, Ca^{+2}, and Zn^{+2}) agreement is quite good at the higher degrees of cross-linking (> 4%). At the lower degrees of crosslinking (1% and 2%), the predicted value of $K_N^{M^T}$ is too small. When the opposite experimental condition is tested, i.e.,the macroion component is divalent (Zn^{+2}, Cd^{+2}, Sr^{+2} and Ca^{+2}) and the trace ion component is univalent (Na^+), agreement is good for the Zn^{+2}, Na^+ system: with Cd^{+2} and Ca^{+2} the predicted $K_M^{Na^T}$+2, while a factor of two too large, parallels the trend in selectivity with cross-linking.

The computed value of $K_{Sr}^{Na^T}$ is too large by a factor of almost 2 at low degrees of crosslinking but converges with the experimentally determined value at the highest degree of crosslinking (>8%). In the divalent-divalent systems the agreement between prediction and experiment is quite good for almost every pair of ions examined. Only for the Zn-Sr^T pair is there considerable

Table II

Selectivity Predictions for Pairs of Ions in Dilute Electrolyte

Trace Ion = Sr^{+2}

% DVB	0.168 M HClO$_4$(22)		0.1 M Ca(ClO$_4$)$_2$(23)		0.1M Zn(ClO$_4$)$_2$(23)		0.1M Cd(ClO$_4$)(23)	
	$K_N^{M^T}$(Pred)	$K_N^{M^T}$(Exp)	$K_N^{M^T}$(Pred)	$K_N^{M^T}$(Exp)	$K_N^{M^T}$(Pred)	$K_N^{M^T}$(Exp)	$K_N^{M^T}$(Pred)	$K_N^{N^T}$(Exp)
1	1.41	3.21	0.90	1.56	0.88	2.83	1.30	2.14
2	1.81	3.51	0.94	1.78	0.99	3.01	1.49	2.41
4	3.06	4.30	1.05	1.72	1.34	3.59	1.80	2.91
8	5.81	6.52	1.13	1.84	2.20	4.85	2.40	3.15
12	9.49	10.88	1.27	1.87	3.30	7.13	2.88	3.99
16	16.28	13.13	1.27	1.83	3.59	11.1	3.20	3.35

Trace Ion = Cd^{+2}

% DVB	0.168M HClO$_4$(22)		0.1 M Cu(ClO$_4$)$_2$(23)		0.1MZn(ClO$_4$)(23)		0.1M Sr(ClO)$_4$(23)	
1	1.22	1.82	1.01	0.76	1.04	1.03	0.99	0.60
2	1.39	1.89	0.99	0.73	0.99	1.14	1.02	0.55
4	2.02	1.97	0.91	0.62	1.12	1.16	0.90	0.53
8	2.81	2.19	0.80	0.56	1.21	1.42	0.74	0.44
12	3.31	2.54	0.72	0.56	1.50	1.80		
16	3.72	2.65	0.71	0.55	1.41	2.24		

Trace Ion = Co^{+2}

% DVB	0.168M HClO$_4$(22)		0.1MCa(ClO$_4$)$_2$(23)		0.1M Zn(ClO$_4$)$_2$(23)		0.1M Sr(ClO$_4$)$_2$(23)	
1	1.27	1.44	1.12	0.75	1.00	1.20	0.89	0.57
2	1.50	1.87	1.04	0.72	0.88	1.08	0.96	0.49
4	1.84	1.82	0.81	0.63	1.00	1.11	0.75	0.47
8	2.56	1.78	0.67	0.52	0.98	1.11	0.57	0.20

Trace Ion = Ca^{+2}

% DVB	0.168M HClO$_4$(22)		0.10M Sr(ClO$_4$)(23)		0.1M Cd(ClO$_4$)(23)		0.1M Cd(ClO$_4$)$_2$(23)	
1	1.26	2.09	0.93	0.70	1.03	1.18	1.03	0.93
2	1.60	2.59	1.05	0.65	1.14	1.26	1.04	0.92
4	2.51	2.97	1.01	0.69	1.22	1.37	1.18	0.89
8	4.10	3.92	0.89	0.61	1.51	1.49	0.87	0.74
12	5.48	6.01	0.79	0.58	1.61	1.88		
16	7.42	7.26	0.74	0.60	1.65	1.40		

Trace Ion = Zn^{+2} Trace Ion = Na^+

% DVB	0.168 M HClO$_4$(22)		0.1 M Ca(ClO$_4$)$_2$(23)		0.1M Zn(ClO$_4$)$_2$(23)		0.1M Sr(ClO$_4$)$_2$(23)	
1	1.17	1.71	0.96	0.81	0.98	0.49	0.96	0.54
2	1.35	1.75	0.88	0.71	0.86	0.38	0.93	0.50
4	1.84	1.68	1.03	0.68	0.91	0.51	0.79	0.61
8	2.15	1.72	1.29	1.24	1.11	0.50	0.68	0.57
12	2.35	1.66	2.22	1.81	1.35	0.65	0.71	0.64
16	2.64	1.50	2.56	2.53	1.51	0.90	0.71	0.71

Trace Ion = Na^+

% DVB	0.1M Ca(ClO$_4$)$_2$(23)	
1	1.01	0.49
2	0.87	0.48
4	0.66	0.42
8	0.81	0.42
12	0.85	0.48
16	0.86	0.49

discrepancy between prediction and experiment. However, the trend in selectivity is predictable with a ratio of approximately three existing between computation and experiment over the complete crosslinking range (1 to 16%).

We have found in many instances that the serious discrepancy between prediction and experiment for the various systems at the lower degrees of crosslinking can be removed by evaluation of $\gamma_M^h T$ at the molality of the macroion component rather than at the molality that the macro-ion form of the trace element corresponds to at the water activity of the equilibrium system. This observation is summarized in Table III.

Table III

Selectivity Predictions at Low Degrees of Crosslinking

System	%DVB	$K_N^{M^T}$ (Pred.)	$K_N^{M^T}$ (Exp)
Zn^T $HClO_4$	1	2.00	1.71
Cd^T, $HClO_4$	1	2.03	1.82
Ca^T, $HClO_4$	1	1.90	2.09
Na^T $Cd(ClO_4)_2$	1	0.49	0.49
Na^T, $Ca(ClO_4)_2$	1	0.48	0.49
Na^T, $Sr(ClO_4)_2$	1	0.39	0.54
Na^T, $Sr(ClO_4)_2$	2	0.42	0.50
Sr^T, $Cd(ClO_4)_2$	1	1.52	2.14
Sr^T, $Cd(ClO_4)_2$	2	1.89	2.41
Sr^T, $Cd(ClO_4)_2$	4	2.47	2.91
Cd^T, $Sr(ClO_4)_2$	1	0.67	0.60
Cd^T, $Sr(ClO_4)_2$	2	0.66	0.55
Cd^T, $Sr(ClO_4)_2$	4	0.52	0.53

From this result we learn that at the lowest counterion concentrations in the resin (< 4% DVB) ion-solvent interaction in the trace-ion component may be best described by employing the molality of the macro-ion component to estimate $\gamma_M^h T$. Apparently only after the concentration of counterions increases does the model we have employed for computation of $\gamma_M^h T$ become most applicable.

A good deal of additional research is required to examine most carefully the concepts we have introduced. There can be no question however, that the predictive quality of the new approach we have developed for the assessment of counterion distribution patterns in ion-exchange resins compares favorably with the less direct methods employed rather successfully in our earlier work and further research is merited.

We believe that fundamental assessment of the factors important in defining counterion distribution in charged polymers (crosslinked and linear) has been resolved with our interpretation of osmotic coefficient data. By our analysis, ion-solvent interactions are believed to contribute most importantly in the polystyrene sulfonate-based resins to their ion-exchange selectivity patterns. The agreement obtained between the prediction of these patterns and their observed distribution without need to resort to a single measurement for calibration of the activity coefficient terms in the polymer provides strong support for the validity of the interpretations made.

ABSTRACT

Successful prediction of ion-exchange selectivity has been accomplished previously by using ion-exchange resin-phase activity coefficients calculated from polyelectrolyte osmotic coefficients with the Gibbs-Duhem equation. Because these computed activity coefficients needed to be defined relative to an activity coefficient arbitrarily assigned at a reference concentration direct calculation of the absolute value of the ion-exchange selectivity coefficient was not possible in these earlier studies without evaluation of the reference activity coefficient ratio by a separate selectivity measurement. This restrictive aspect is circumvented by cancellation of the equivalent electrostatic contribution (ion-ion interaction) to the resin phase activity coefficients of pairs of ions exchanging between simple electrolyte and the resin. Resolution of the contribution of ion-solvent interaction to the mean activity coefficient of each ion in this way has permitted estimate of selectivity coefficients directly without the reference activity coefficient ratio calibration step. The relative success of this approach supports strongly the concept that ion-solvent interaction contributes most importantly to the ion-exchange selectivity patterns encounted with the DVB crosslinked polystyrene sulfonate resins.

Literature Cited

1.   Gregor, H. P. J. Am. Chem. Soc. (1948), 70, 1293.

2.   Gregor, H. P. ibid (1951), 73, 642.

3.   Ekedahl, E.; Hogfeldt, E., Sillen, L. G. Acta. Chim. Scand. (1950), 4, 828.

4.   Hogfeldt, E.; Ekedahl, E.; Sillen, L. G. ibid (1950), 4, 556.

5.   Bonner, O. D.; Argersinger, W.J. Jr.; Davidson,A.W. J. Amer. Chem. Soc. (1952), 74, 1044.

6. Glueckauf, E. Proc. Roy. Soc. Ser. A. (1952), 214, 207.

7. Duncan, J. F. Aust. J. Chem (1955), 8, 293.

8. Soldano, B.; Chestnut, D., J. Am. Chem. Soc (1955), 77, 1334.

9. Myers, G. E.; Boyd, G. E., J. Phys. Chem. (1956), 60, 521.

10. Boyd, G. E.; Lindenbaum, S.; Myers, G. E., ibid (1961), 65, 577.

11. Feitelson, J. ibid (1962), 66, 1295.

12. Marinsky, J. A. Ibid (1967), 71, 1572.

13. Helfferich, F., "Ion Exchange", McGraw-Hill, New York (1962).

14. Donnan, F. G.; Guggenheim, E. A. Z. Physik Chem. (Leipzig) (1932), 162A 356.

15. Donnan, F. G. ibid, (1934), 168A, 369.

16. Bukata, S.; Marinsky, J. A. J. Phys. Chem. (1964), 68, 994.

17. Robinson, R. A.; Stokes, R. H. "Electrolyte Solutions", 2nd ed. Butterworth Scientific Publications, London (1959).

18. Reddy, M. M.; Marinsky, J. A.; Sarker, A. J. Phys. Chem. (1970), 74, 3891.

19. Soldano, B.; Larson, Q. V. J. Am. Chem. Soc. (1955), 77, 1331.

20. Mukherjee, P. J. Phys. Chem. (1961) 65, 740.

21. Soldano, B.; Larson, Q. V.; Myers, G. S. ibid, (1955), 77, 1339.

22. Reddy, M. M.; Marinsky J. A. J. Macromol. Sci-Phys., B5(1), (1971),   135.

23. Reddy, M. M.; Amdur, S.; Marinsky, J. A. J. Am. Chem. Soc (1972), 94, 4087.

24. Marinsky, J. A.; Reddy, M. M.; Amdur, S., J. Phys. Chem. (1973), 77, 2128.

25. Yang, R.; Marinsky, J. A. J. Phys. Chem. (1977), 83, 2737.

26. Manning, G. S. J. Chem. Phys. (1969), 51, 924, 3249.

27. Boyd, G. E.; Bunzyl, K. J. Am. Chem. Soc. (1974), 96, 2054.

28. Baldwin, R. "Metal Ion Complexation by Sodium Polystrene Sulfonate and Sodium Polyphosphate and Effects of Complexation on Ion-Exchange Selectivity", Ph.D. Thesis, State University of New York at Buffalo, (1978).

29. Travers, L.; Marinsky, J. A. J. Polymer Sci. Symposium, No. 47 (1974), 285.

30. Linderstrøm-Lang, Compt. Ren. Trav. Ab. Carlsberg, (1924), 15 No. 7.

31. Scatchard, G.; Ann. N. Y. Acad. Sci., (1949), 51, 660.

32. Tanford, C. J. Am. Chem. Soc. (1950), 72, 441.

33. Arnold, R.; Overbeek, J. Th. G. Rec. Trav. Chim. Bas, (1950), 69, 192.

34. Marinsky, J. A.; Hogfeldt, E. Chemica Scripta, (1976), 9, 233.

35. Boyd, G. E.; Myers, G. E.; Lindenbaum, S, J. Phys. Chem. (1974), 78, 110.

RECEIVED January 3, 1980.

Fluid Exudation and the Load–Deformation Properties of Articular Cartilage During Compression

HAROLD LIPSHITZ

Department of Orthopedic Surgery, Harvard Medical School, Boston, MA 02115

Articular cartilage is a tissue that covers the ends of diarthrodial joints. As a polymeric material, it exemplifies an especially interesting role for water in polymers. Biologically, its main physiological function is to act as a bearing material. As such it is remarkably wear resistant. Except in pathological conditions it is able to withstand the multitude of stresses imposed on it during an individual's lifetime without significant wear despite the fact that it has relatively little reparative capability (1). Its ability to sustain high loads and rapidly achieve stress relaxation are thought to be factors rendering it so wear resistant. These studies were undertaken as part of an effort to understand the mechanisms of its load-deformation and stress relaxation properties. Ultimately it is hoped to relate these mechanisms to the unique chemical and ultrastructural makeup of the tissue.

Cartilage consists, in addition to cells and inorganic ions, of a network of entangled high molecular weight polymers (predominantly ionomers), in various states of aggregation, that is swollen with water ($\sim$75–80% water). Its structure, composition and properties have been extensively reviewed (2,3,4). When the tissue is compressed its interstitial fluid is exuded (5). This property, together with the factors stated below, make its load-deformation behavior extremely complicated.

As known, the macroscopic load-deformation or stress-strain and stress relaxation characteristics of any polymeric material, during and following deformation, are a consequence of the mean molecular motions of its chains (6). These properties are additionally affected in the case of cartilage by the following: (1) many of the tissue's macromolecules are associated in a variety of fibrous and other arrays (7); (2) the tissue is both inhomogeneous and anisotropic; its composition and mechanical properties vary as a function of depth from its surface (8,9,10); and (3) its intermolecular crosslink densities vary with depth (9). In addition, it has been proposed that its load-deformation relationships and load dissipation rates are governed, in part, by the resistance to fluid flow within and out of the tissue

0-8412-0559-0/80/47-127-403$07.00/0
© 1980 American Chemical Society

during and following deformation (11-15). According to this
theory, unlike most polymeric materials, the tissue's mechanical
properties are governed not only by the "stiffness" of its chains,
the magnitudes of the activation energies between its various
macromolecular conformations and the internal frictional resis-
tances to macromolecular segmental motion, characteristic of all
viscoelastic polymeric materials, but also by fluid transport
phenomena within and out of the tissue. At present, however, the
relationships between fluid flow and the overall mechanical
properties of the tissue are not clearly elucidated. It was the
purpose of this study to assess the relative contribution of
fluid exudation on some load-deformation characteristics of the
tissue during compression.
    This paper will describe: (1) the load to strain relation-
ships of the tissue during confined and unconfined compression
against platens of different porosity; (2) their strain rate
dependency; (3) some aspects of the tissue's load-dissipation
characteristics; and (4) the time invariant or asymptotic load
to strain relationship of cartilage following relaxation.

Experimental Procedures

Source and Form of the Cartilage. The measurements were
made on circular cylindrical plugs of cartilage and underlying
bone that were cut from the parapatellar region of the medial
femoral condyles of adult steers (approx. 2 years old). The
techniques and tools used for their preparation are described
elsewhere (16). The plugs, (2.5 mm in diameter), were prepared
in a manner such that their surfaces were both flat and perpendicular to the axes of the cylindrical shafts of underlying bone.
The small size of the specimens facilitated the preparation of
flat plugs from the predominantly curved condylar surface. A
test specimen is pictured in Figure 1.

Apparatus and Procedures. The apparatus used was designed
and built for these studies. The instrument consists, in
essence, of a piston moving through a hole in a rigidly mounted
bar and a porous platen attached to a load cell (model U231, BLH,
Inc., Waltham, MA) that is securely mounted over the bar (Figures
2 and 3). The movement of the piston was precisely controlled by
a micrometer screw that was driven by a variable speed motor.
The specimens, which were placed in the hole of the bar, were
compressed by the piston against the platen. Displacement in the
axial direction was measured by a suitably mounted linear voltage
displacement transducer (LVDT) whose core was attached to the
piston (Figure 2). The calibrated output potentials of the LVDT
and load cell were monitored by a fast responding dual channel
recorder.
    The hole in the bar, (the diameter of which was only 0.01 mm
larger than that of the plug), served as a confining chamber to

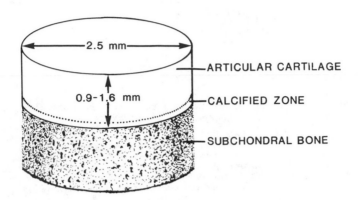

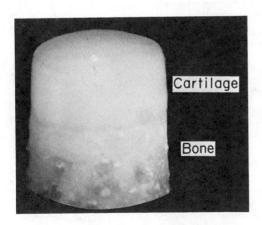

*Figure 1A. Schematic of a cylindrical plug of cartilage and underlying bone show-
ing the average dimensions of the test specimen*

Figure 1B. Photograph of plug (×16)

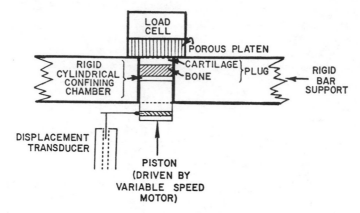

Figure 2A.   Schematic of the apparatus used to study the confined compression and relaxation of cartilage

The tissue was confined within a hole of the bar, and its surface was flush with the bar's surface. The porous platin (attached to the load cell) was positioned over the hole against the surface of the bar. (Since the cartilage is many times more compliant than the underlying bone, the bone's deformation and relaxation were considered small relative to the overall measurements. This was confirmed by control experiments run on cylindrical specimens of underlying bone with the cartilage removed.)

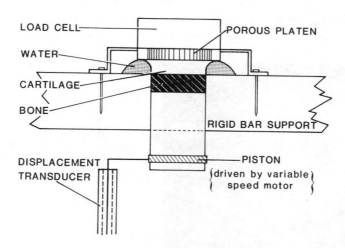

Figure 2B.   Schematic of the experimental arrangement used for the unconfined compression of the tissue

The platen was positioned above the surface of the bar. The bony end of the plug was within the hole of the bar, and the cartilage was above the bar's surface. A few drops of water placed on the surface of the bar next to the tissue, as seen, precluded its dehydration during the experiment.

restrict the lateral expansion of the tissue during compression
in the majority of experiments. In this experimental configuration the strain field was presumed to be predominantly uniaxial
while the tissue was compressed triaxially. The volume change
corresponding to displacement in the axial direction was obviously somewhat greater in this case than when the tissue was
compressed to the same extent while unconfined and the relationship between displacement and volume change could be determined.
This was not the case when the tissue was compressed while unconfined.

In studies where the cartilage was not confined, the hole
in the bar served to maintain the plug in an upright position
while being compressed. In this series, only the bony end of
the plugs were in the hole; the cartilage was always above the
level of the bar (Figure 2b). For these studies the tissue was
surrounded by a small reservoir of water to preclude its dehydration. The confining chamber (i.e., the bar) was separated from
the platen (load cell) in the apparatus to assure that frictional
forces between the plug and chamber during compression were not
detectable by the load cell.

The experimental procedure consisted of the following.
Prior to the start of each experiment, the plugs were equilibrated
with 0.03M phosphate buffer (pH 7.4) at 4°C for at least 2 hours.
This time period was found to be sufficient for the cartilage
specimens to attain their equilibrium swelling ratios. The
buffer contained, in addition to the dissolved salts, 1×10^{-4} M
phenylmethylsulfonylfluoride (PMSF). This compound inhibited
degradative enzyme reactions from occurring within the tissue
during the experimental period (17). All experiments were carried
out at room temperature. The tissue's load-deformation and load
dissipation characteristics were found to be insensitive to
temperature variations in this small range (23-27°C) (18).

The thickness of the fully swollen cartilage (i.e. the
distance between its surface and calcified layer (Figure 1) were
measured by viewing the sides of plugs through a dissection
microscope equipped with a calibrated grid. Sometimes a specimen's thickness on one side differed from that on the other by
about 3-5%. In those cases, the mean value of the two was taken
as its initial thickness.

The underlying bone of each plug was trimmed (using a
special apparatus) so that its surface was flat and between 1
to 1.5 mm thick. This enabled the plugs to rest squarely on the
surface of the piston and minimized friction between the plug
and chamber during compression. After the plugs were placed in
the hole of the bar, the bar and load cell assembly were firmly
secured to the apparatus, such that the load on the bar was
approximately 4.5 kg. (The load cell used had a capacity of
9.1kg). This was achieved by judiciously tightening the screws
securing the load cell to the apparatus (Figure 3). The output
potentials of the load cell, corresponding to these loads, were
electronically zeroed before compression was started. However,

Figure 3.   The apparatus

A, load cell; B, speed control for the motor drive; C, support bar (a hole in this served as the confining chamber); D, support disc used to secure the load cell to the apparatus. (This initial load on the load cell—i.e., before compression—was regulated by adjusting the tension in these screws.); E, displacement transducer; F, motor-driven micrometer screw; G, power supplies for the load cell and the displacement transducer; H, recorder.

as will subsequently be explained, it was necessary to multiply
the loads (as measured by the load cell) by a correction factor
to obtain the true loads on the tissue.

To ascertain the point where compression of the cartilage
began, it was necessary to be able to position the tissue against
the platen without compressing it. This was accomplished as
follows. The gain was increased so that as little as 1.5 gms
was detectable. The piston was then raised till the first faint
signal could be seen, whereupon the motor drive was stopped.
This was taken as the position where the surface of the cartilage
was just making contact with the platen without being compressed.
The speed setting for the motor drive was then adjusted to com-
press the tissue at a desired strain rate and the experiment was
begun. At selected strains the motor drive was stopped and the
dissipation of the load with time was followed until a time
invariant value, characteristic for a given strain, was attained.
The tissue was then compressed to a new strain and the procedure
repeated.

Because the cartilage, even when swollen, is very thin
(approx. 1-1.5 mm thick), the compliance of the load cell had to
be taken into account in determining the axial displacement of
the tissue during compression. This was done by subtracting the
displacement of the platen-load cell at each load (as determined
from the displacement-load curve of the load cell) from that
measured by the displacement transducer. Likewise it was
necessary to add an appropriate displacement (as calculated)
during the relaxation phase of the experiment. With the particular load cell used in our apparatus, these corrections ranged
from 3 to 5% of the displacement at peak loads, to negligible
amounts at low loads.

Of further note, to assure that the tissue remained confined
during compression in that experimental configuration, it was
necessary that the bar be more compliant than the load cell. If
this condition was not satisfied, then compression of the tissue
against the platen (load cell) would result in a gap between the
bar and platen into which the tissue would spread.

As stated, the loads measured by the load cell had to be
multiplied by a correction factor to determine the true loads on
the tissue. The reasons are as follows. The securing of the
load cell to the apparatus caused a small deflection of the bar.
Clearly, at the start of an experiment (i.e. under the static
conditions prior to compression) the force of the load cell on
the bar (due to its weight and the tension in the screws) was
necessarily exactly equal to the force of the bar against the
load cell. The output potentials of the load cell due to this
were, as stated, electronically zeroed prior to compression.
When, however, the tissue was compressed against the platen, the
sensor of the load cell was necessarily displaced. The bar,
being more compliant than the load cell, moved with the sensor
and consequently reduced the initial deflection, and hence the

force of the bar, on the load cell. Since the force (as measured by the load cell) was equal to the <u>sum</u> of the forces due to the bar and those of the tissue as it compressed against it (all other forces remaining reasonably constant) the load on the tissue was consequently equal to that measured by the load cell plus the difference between the upward force of the bar at the start of the experiment and that at any time thereafter. Therefore to determine the true load on the tissue at any instant, it was necessary to correct the measured load by adding an incremental value.

The magnitude of the correction at any instant was determined as follows. The force-displacement curve of the bar (as secured in our apparatus in the region where it interfaced with the platen) was measured. The bar's displacement was measured as a function of load with a cathetometer. The load-displacement curve was, as expected, linear. Since the bar was more compliant than the load cell, its displacement during and following compression was necessarily equal to that of the load cell. Hence, changes during a run of the bar's displacement could be determined from the load-displacement relationship of the load cell.

The true load on the tissue was computed by using the equation:

$$[1] \qquad L_t = L_c \left(1 + \frac{k}{k'}\right)$$

where L_t is the actual load on the tissue, L_c is the load measured by the load cell and k and k' are the "spring" constants of the bar and load cell respectively. With our apparatus, the numerical value of the correction factor $(1 + \frac{k}{k'})$ was found to be slightly less than 2.

The derivation of this equation is rather straightforward. As explained above during an experiment $L_t = L_c + \Delta L_B$ where ΔL_B is the change of force on the load cell due to changes in the deflection of the bar. But $\Delta L_B = kD$ where k is the spring constant of the bar and D is its displacement, as was determined experimentally. Since displacement of the bar was necessarily equal to the displacement of the load cell, $D = \frac{L_c}{k'}$. Thus $L_t = L_c + \frac{k}{k'}L_c$ and $L_t = L_c(1 + \frac{k}{k'})$.

The Porous Platens. Since the interstitial fluid of cartilage is exuded during compression (5), its measured mechanical properties will vary if any impediments to flow are imposed by the experimental apparatus (i.e. a resistance above that inherent to the tissue). Measurement of the tissue's real mechanical properties at significant strains, would require that it be compressed against completely free draining platens, (i.e. ones that conceivably have zero or negligible resistance to flow) that cause no distortion to its surface during compression. Then the confined compression of the tissue would presumably result in predominantly uniaxial flow fields with little lateral flow of interstitial fluid.

Unfortunately, it was not possible to make such platens even though they all appeared, on inspection, to be free draining, i.e., water drained freely through them. This criterion was, however, insufficient to assure that the platens did not impede the flow of water from the tissue when compressed against them. Even the obtaining of a platen that was as free draining as possible, with the constraints that it be: (1) sufficiently stiff to sustain the pressures imposed on it without deforming, and (2) have a flat surface, proved very difficult. If the holes were made very small, capillary effects precluded its being free draining. On the other hand, because of the high compliance of the cartilage larger holes resulted in its protruding into them during compression. This effect necessitated a more complicated consideration of the tissue's average strain during compression, (as will be explained below) and precluded a simple, clear cut assessment of the average stress, normal to its surface, even during confined compression. Other practical limitations to making platens of greater porosity included machining difficulties (despite the use of the most advanced techniques) and the fact that a minimum distance between holes was required. If the space was too little, the walls cut into the tissue during compression.

Since the obtaining of a platen that approached ideal properties was not possible, it was necessary to assess the variance of the measured mechanical properties of the cartilage as a function of the porosity and size of holes of the platens. The porosity is defined here as the ratio of the total area of the holes, A_h, within the contact area, A_c, made between the tissue and platen (i.e. porosity $\equiv A_h/A_c$).

Five platens 2.5 mm thick were made from type 302 stainless steel. One platen was solid (i.e. it had no holes), three had ratios of A_h/A_c that were respectively 0.20, 0.35, and 0.45 where all holes were of diameter 0.55 mm and one platen had a value of A_h/A_c equal to 0.35 but holes of diameter 0.31 mm.

<u>The Measurement of the Extent of Penetration of the Tissue into the Holes of the Platen During Compression</u>. The extent of penetration into the holes during and following compression was measured using the apparatus that is pictured schematically in Figure 4. A stiff rod polished flat at its end, 0.25 mm in diameter, was attached to the core of an LVDT displacement transducer. The barrel of the displacement transducer was secured and positioned precisely over the platen. The core of the transducer and attached rod were made practically weightless by counterbalancing it over a "frictionless" pulley (Figure 4). (The rod and core assembly had an "effective" weight of a few milligrams to assure that it made contact with the tissue throughout the experiment.) The rod, when inserted into a hole at the center of each of the platens, was so positioned that it hung free (i.e. it did not touch the walls of the hole).

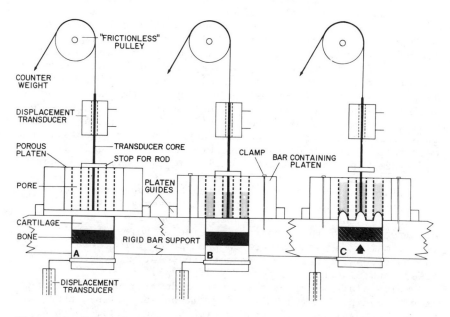

Figure 4A. Schematic of the apparatus used to measure the penetration of the tissue into the holes of the platen during and after confined compression.

Section A: the porous platen is shown secured within a second bar, its surface flush with the surface of that bar; the surface of the rod is flush with the surface of the platen and the surface of the tissue is flush with the surface of the bar it is contained in. Section B: the assembly is clamped to the test bar so that the tissue is in contact with the platen without being compressed. Section C: the tissue penetrates the hole of the platen during compression, lifting up the rod connected to the core of the displacement transducer.

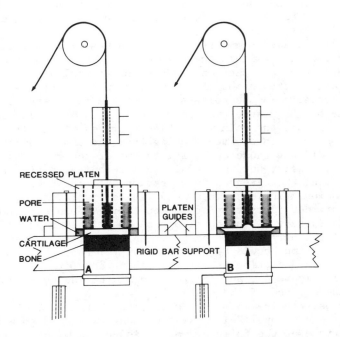

Figure 4B.   Schematic of the apparatus used to measure the tissue's penetration during unconfined compression. In this case, the porous platen was receded into its contained bar, and the tissue was raised above the surface of the bar it was contained in.

On compression, the extent of penetration was followed by measuring the displacement of the rod (as monitored by the cali-brated output potentials of the LVDT) as a function of time and the overall compression of the tissue in the axial direction. By overall compression is meant the average displacement of the "plane" of the top of the calcified zone relative to the surface of the platen (Figure 1). This was monitored by the displacement transducer attached to the piston (Figure 4).

To preclude the possibility that water exuding from the tissue during compression might have artifactually caused dis-placement of the rod, the following experiments were done. The tissue was compressed a given amount and its penetration into the holes after relaxation, (see results) as measured by the rod's displacement, was noted. The rod was then removed from the platen and water within the hole was sucked out (by placing a tube attached to a suction flask under partial vacuum over the hole for a short period of time). The position of the rod after being replaced in the hole (such that its tip rested on the surface of the tissue) was found to be identical with that of the rod before the hole was subjected to suction. It was there-fore concluded that water exuding from the tissue during comp-ression did not cause any displacement of the rod.

There were, however, additional technical difficulties in correlating the extent of penetration into the holes with overall compression. The distances measured were very small. It was therefore necessary that at the beginning of an experiment that (1) the surface of the cartilage be positioned exactly flush with that of the bar so that the position of initial contact (prior to compression) could be ascertained, (2) that the porous platen be positioned precisely over the tissue and firmly secured to the bar (Figure 4), and (3) the end of the rod be exactly flush with the surface of the platen.

These requirements were satisfied as follows. Light section microscopy (Zeiss, West Germany) was used to assure that the surface of the fully hydrated tissue was within a micron or less of the surface of the bar, as the tissue was raised by the piston. The porous platen was precisely positioned over the tissue and firmly attached to the bar by properly securing the test platen in a hole of a second bar (Figure 4) (such that the surface was flush with that bar) and clamping the entire assembly (with the aid of guides) to the original bar with C clamps. To assure that the rod's end was flush with the platen surface, a plastic ring attached to the rod, precisely 2.5 mm from its end was used as a stop (Figure 4). At periodic intervals, it was necessary to add a few drops of water to the tissue through the other holes of the platens to preclude its dehydration in this open system.

The technique was slightly modified to measure the extent of penetration during unconfined compression. For these measure-ments the platen was receded into the hole of the bar it was

contained in. The tissue after being positioned flush with the
surface of the bar it was contained in, was raised by the micrometer screw the precise distance that the platen was receded into
the second bar, such that its surface just touched the platen
without being compressed. The rest of the procedure was the same
as that used for confined compression.

Determination of the Average Strain of the Tissue in the z
or Axial Direction. To assess the average strain of the tissue
during compression in the z or axial direction, it was necessary
that the extent of its penetration into the holes of the platen
be taken into account (see results section). Under these conditions, the average strain ε_z could be described by the equation:

$$[2] \qquad \varepsilon_z = \frac{z_0 - [(1 - \alpha)z + \alpha(\langle z' \rangle_{\dot{\varepsilon}}(\varepsilon) + z)]}{z_0}$$

where z_0 is the tissue's thickness when fully hydrated, prior to
compression (defined as the distance between the "plane" of the
top of the calcified zone to the surface of the cartilage), z is
the distance from the same reference point to the surface of the
platen, during compression, $\langle z' \rangle_{\dot{\varepsilon}}(\varepsilon)$ is the average height or
distance between the surface of the platen and the surface of the
tissue penetrating the hole for a strain, ε (averaged over the
area of the hole), when compressed at a strain rate, $\dot{\varepsilon}$, and α is
the fraction of the total contact area between the tissue and
platen that is hole (i.e. A_h/A_c).

The average strain during compression is described here in
terms of changes in relative thickness. For these experimental
conditions, the thickness of the tissue is at any time equal to
an average of the fraction under the solid portion of the platen
and that under a hole. As such the relative changes in thickness,
or strain, are described by equation [2].

It was necessary, however, to be able to evaluate $\langle z' \rangle_{\dot{\varepsilon}}(\varepsilon)$.
To do this, we assumed the tissue penetrating the hole to have
the shape of an inverted circular parabaloid whose base area had
a radius of 0.275 mm (the actual size of the holes).

The mean height of such a parabaloid, $\langle z' \rangle_{\dot{\varepsilon}}(\varepsilon)$, is equal to:

$$[3] \qquad \langle z' \rangle_{\dot{\varepsilon}}(\varepsilon) = \frac{2\pi \int_0^{0.275} Z''(r) r \, dr}{2\pi \int_0^{0.275} r \, dr}$$

where $Z''(r)$ is the variance of the height of the parabaloid with
the radius of the hole, r.

For an inverted circular parabaloid, $Z'' = p - \beta r^2$, where p is the distance between the surface of the platen and the peak of the parabaloid. p was experimentally shown in this work to be equal to $1.15(z_0 - z)$ for compression rates of around 0.30 and 0.17 mm/sec., and $(z_0 - z)$ for compression rates of around 0.05 mm/sec. (figs. 5 and 6). For holes of radius 0.275 mm, β is equal to 13.2p. Thus, $Z''(r) = k[p - 13.2pr^2] = k(z_0 - z)[1-13.2 r^2]$ where k = 1.15 for the higher compression rates and 1 for the slowest compression rate. Thus:

$$\langle Z' \rangle_{\varepsilon} = \frac{k(z_0 - z) \int_0^{0.275} (1 - 13.2r^2) r\,dr}{\int_0^{0.275} r\,dr} = [z_0 - z](0.501)k$$

and

$$\varepsilon_z = \frac{z_0 - [(1 - \alpha)z + \alpha[(z_0 - z)0.501k + z]}{z_0}$$

[4]

For our most porous platen (i.e. $\alpha = 0.45$)

[5]
$$\varepsilon_z = \frac{z_0 - z}{z_0} (1 - 0.225k)$$

Results

The Extent of Penetration into the Holes of the Platens.
The extent to which the tissue penetrated the holes of the platens during confined compression are plotted in Figures 5 and 6 as functions of overall compression (i.e. the displacement of the top of the calcified layer (Fig. 1) relative to the surface of the platen). As seen, penetration was approximately a linear function of overall compression (for the compression rates studied). Slight changes in slope were observed at around 0.15 to 0.18 mm displacement. Compression at slower rates resulted in less penetration. The extent of penetration was independent of the porosity of the platens (Fig. 6) and the tissue's initial thickness. The penetration was thus a function of the extent of compression or displacement but not strain.

When compression ceased (i.e. during relaxation), the tissue receded in the hole until an asymptotic penetration was attained after around 100 seconds (Fig. 7).

Unconfined compression of the tissue resulted in the tissue penetrating the holes of the platen to a considerably lesser

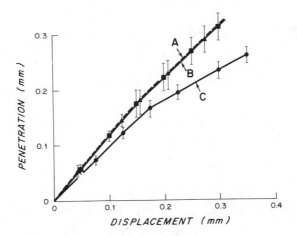

Figure 5. Penetration of cartilage into a platen ($A_h/A_c = 0.45$) during its confined compression (at three compression rates) as a function of overall compression (displacement). Compression rates (mm/sec): curve A = 0.30; curve B = 0.17; curve C = 0.05. (Mean values of 8–10 runs ± 1 SD.)

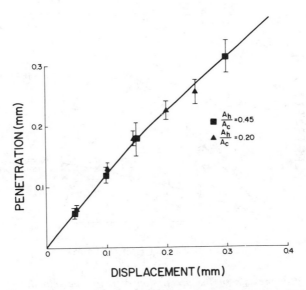

Figure 6. Penetration of tissue into platens of different porosity during compression. Note that platen porosity did not affect penetration. (Mean values of 9 runs ± 1 SD.)

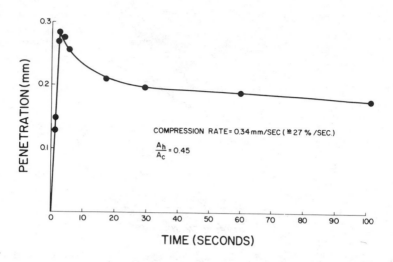

Figure 7.   *Changes in penetration with time after cessation of compression. (Mean values of 5 runs.)*

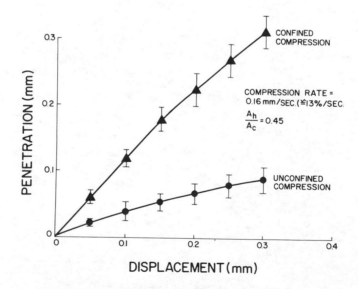

Figure 8.   *Compression of the penetration of the tissue into the platens during confined and unconfined compression. (Mean values of 10 runs ± 1 SD.)*

extent for the same axial displacement (Fig. 8).

The Load-Strain Relationships During Confined and Unconfined
Compression. As stated, since the tissue penetrated the pores of
the platen during compression, the stress field at its surface
was extremely complex and presently indeterminate. We could
therefore only describe its load-strain and not its stress-strain
characteristics. In Figure 9, typical curves of load versus time
during compression and relaxation are shown. During the compressive phase, the load increased at increasing rates until compression ceased. Thereafter a sharp and rapid dissipation of the
load was observed. This decreased at decreasing rates with time
until time invariant values, characteristic for a given strain
(but independent of the strain rate) were attained. During
unconfined compression, the loads increased to considerably lesser
extents for the same axial displacements (Fig. 10). Furthermore,
the load dissipation rates were considerably slower. Although
the tissue cannot be modeled as a simple Maxwell body (15) considering the time elapsed for the load to reach $1/e$ of its maximum
value (as an arbitrary point) as a measure of the average stress
relaxation rates,following compression of the tissue to a given
strain, it was found that this time was generally a factor of
2 to 3 times greater, after unconfined compression than confined
compression.

The load-strain relationships during confined compression
against platens of porosities (i.e. A_h/A_c) 0.20, 0.35 and 0.45
are seen in Figures 11, 12 and 13. As might be expected, compression of the cartilage against platens of lower porosity resulted
in considerably higher load to strain ratios (Fig. 14) and correspondingly slower load dissipation rates. Furthermore, against
platens of lower porosity, the load-strain curves were somewhat
more strain rate dependent, than when the tissue was compressed
against the most porous platen ($A_h/A_c = 0.45$). Against this
platen, the load-strain curves had very weak, if any, strain rate
dependence. The load-strain curves of the cartilage against
platens of the same porosity (0.35) but different hole size were
identical (Fig. 14). Thus, if the holes were sufficiently large
to preclude capillary effects, hole size did not appear to be a
critical parameter.

A plot of the mean loads at each strain (obtained from the
load-strain curves against each of the platens) against the porosities of the platens (Fig. 15) resulted in a family of linear
curves that extrapolated to approximately the same point at
around 65% porosity. Uncertainties, due to experimental scatter,
preclude very precise extrapolations, but the data indicate that
the confined compression of the tissue against platens of at
least 65% porosity would result in very low loads at each strain
(perhaps approaching the time independent load-strain curve)
(Fig. 16). The time invariant load-strain relationship (i.e.
the asymptotic load at each strain following the relaxation

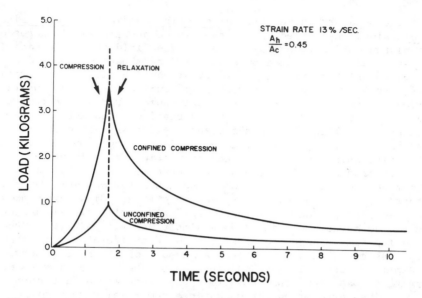

Figure 9.   *Typical curves of the changes in load during compression and relaxation of cartilage (for confined and unconfined compression). Note the considerably lower loads attained during unconfined compression.*

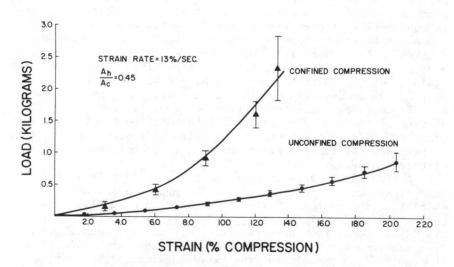

Figure 10.   *Comparison of the load–strain curves of cartilage during confined and unconfined compression. (Mean values of 12 runs ± 1 SD.)*

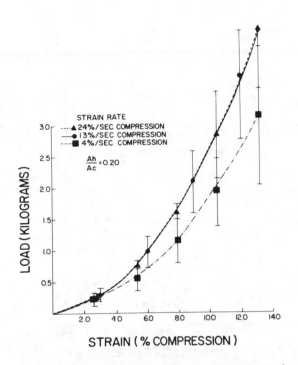

Figure 11.   Load–strain curves, at three strain rates, of cartilage when compressed against the 20% porosity platen. (Mean values of 8–10 runs ± 1 SD.)

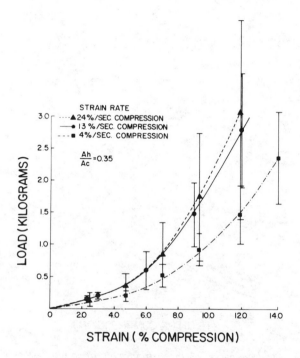

Figure 12.   *Load–strain curves of cartilage, at three strain rates, when compressed against the 35% porosity platen. (Mean values of 12–15 runs ± 1 SD.)*

of the tissue after its confined compression) is seen in Figure
16. The curve is practically linear up to about 30% compression,
whereupon there is a rapid upsweep. This relationship was found
to be independent of the porosity of the platens against which
the tissue was compressed and the strain rate it was compressed
at.

Discussion

These studies point to a rather unique function of water in
cartilage. The results clearly demonstrate that the measured
mechanical properties were a function of the impedance to flow
from the tissue during and following compression as evidenced
by the lower load-strain relationships and the faster relaxation
rates found when compressed against platens of greater porosity.
Further, the data illustrate the difficulty, if not impossibility,
of obtaining measurements of time dependent bulk mechanical properties of cartilage in compression (at significant strains) that
are inherent to the tissue and independent of experimental configuration. The same can probably be said of permeability measurements.

The loads at each strain, obtained from the load-strain
curves against the different platens, when plotted against porosity, all extrapolated to approximately the same porosity at low
loads (Fig. 15). This strongly suggests that the confined compression of the tissue against platens of at least 65% porosity,
to a strain of 0.15 and possibly more would result in very low
loads. This indicates that the experimental load-strain curves
were primarily a function of the resistance to flow from the
tissue due to the finite porosities of the platens and not to
the viscoelastic properties of the matrix. However, of more
importance, the results indicate that the tissue is probably
very permeable to fluid flow normal to its surface (i.e., it
offers little resistance to the flow of water in that direction)
at the pressure gradients and flow rates generated during compression. It is therefore contended that fluid contained within a
region of the tissue opposite that of a hole flowed freely from
the tissue on compression.

This contention is consistent with 1) the decreasing moduli
at the strains measured, when the tissue was compressed against
platens of greater porosity; 2) the extremely weak (if any) strain
rate dependence of the load-strain curves, particularly when compressed against the 45% porosity platen (Fig. 13); 3) the
apparent independence of the load-strain curves on the size of
the holes (Fig. 14) -- they being a function of the porosities of
the platens only; and 4) the reported weak dependence of the
permeability coefficients on strain, for given pressure gradients,
when the cartilage was interfaced against porous discs (19).

The contention is, however, inconsistent with the low values
that have been reported for the permeability coefficients of

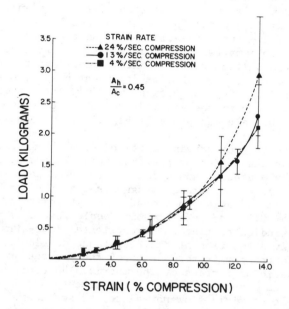

Figure 13.   Load–strain curves of cartilage, at three strain rates, when compressed
against the 45% porosity platen. Note little, if any, strain-rate dependence. (Mean
values of 10 runs ± 1 SD.)

cartilage by Mansour and Mow (19) and others (20,21). These values would indicate the opposite (i.e. a high resistance to flow normal to the tissue's surface). Those studies were, however, carried out with slices of cartilage interfaced against porous platens. Their measured permeabilities may therefore reflect the combined resistance to flow of the tissue in contact with porous platens and not the tissue itself, and may thus be artifactual. Furthermore, the measurements were obtained under steady state conditions of flow that are not necessarily applicable to the non-steady state conditions occurring during the compression and relaxation of the tissue.

The data would also seem to suggest that the tissue is anisotropic with respect to interstitial fluid flow; that is, it may be considerably more permeable normal to its surface than tangential to it. This contention is supported by 1) the considerably slower relaxation rates of the cartilage following its unconfined compression to approximately the same volume change (at the same strain rates), as compared to when the tissue was compressed while confined (Fig. 9), and 2) the rather steep dependence of the load-strain curves on the porosities of the platens. Were the tissue equally permeable in all directions, the relaxation rates following unconfined compression would be expected to be greater than those following confined compression. However, the load- or stress-strain curves of any material are a function of the extent of relaxation occurring during deformation. It is therefore possible that greater relaxation occurs during unconfined compression which, in turn resulted in the the lower load-strain relationships and the slower relaxation rates.

Fluid exudation from the tissue during and following compression occurs as a result of 1) pressure gradients between the deeper regions and surface of the tissue, and 2) transitory internal pressures within the tissue that are consequent to the slow relaxation rates of the matrix relative to the rates of compressive strain. The former arise from the fact that the calcified zone is essentially impermeable to the flow of fluid while the tissue was compressed against semi-porous platens, and, relaxation rates from the deeper regions of the tissue may be greater than those at the surface,   thereby generating pressure gradients (10).

The tissue has been modeled from a continuum mechanics point of view by Mow and students (11,12,13). The model treats cartilage as a biphasic material (the "solid"organic matrix was assumed, for these purposes, to be one "phase" and the water the other). The "solid" organic matrix was further assumed to behave as a single Kelvin-Voigt body whose viscoelastic properties are attenuated by the frictional resistance to fluid flow from the tissue. Attempts were made to use this model to explain the load-strain and load dissipation properties of the tissue for the experimental configuration described in this paper (15).

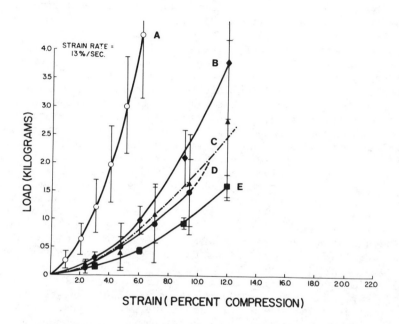

Figure 14.   Load–strain curves·of the tissue vs. each of the platens.

Curve A = solid platen (i.e. no holes); curve B = (A_h/A_c = 0.20: hole diameter 0.55 mm); curve C = (A_h/A_c = 0.35: hole diameter 0.55 mm); curve D = (A_h/A_c = 0.35: hole diameter 0.31 mm); curve E = (A_h/A_c = 0.45: hole diameter 0.55 mm). The shape of curve A was, no doubt, a function of the experimental configuration in that the seepage of fluid out of the chamber during the compression of the tissue was not prevented. (Mean values of 9–15 runs ± 1 SD.)

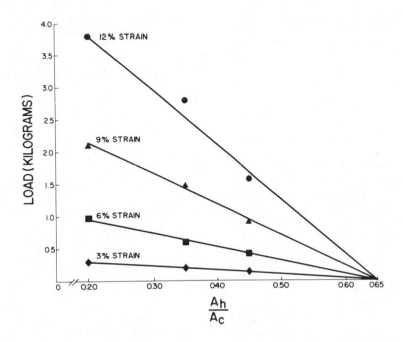

Figure 15.   Loads at each strain plotted as a function of the the porosities of the platens.   (Each point is the mean value of 9–15 runs.)

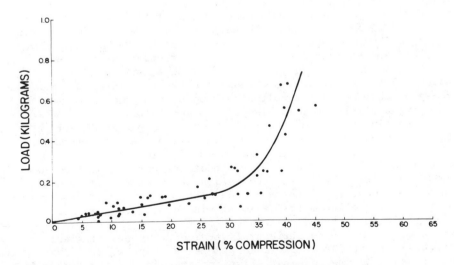

Figure 16.   Load–strain relationship of cartilage after confined compression and relaxation at each strain

Series solutions were obtained for linear partial differential
equations that shared the same boundary conditions and represented
extreme values of the non-linear constitutive equation, developed
by these authors, to describe the time dependent deformation of
the tissue. The solutions were indeed shown to bound the experimental load decay curve (15). However, at the time the work was
done it was not appreciated that the experimental load-strain and
load-decay curves were a function of the attenuation of flow
from the tissue by the porous platens and hence not sole functions
of fundamental properties of the tissue. Furthermore, the
bounded solutions for the compressive phase generated theoretical
load-strain curves that were convex in shape, when, in fact, the
experimental curves were always concave (Figs. 10-13). To generate concave shaped curves from these equations, it was necessary
to assume a strong functionality (presumably exponential) between
the permeability of the tissue and strain (15). However, this
work indicates that if load-strain curves were obtained against
completely free draining platens (i.e., platens of at least 65%
porosity) where the relationships would be reflective of the
tissue only, no such dependency would be obtained. Leaving aside
questions of the meaningfulness of describing the viscoelastic
properties of an inhomogeneous, anisotropic organic matrix by a
single Kelvin-Voigt model, the theory requires that there be
significant resistance to flow, inherent to the tissue, normal
to its surface. Perhaps this should be reexamined. It is possible that the unique mechanical properties of the tissue in its
normal environment can be explained in terms of an anisotropic
nature to the resistance to flow.

That fluid continues to flow after compression has ceased
(i.e., during relaxation) is evidenced by its electromechanical
properties (22). When the tissue is compressed, an electric
potential difference occurs between the surface and the deepest
regions. This potential difference was attributed to the flow
of fluid and entrained counterions with respect to the net
negatively charged solid matrix (22). That is, the observed
potential is essentially a streaming current potential. The
decay of this potential following compression was found to
exactly parallel the load dissipation curves with time (22).

The time invariant load-strain curves appear to be an
inherent property of the tissue. The values were independent
of the porosity of the platens and the strain rates that the
tissue was compressed at. Up to about 30% compression the data
could be fitted, by linear regression, to a straight line. As
such, the curve represents the equilibrium load-strain relationship of the swollen matrix. It should be pointed out, however,
that while the strains calculated for this and the other curves
are probably close to the actual strains, they nevertheless are
approximations, due to the assumptions of shape of the penetrating tissue and the possible slight compression of the tissue by
the rod.

Abstract

Some load-strain and load dissipation characteristics of articular cartilage during and following its confined and unconfined compression were studied. Using apparatus that was designed and built for these studies, the tissue was compressed against platens of different porosity at various strain rates. The tissue penetrated the pores of the platen during compression. This had to be taken into account in computing the strain.

During compression, the loads increased at increasing rates with strain until compression ceased. Thereafter a sharp and rapid dissipation of the load was observed. This decreased at decreasing rates with time until time invariant values, characteristic for a given strain, but independent of strain rate, were attained. During unconfined compression the loads increased to lesser extents for the same axial displacements, and relaxation was considerably slower. Against our most porous platen, the load-strain curves had little, if any, strain rate dependence. Plots of the mean loads at each strain against the porosities of the platens resulted in a family of linear curves that extrapolated to approximately the same porosity at low loads.

The data is interpreted to indicate that the tissue is very permeable to the flow of fluid, normal to its surface, at the pressure gradients and flow rates generated during compression. It is suggested that the unique mechanical properties of cartilage in its normal environment may be due to an anisotropic nature of the resistance to flow.

Acknowledgment

This work was supported in part by funds from NIH grant AM 15671 and the New England Peabody Home for Crippled Children, Inc. The author also wishes to thank Mr. Robert Etheredge, III, for his invaluable help.

Literature Cited

1. Sokoloff, L., "The Biology of Degenerative Joint Disease"; University of Chicago Press: Chicago, IL, 1969.
2. Freeman, M.A.R., Ed. "Adult Articular Cartilage"; Grune and Stratton: New York, NY, 1974.
3. Serafini-Fracassini, A.; Smith, J.W., "The Structure and Biochemistry of Cartilage"; Churchill Livingstone: Edinburgh, 1974.
4. Barrett, A.J. in "Comprehensive Biochemistry", Florkin, M. and Stolz, E.H., Eds. Elsevier Publishing Co.: Amsterdam, 1968; pp. 431-456.
5. Edwards, J. Proc. Inst. Mech. Eng. 1967, 181 (pt 3J), 16-24.
6. Ferry, J.D. "Viscoelastic Properties of Polymers"; John Wiley & Sons: New York, NY, 1970.

7. Millington, P.F.; Gibson, T.; Evans, J.H.; Barbanel, J.C., in
    "Advances in Biomedical Engineering", Kenedi, R.M., Ed.
    Academic Press: New York, NY, 1971; pp. 189-248.
 8. Kempson, G.E.; Freeman, M.A.R.; Swanson, S.A.V., Nature,
    1968, 220, 1127.
 9. Lipshitz, H.; Etheredge, R., III; Glimcher, M.J., J. Bone
    Joint Surg., 1976, 58A, 1149-1156.
10. Lipshitz, H.; Etheredge, R.; Glimcher, M.J., Proc. 176 ACS
    Meeting, Miami, FL, 1978.
11. Torzilli, P.A.; Mow, V.C., J. Biomech., 1976, 9, 541.
12. Torzilli, P.A.; Mow, V.C., J. Biomech., 1976, 9, 587.
13. Mow, V.C.; Mansour, J.M., J. Biomech., 1977, 10, 31.
14. Lipshitz, H.; Mow, V.C.; Torzilli, P.A.; Eisenfeld, J.;
    Glimcher, M.J., Proc. VII Int. Congr. Rheology, Klason, C.
    and Kubat, J., Eds., 1976; pp. 198-199.
15. Eisenfeld, J.; Mow, V.C.; Lipshitz, H., Math. Biosci., 1978,
    39, 97-111.
16. Lipshitz, H.; Glimcher, M.J., J. Biomech., 1974, 7, 293-295.
17. Lipshitz, H.; Glimcher, M.J., Wear, 1979, 52, 297-339; cf.
    p. 299.
18. Lipshitz, H., "Consideration of the load-deformation and
    load-dissipation characteristics of articular cartilage in
    compression." (in preparation).
19. Mansour, J.M.; Mow, V.C., J. Bone Joint Surg., 1976, 58A,
    509.
20. Maroudas, A.; Muir, H.; Wingham, J., Biochim. Biophys. Acta
    1969, 177, 492.
21. Maroudas, A., Biorheology, 1975, 12, 233.
22. Grodzinsky, A.J.; Lipshitz, H.; Glimcher, M.J., Nature, 1978,
    275, 448-450.

RECEIVED January 17, 1980.

SYNTHETIC POLYMERS:
WATER INTERACTIONS

Water in Nylon

HOWARD W. STARKWEATHER, JR.

E. I. du Pont de Nemours and Co., Central Research and Development Department,
Wilmington, DE 19898

The sorption of water by nylon has a major effect on properties of engineering and scientific importance. In molded 66 nylon
at room temperature, the modulus decreases by about a factor of
five, the yield stress decreases by more than half, and there are
major increases in the elongation and energy to break as the water
content is increased from dryness to saturation (1). Thus,
reported properties of nylon are frequently those of a mixture of
nylon and water. It is important to specify the water content or
the relative humidity with which the polymer is in equilibrium.

The changes in properties due to absorbed water closely
parallel those which occur as the temperature is increased. In
such systems, the time-temperature superposition which is familiar
in studies of viscoelasticity can be extended to a time-tempera-
ture-humidity superposition (2-4).

Thermodynamic Properties

A sorption isotherm at $23^\circ C$ for an extruded film of 66 nylon
which had been annealed at $250^\circ C$ (5) is shown in Figure 1. The
film was 0.010" (0.25 mm) thick and was 57% crystalline. For this
sample, the isotherm had an upward curvature which became more
pronounced at higher humidities. For less fully annealed samples,
sorption isotherms have been reported which have a downward curvature at low humidities (6). This indicates that polymer-water
contacts are strongly preferred.

The upward curvature of the isotherm is indicative of clustering of water molecules. According to a formula derived by Zimm
(7), the number of water molecules in the neighborhood of a given
water molecule in excess of the mean concentration of water is
given by

$$c_1 G_{11} = (1-\phi_1) \left(\frac{\partial \ln \phi_1}{\partial \ln a_1} \right)_{p,T} - 1 \qquad (1)$$

0-8412-0559-0/80/47-127-433$05.00/0
© 1980 American Chemical Society

where c_1, ϕ_1, and a_1 are the molar concentration, volume fraction, and activity of water, and G_{11} is the cluster integral. Therefore, the apparent number of water molecules in a cluster is

$$c_1 G_{11} + 1 = (1 - \phi_1) \left(\frac{\partial \ln \phi_1}{\partial \ln a_1} \right)_{p,T} \qquad (2)$$

One must be cautious in equating this statistical quantity with physical clusters. For example, the fact that a polymer chain is a cluster of volume elements means that small molecules mixed with it must be clustered in a complementary way. The application of Equation 2 to the Flory-Huggins equation for a θ solution leads to an apparent cluster size of $(1 - \phi_1)^{-1}$ (8). This effect will be very small at the levels of ϕ_1 for water in nylon.

It is interesting to consider the number and size of the clusters in relation to the polymer structure (9). If m is the number of water molecules per amide group in the amorphous regions, the corresponding number of clusters will be

$$\frac{m}{c_1 G_{11} + 1}$$

Figure 2 is a plot of the number of clusters per amorphous amide vs. the number of water molecules per amorphous amide. The sloping lines correspond to various average numbers of water molecules per cluster. The water is almost unclustered up to a concentration of one water molecule per two amide groups. Puffr and Sebenda (10) considered that this level which probably corresponds to water molecules hydrogen bonded to the oxygen atoms of two amide groups represents the first and most strongly bonded type of absorbed water. Their view has been widely supported by later workers. As the concentration of water is increased, the number of clusters remains within a narrow range. Decreases in the number of clusters may occur when newly absorbed water forms bridges between previously uncorrelated clusters. Near saturation, the apparent average cluster size is about three water molecules. This is close to Puffr and Sebenda's second stage of water absorption which is three molecules per two amide groups in the amorphous regions. The second and third water molecules were said to be bonded between the carbonyl of one amide group and the NH of another. A similar treatment (9) of Bull's data on nylon fibers (6) also indicated that clustering begins at about one water molecule per two amide groups in the amorphous regions. The apparent number of clusters declined as more water was absorbed, and the apparent cluster size was about three water molecules near saturation.

The volume of a mixture of nylon and water is less than the sum of the volumes of the components (5). The partial specific volume of the first water to be absorbed by dry nylon is about

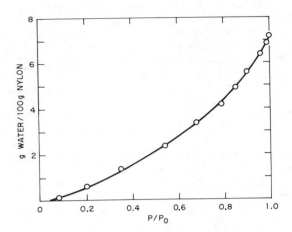

Figure 1.   Sorption isotherm for water in 66 nylon at 23°C

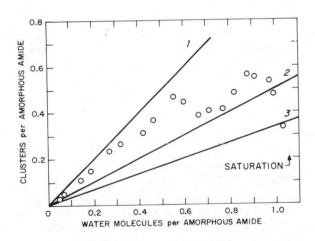

Figure 2.   Clustering of water in 66 nylon

0.5 g/cc as shown in Figure 3. This is attributed to the displacement of amide-amide bonds associated with highly strained chain conformations and poor packing. At a water concentration near one molecule per two amide groups in the amorphous regions, the partial specific volume is about 0.85 cc/g. Above 90% R.H., this quantity rises again and reaches a value of 1.2 cc/g near saturation. This indicates that the last water to be absorbed when there is maximum clustering is packed much less favorably than that absorbed earlier.

Between 35 and 100% R.H., the sorption isotherm is closely approximated by the Flory-Huggins equation with an interaction parameter, χ, of 1.46 ± .02. Since χ is largely an internal energy parameter, the energy term in the Flory-Huggins equation is approximately $RT\chi(1-\phi_1)^2$, but this does not include the effect of changes in volume. The partial molar heat of sorption at constant pressure is

$$(\Delta \bar{H}_1)_p = (\Delta \bar{E}_1)_v + T(\partial P/\partial T)_v \Delta \bar{V}_1 \qquad (3)$$

$$= RTX(1-\phi_1)^2 + T(\alpha/\beta)\Delta \bar{V}_1 \qquad (4)$$

where α is the coefficient of thermal expansion, β is the compressibility, and $\Delta \bar{V}_1$, is the partial molar change of volume for the total system of nylon plus water. The subscripts, 1, refer to water. It has been found that Equation 4 agrees with experimental data quite well (5). The heat of sorption declines from several kcal/mole at low humidities to zero at saturation. Thus the expansion which occurs above 90% R.H. is just enough to satisfy the condition that

$$\Delta \bar{F}_1 = \Delta \bar{H}_1 = \Delta \bar{S}_1 = 0 \qquad (5)$$

when $P = P_o$.

Viscoelastic Properties

There are three major viscoelastic relaxations in 66 nylon (11). The α-relaxation occurs at 65°C in the dry polymer and reflects the motion of fairly long chain segments in the amorphous regions. Boyd (12) has estimated that these segments contain about 15 amide groups. The β-relaxation at -50°C has been attributed to the motion of labile amide groups. It may be absent in very dry, annealed samples (1). The γ-relaxation at -120°C is similar to the γ-relaxation in polyethylene and has been attributed to the motion of short polymethylene segments. Since it is dielectrically active (13,14) the motions must involve some amide groups as well.

The temperatures of these relaxations are plotted against relative humidity in Figure 4 and percent water in Figure 5. These data were obtained with the Du Pont Dynamic Mechanical

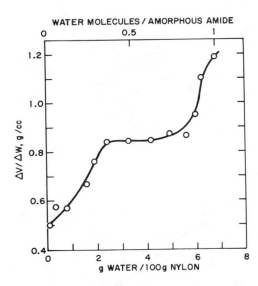

Figure 3.   *Partial specific volume of water absorbed in 66 nylon*

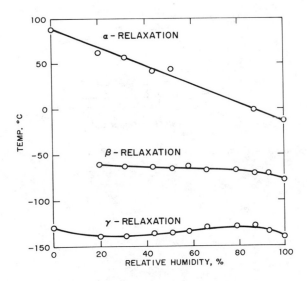

Figure 4.   *Temperature of peaks in loss modulus vs. relative humidity*

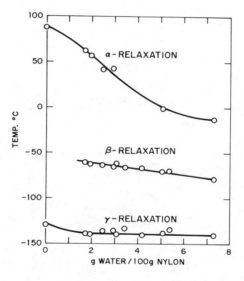

Figure 5.   Temperature of peaks in loss modulus vs. present water

Analyzer on a sample of roll-oriented nylon tape. The
temperature of the α-relaxation declines by about 90°C
between dryness and saturation (1,15,16). The variation appears to be linear with relative humidity, i.e.,
the activity or chemical potential of the water. The
relaxation passes room temperature near 35% R.H., and
this is why the water content of nylon has such a large
effect on properties of engineering importance. As
mentioned earlier, time, temperature, and humidity can
be treated as complementary variables. (2-4)

The activation energy for the α-relaxation decreases from 46 kcal/mole in the dry polymer to 18 kcal/
mole at 8.7% water with most of the decrease coming
below 0.88% water. (12) The increase in the dielectric
constant associated with the α-relaxation is greater
when water is present. This indicates that water molecules are bonded to the amide groups and participate in
the motion of chain segments.

The temperature of the β-relaxation decreases linearly with the concentration of water, not the relative
humidity. The height of a dynamic mechanical loss peak
has a maximum value at one water per two amide groups
in the amorphous regions. (17) The temperature of the
γ-relaxation decreases slightly when the first water is
added and remains almost independent of water content.
The height of the loss peak decreases with increasing
water, especially below a concentration of one molecule
per two amide groups in the amorphous regions. (13,15,
17)

At temperatures below the α-relaxation but above
the γ-relaxation, the modulus is increased by the
presence of water. (1,15,18) This is part of a familiar pattern of antiplasticization. The temperature
of a primary relaxation is reduced, the strength of a
secondary relaxation is reduced, and the modulus between them is increased.

Conclusions

There is a good deal of evidence to support the
suggestion of Puffr and Sebenda (10) that the first to
be absorbed and most tightly bound water is hydrogen
bonded to the oxygen atoms of two amide groups in the
amorphous regions. At this point clustering begins,
but all of the water is bonded to the amide groups,
and there is no evidence for freezable liquid water.

At low temperatures, water forms mechanically
stable bridges between amide groups. This is reflected
in a partial suppression of the γ-relaxation and an
increase in the modulus between the γ- and α-relaxa-

tions. At higher temperatures, water facilitates
motion in the amorphous regions and shifts the α-
relaxation progressively to lower temperatures. This
is accompanied by major increases in ductility and
toughness.

Literature Cited

1. Starkweather, H. W., Chapter 9 in "Nylon Plastics"
    Kohan, M. I., Ed.; Wiley: New York, 1973.
2. Quistwater, J. M. R. and Dunell, B. A., J. Polym.
    Sci., 1958, 28, 309.
3. Quistwater, J. M. R. and Dunell, B. A., J. Appl.
    Polym. Sci., 1959, 1, 267.
4. Onogi, S., Sasaguri, K., Adachi, T. and Ogihara,
    S., J. Polym. Sci., 1962, 58, 1.
5. Starkweather, H. W., J. Appl. Polym. Sci., 1959,
    2, 129.
6. Bull, H. B., J. Am. Chem. Soc., 1944, 66, 1499.
7. Zimm, B. H., J. Chem. Phys., 1953, 21, 934.
8. Starkweather, H. W., Chapter 3 in "Structure-
    Solubility Relationships in Polymers", Harris,
    F. W. and Seymour, R. B., Ed. Academic Press:
    New York, 1977.
9. Starkweather, H. W., Macromolecules, 1975, 8, 476.
10. Puffr, R. and Sebenda, J., J. Polym. Sci., Part C,
    1967, 16, 79.
11. Schmieder, K. and Wolf, K., Kolloid Z., 1953, 134,
    149.
12. Boyd, R. H., J. Chem. Phys, 1959, 30, 1276.
13. Dahl, W. V. and Muller, F. H., Z. Elektrochem.,
    1961, 65, 652.
14. Curtis, A. J., J. Res. NBS, 1961, 65A, 185.
15. Woodward, A. E., Crissman, J. M., and Sauer, J. A.
    J. Polym. Sci., 1960, 44, 23.
16. Prevorsek, D. C., Butler, R. H., and Reimschussel,
    H. K., J. Polym. Sci., 1971, 9, 867.
17. Kolarik, J. and Janacek, J., J. Polym. Sci., Part
    C, 1967, 16, 441.
18. Starkweather, H. W., J. Macromol. Sci. Phys., 1969
    B3(4), 727.

RECEIVED January 17, 1980.

Clustering of Water in Polymers

GEORGE L. BROWN

Mobil Chemical Co., Edison, NJ 08817

The perception that water sorbed in polymers can be clustered has a dual origin. The earliest inference of clustering was based on the development of opacity ("blushing") in films or coatings exposed to water. The occurrence of blushing is particularly prevalent in coatings or films, such as those produced from emulsion polymers, which contain water soluble impurities (emulsifiers and initiator residues). This form of clustering is not the same as that which can occur in homogeneous water swollen polymer films. It is, however, important to bear in mind because the presence of unrecognized traces of water soluble impurities may result in substantial errors in the interpretation of sorption measurements.

In binary solutions – in the case under consideration, water sorbed in a polymer – non-random mixing is also described as clustering. For this case, the cluster size is rarely, if ever, sufficient to produce visual opacity, and the evidence for the phenomenon is found in peculiarities of the sorption isotherm.

If the two cases cited – a polymer containing polymer insoluble, water soluble components, and a homogeneous polymer – represent two ends of a spectrum, a variety of intermediate cases can be imagined which represent varying degrees of heterogeneity (graft or block copolymers, ionomers, non-random copolymers).

Thus it is clear that the term "clustering" can be used in a variety of situations, and is not precisely defined. In the present case, it will be used in the context of the statistical thermodynamic treatment of binary solutions developed by Zimm (1) and Zimm and Lundberg (2), which provides a calculation of a cluster integral, and which can be extended to specify a cluster size for each component. In describing the role of the "clustering" theory in relationship to previously developed solution theories, such as the widely used Flory-Huggins theory (3), Zimm and Lundberg point out that "Our considerations are not intended as a replacement for the previous theories, but as

0-8412-0559-0/80/47-127-441$05.00/0
© 1980 American Chemical Society

an adjunct thereto, interpreting experimental data in molecular terms." The analysis which will be described is in accord with the complementarity of the two approaches. Water sorption of non-polar polymers (polyolefins, polystyrene) is very low, and is probably adequately described by application of conventional solution thermodynamics, i.e., The Flory-Huggins theory. For mixtures of water and polar polymers, the increase in water sorption with increasing humidity is much greater than that which would be predicted from this theory. This is not surprising, because the theory is based on an assumption of random mixing, and does not contemplate the specific associations possible between water and polar groups in the polymer and between water molecules which can lead to non-random mixing. A method will be described which combines conventional solution theory and cluster theory and provides an interpretation for the sorption of water by certain polar polymers. The total water sorbed by a polymer will be viewed as a sum of an amount contributed by normal, random mixing, plus an increment due to association or clustering of the water.

Mathematically, the sorption data for the amorphous acrylic polymers which will be considered here cannot be correlated by a single parameter sorption isotherm. In the Flory-Huggins theory, the parameter is the interaction parameter, which for the simplest possible case characterizes the enthalpy of mixing which results from the intermolecular bonding mismatch between polymer and water. A two parameter sorption isotherm provides an excellent vehicle for data treatment. The two parameters can be identified as the interaction parameter and a clustering parameter.

Methods for Evaluation of Experimental Data. A variety of solution and sorption isotherms have been developed to account for water-polymer mixtures, and this has been extensively summarized by Barrie (4). However, few studies have been performed where simple, systematic changes in polymer composition are involved. An examination of published experimental data on amorphous acrylate polymers indicates that a rather simple, consistent isotherm is applicable, which allows some interesting conclusions as to the influence of polar content and glass temperature on the cluster size.

Experimental data on water sorption by polymers is usually presented based on some form of Henry's Law. That is, a measure of the quantity of water sorbed (weight, volume, moles) per unit quantity of substrate or substrate plus water is plotted against the activity of water in the surrounding vapor. Since data for water rarely display linearity over any appreciable range of sorption, a search was made for a function which would provide better linearity - particularly for high humidity sorption measurements. From a practical standpoint, accurate interpolation and extrapolation for samples exposed to high humidity is very important in predicting behavior of coatings and barrier films.

For a large number of polymers, a plot of the reciprocal of water sorption against the reciprocal of the partial pressure of water exhibits near-linear behavior. With the expectation that the volume fraction of water will be the expression of water concentration relevant to basic polymer solution thermodynamics, this measure has been used. However, a variety of other expressions of concentration can be used (weight or volume per unit weight or volume of polymer, weight percent, etc.) and substitution of these will alter the slope and intercept, but not the linearity of the plot.

The use and interpretation of the technique are exemplified in Figures 1 and 2 for poly(ethyl methacrylate) sorption data of Williams, Hopfenberg, and Stannett (5). The upper curve in Figure 1 shows experimental data plotted in the conventional manner. The lower curve is the Henry's Law limiting isotherm, tangent to the experimental curve at the origin. The lower line in Figure 2 shows the plot of reciprocal quantities ($1/\phi$ vs $1/P$). It shows excellent agreement with the linear equation

$$\frac{1}{\phi} = \frac{k_1}{P} - k_2 \tag{1}$$

with $k_1 = 167.9$, and $k_2 = 104.7$ with a correlation coefficient greater than 0.999. An interpretation of Equation (1) is that it represents Henry's Law sorption for $k_2 = 0$, a Langmuir or attentuated type of isotherm where k_2 is negative, and an enhanced or "clustered" isotherm where k_2 is positive. The limiting, infinite dilution isotherm, as P approaches zero, is given by the inverse Henry's Law expression

$$\frac{1}{\phi_H} = \frac{k_1}{P} \tag{2}$$

Using the value of 168 for k_1, this equation is plotted as the upper line in Figure 2. Graphically, this is done by drawing a line parallel to the experimental data passing through the origin. The inverse of this relationship is the Henry's Law isotherm shown in Figure 1.

The postulate which will be pursued is that water sorption at infinite dilution of the water is normal in behavior, representing the true interaction of water and polymer molecules. That is, if water sorption occurs on two types of sites, a polymer site and a polymer-water site, the influence of the former will predominate as P and the amount of sorbed water simultaneously approach zero. A plot of ϕ vs P with a slope of k_1 (the inverse of Equation 2), as shown in Figure 1, is tangent to the experimental isotherm at the origin. The analysis outlined allows a unique specification of χ, the interaction parameter, through use of the limiting (Henry's Law) approximation of the Flory-Huggins theory (3). That is

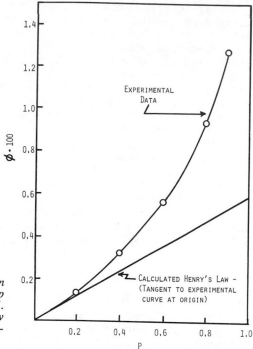

Figure 1. Conventional plot of sorption of water by poly(ethyl methacrylate): top curve, experimental data from Ref. 5. Bottom curve, calculated Henry's Law isotherm based on extension of low relative pressure sorption

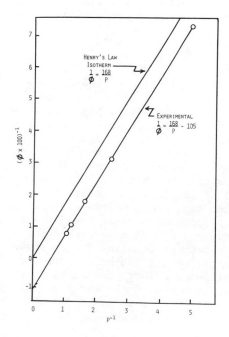

Figure 2. Inverse plot of sorption of water by poly(ethyl methacrylate): top curve, calculated Henry's Law isotherm; Bottom curve, experimental data

$$P \stackrel{\sim}{=} \phi e^{(1 + \chi)} \stackrel{\sim}{=} k_1 \phi \qquad (3)$$

and $\chi \stackrel{\sim}{=} \ln k_1 - 1$

At any relative pressure, the experimental sorption can be compared to that predicted for the Henry's Law isotherm to provide a ratio, N_e, which will be termed the "enhancement number", calculated from Equations (1) and (2) as follows:

$$N_e = \frac{\phi}{\phi_H} = \frac{k_1}{k_1 - k_2 P} = 1 + k_2 \phi \qquad (4)$$

The enhancement number is a measure, then, of the extent to which the sorption of water is increased by the abnormalities of the process which result from non-random mixing. Another index is available from the calculation of the cluster size, based on the Zimm-Lundberg concepts.

In his initial paper, Zimm (1) provides for the calculation of a cluster integral from the solution isotherm. Neglecting a small term involving compressibility, for component 1 in a binary mixture, a cluster integral, G_{11}, is calculated from:

$$\frac{G_{11}}{v_1} = -(1-\phi_1) \left[\frac{\partial \left(\frac{a_1}{\phi_1} \right)}{\partial a_1} \right] - 1 \qquad (5)$$

where v_1 is the molecular volume, ϕ_1 the volume fraction and a_1 the activity of component 1. For a random solution, the activity is proportional to the volume fraction, and

$$\frac{G_{11}}{v_1} = -1 \qquad (6)$$

The extent of clustering in solutions is indicated by the extent to which G_{11}/v_1 exceeds minus one.

A more useful index has been provided by Zimm and Lundberg (2), in the quantity $\phi_1 G_{11}/v_1$, which is "the number of type 1 molecules in excess of the mean concentration of type 1 molecules in the neighborhood of a given type 1 molecule." From Equation (6), this quantity is obviously $-\phi$ for a random solution.

For many purposes, a calculation of the average number of solvent molecules in a cluster is useful. We shall denote this as N_c, the cluster number, and Starkweather (6) has suggested that this is provided by the expression:

$$N_c = \frac{\phi_1 G_{11}}{v_1} + 1 \qquad (7)$$

For a random solution (an "ideal" solution in this context) the activity coefficient, a_1/ϕ_1, is invariant with concentration, and from a combination of Equations (7) and (5), this would give for such a solution:

$$N_c = 1-\phi \tag{8}$$

However, there should be no clustering for this case, and the cluster number should be one.

We believe that the difficulty arises from the peculiarity of the definition of the quantity $\phi_1 G_{11}/v_1$, which was provided above, which specifies the excess solvent molecules in the neighborhood of (but not including) the central solvent molecule. The cluster number should then be specified as

$$N_c = \phi_1 \left(\frac{G_{11}}{v_1} + 1 \right) + 1 \tag{9a}$$

$$N_c = -\phi_1(1-\phi_1) \left[\frac{\partial \left(\frac{a_1}{\phi_1} \right)}{\partial a_1} \right] + 1 \tag{9b}$$

Values of N_c calculated from Equation (9a) or (9b) exceed those calculated on the basis of Equation (7) by the quantity ϕ_1. This difference is negligible for the cases which will be treated here, but could be appreciable if the method is applied to the determination of cluster number for dilute aqueous polymer solutions, for example. These corrected equations provide a cluster number of one for an "ideal" solution.

Using the partial pressure of water, P, as an adequate approximation to the activity, we see from Equation (1), that for the isotherm under consideration here, the derivative within the brackets in Equation (9b) is equal to $-k_2$. The cluster number is, for this case, given by:

$$N_c = 1 + k_2\phi - k_2\phi^2 \tag{10}$$

and comparing this to the enhancement number from Equation (4):

$$N_c = N_e - k_2\phi^2 \tag{11}$$

Evaluation and Interpretation of Data. Sorption data on four polymers reported in the literature have been examined. Data on poly(ethyl methacrylate) from Williams et al (5) were specified as volume fraction of water sorbed as a function of partial pressure. In an earlier publication of Stannett and Williams (7) the glass temperature of the polymer was given as 65°C. Sorption and desorption measurements showed a very close concordance. Data on poly(methyl methacrylate) by Brauer and Sweeney (8) were transformed from gravimetric to volume fraction using a density of 1.19. The glass temperature of this polymer is 105°C. The authors report that "thin specimens reach a steady state value within a short time." Data for poly(methyl acrylate) and a copolymer of methyl acrylate and acrylic acid, 89/11, are from Hughes and Fordyce (9). Densities used to convert gravimetric data to volume fraction were 1.22 and 1.24, the latter based on the assumption of volume additivity using a density of 1.4 for polyacrylic acid, kindly furnished by D. B. Fordyce. The glass temperature of polymethyl acrylate is 6°C. The sorption results were reported to be equilibrium values. Measurements on poly(ethyl methacrylate) were made at 25°C, and on the remaining three at 30°C.

Data on the four polymers, calculated from the equations of the previous section, are presented in Table I. In descending order, the series represents increasing hydrophilicity, with ester groups per unit volume increasing from ethyl methacrylate to methyl acrylate, and with carboxyl groups inserted in the fourth polymer. This is reflected in terms of an orderly decrease in χ and increase in saturation sorption. However, no trend in the clustering can be discerned. Thus this phenomenon appears to be related to the nature of the water molecule and quite probably to the hydrogen bonding propensity of the polar groups on the polymer, but not to any appreciable extent to the concentration of polar groups.

A number of investigators have suggested that effects of initial water sorption on polymer mechanical properties ("plasticization") would facilitate further accommodation of water. The first two polymers in Table I are glassy, and the third rubbery, and there is no evidence of any mechanistic difference in sorption. Although relaxation effects will be expected to influence rate of sorption for these amorphous polymers, influence on equilibrium sorption appears negligible. Viewed in another manner, water is certainly capable of plasticizing polymers, but organic penetrants which sorb in a normal fashion also do so. It is therefore questionable to assign the unusual nature of water sorption to plasticization.

The analysis produces a unique value for χ, which should relate to the fundamental interaction of water and the polymer. An alternative view, which has been taken by many authors, is to calculate χ for each experimental point. For poly(ethyl methacrylate), for example, Williams et al (5) show a decreasing χ with increasing relative pressure. Their values can be extrapolated to zero pressure, to give $\chi = 4.1$, the value derived here.

Two independent methods have been utilized to examine the nature of the sorption isotherm. An analysis of the experimental isotherm compared to an extrapolation of the infinite dilution behavior allows calculation of an enhancement number for any of the polymers at any given partial pressure. Calculation of a cluster number based on an independent method shows very close concordance with the enhancement number, providing strong support for the postulate that associated groups of water molecules sorb in the polymer, and account for the anomolous sorption.

Table I

Polymer of	k_1	k_2	χ	Values at P = 1		
				ϕ	N_e	N_c
Ethyl Methacrylate	168	105	4.1	0.016	2.7	2.6
Methyl Methacrylate	119	80	3.8	0.025	3.0	2.9
Methyl Acrylate	86	53	3.5	0.030	2.6	2.5
Methyl Acrylate-Acrylic Acid 89/11	33	22	2.5	0.091	3.0	2.8

The sorption isotherm utilized:

$$\frac{1}{\phi} = \frac{k_1}{P} - k_2 \tag{1}$$

can in a sense be derived analogously to the Langmuir isotherm, by postulating a forward rate in which additional sorption sites are made available by sorbed water molecules, and a rate of desorption proportional to total water content.

Alternatively, a Henry's Law sorption can be written for the polymer component and the water component, each term multiplied by the volume of the component, and the contributions summed to give:

$$\phi = k'P(1-\phi) + k'P\phi \tag{8}$$
$$= k'P + (k''-k')P\phi$$

which is similar to Equation (1) when rearranged in the form:

$$\phi = \frac{P}{k_1} + \frac{k_2}{k_1}P\phi \tag{9}$$

At best, the isotherm is a limiting form which could be true only for low sorption. When rewritten as

$$P = \frac{k_1\phi}{1+k_2\phi} \tag{10}$$

We see that it becomes Henry's Law for $k_2 = 0$, and speculate that the substitution of the Flory-Huggins isotherm for $k_1\phi$ to give:

$$P = \frac{\phi e^{(1-\phi)+(1-\phi)^2\chi}}{1+k_2\phi} \tag{11}$$

may provide more general applicability.

Since the "base" isotherm used here utilizes the volume fraction of water as the measure of concentration, for convenience the enhancement has been based on volume fractions. For highly hydrophilic polymers, enhancements in the range reported in Table I would result in volume fractions of water greater than unity. This problem could be eliminated by treating the water sorption in the base isotherm on a volume fraction basis, as outlined, but treating the clustering on the basis used for absorption processes, considering the volume added by clustering to a unit volume of polymer containing unclustered water.

Incorporation of the two suggested improvements leads to a more complicated process for identifying the interaction parmeter, the "base" isotherm, and the extent of enhancement, but we suggest that the simple, coherent picture which emerges for the systems reported here justifies extension to develop the more general treatment.

The qualitative mechanism suggested by the data and its interpretation is that at low relative pressures, water is distributed throughout the polymer, but probably preferentially where hydrogen bonding is possible. At higher pressures, chains of water on the hydrogen bonding sites predominate. The initial process can be described in terms of conventional solution theories and the enhancement process can be viewed as one of occupancy of sites – analogous to that found in adsorption processes.

REFERENCES

(1) Zimm, B.H. J. Chem. Phys., 1953, 21, 934-935

(2) Zimm, B.H.; Lundberg, J.L. J. Phys. Chem., 1956, 60, 425-428

(3) Flory, P.J. "Principles of Polymer Chemistry"; Cornell University Press, Ithaca, N.Y., 1953; p. 512-514

(4) Barrie, J.A. In "Diffusion in Polymers", Crank, J.; Park, G.S., Ed.; Academic Press: London and New York, 1968; Chapter 8

(5) Williams, J.L.; Hopfenberg, H.B.; Stannett, V.T. Polymer Preprints, 1968, 9(2), 1503-1510

(6) Starkweather, H.W. Polymer Letters, 1963, 1, 133-138

(7) Stannett, V; Williams, J.L. J. Polymer Sci., 1966, C-10, 45

(8) Brauer, G.M.; Sweeney, WT. Modern Plastics, May 1955, 32, 138

(9) Thompson Hughes, L.J.; Fordyce, D.B. J. Polymer Sci., 1956, 22, 509-526

RECEIVED January 4, 1980.

Water Sorption and Its Effect on a Polymer's Dielectric Behavior

G. E. JOHNSON, H. E. BAIR, S. MATSUOKA, E. W. ANDERSON, and J. E. SCOTT[1]

Bell Laboratories, Murray Hill, NJ 07974

Recently we reported a way to measure the amount of water which has associated to form microscopic water-filled cavities (clusters) in polyethylene at a level of 10ppm and greater (1,2). By combining this calorimetric technique with a coulometric method, it was possible to differentiate between clustered water and the total water sorbed by the polyethylene. It was found that clusters are formed when polyethylene is saturated with water at an elevated temperature and is rapidly cooled to room temperature. During cooling the solubility of water in polymers is lowered and some water condenses in the form of microscopic water-filled cavities, providing the internal pressure which is generated by the excess water exceeds the strength of the polymer. Figure 1 shows 2-micron clusters formed in polyethylene quenched from the melt in the presence of water.

In sorption studies of polycarbonate (3) it was learned that this polymer absorbs water in two stages. In the initial period of absorption at an elevated temperature, but below T_g, all of the water was found in an unassociated state when cooled to room temperature. In the second stage at later times, most of the water gained by the polymer was identified in a separate liquid phase (clustered water). In addition after the polymer was saturated with water at a temperature above T_g and cooled, its solubility was lowered and water condensed in the form of microscopic water filled cavities. Below T_g the clusters were formed only after the polycarbonate's strength ($M_w = 26,600$) was decreased by hydrolysis whereas above T_g clusters were formed without degradation.

The dielectric loss behavior of both polyethylene's γ-transition and polycarbonate's β-transition was enhanced by the presence of unassociated water. The area under the associated loss peak was found to increase in direct proportion to the concentration of unassociated water. In addition a secondary dielectric loss peak associated with frozen clustered water occurred in polycarbonate about 40°C below its β-transition. Liquid clustered water at

[1] Current address: Prairie View A&M, Prairie View, TX 77445

0-8412-0559-0/80/47-127-451$05.00
© 1980 American Chemical Society

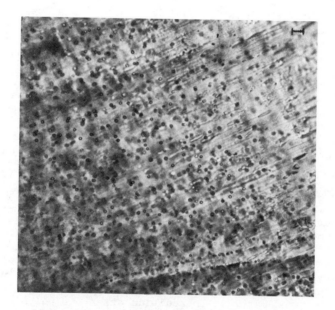

Figure 1.   Two-micron water clusters in polyethylene

23°C yielded a loss mechanism in the mHz region in polyethylene and in the kHz region for polycarbonate that was interpreted as a Maxwell-Wagner effect.

Early investigations of the effect of water on the low-temperature relaxations of several aromatic polymers including polycarbonate, polyamides, and a polyurethane have shown several low-temperature anomalies (4,5). In the case of a water-saturated polysulfone polymer which exhibited a doublet in its β-loss process, Jackson suggested that the secondary peak may be due to water-filled cavities. In this work we have employed the DSC technique for water cluster analysis along with the total water content measurements to elucidate the water sorption behavior of polysulfone and poly(vinyl acetate) (PVAc). In addition DSC and dielectric methods were used cooperatively to understand the T_g behavior of polymers in the presence of water.

Experimental

Water Analysis. The water gained by a sample was measured on a duPont 26-321A moisture analyzer. This instrument uses a coulometric technique to measure the total amount of water in a sample. Samples were heated for 15 minutes above their T_g to drive off the water.

The determination of clustered water was done calorimetrically (1) using a differential scanning calorimeter (Perkin Elmer DSC-2). Samples were placed into a nitrogen-flushed dry box before they entered the DSC sample holder. All experimental runs were made at 20°C/min. All reported values of water content in this paper are in weight percent as determined both calorimetrically and coulometrically.

When about 20 milligrams of sample were placed in the DSC-2 and cooled at 20°/min. from room temperature to -140°C the onset of crystallization of the clustered water was normally detected near -40°C and proceeded at a maximum crystallization rate at -50°C. We believe that the large undercooling is due to the microscopic size of the clusters and the absence of heterogeneities in the water. When the sample was reheated from -120°C to room temperature a first order transition was detected near 0°C as expected.

Dielectric Measurements. Dielectric measurements on polysulfone were conducted at 10^2, 10^3, and 10^4 Hz. The data were obtained by combining a Princeton Applied Research 124 lock-in amplifier and a General Radio 1615A capacitance bridge. The bridge was connected to a Balsbaugh LD3 research cell inside a test chamber. After the test chamber was equilibrated at -160°C for one hour, measurements were made at ten to twenty degree intervals with a fifteen minute waiting period between each discrete change in temperature until room temperature was reached.

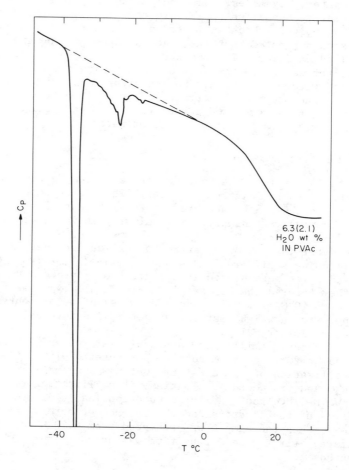

Figure 2. DSC cooling curve for PVAc containing 6.3% total water (2.1% clustered)

Dielectric measurements on poly(vinyl acetate) were obtained utilizing a Fourier transform dielectric spectrometer developed in our laboratory (6). A voltage step pulse was applied to the sample and the time dependent integrated current response, Q(t), was collected by computer. The frequency dependent dielectric properties, ε' and ε'' were then obtained from the Fourier transform of the integrated current. For this study frequency dependent dielectric data in the 10^0 to 10^4Hz range were obtained isothermally. A sample cell with low thermal mass and copper screening allowed rapid equilibration between temperatures. This allowed the frequency dependence at several temperatures to be measured while minimal water was lost from the samples.

Results and Discussion

Determination of Unassociated and Clustered Water. Compression molded samples of polysulfone were immersed in water at 100°C and below until they came to equilibrium. Above 100°C an autoclave was used. As with polycarbonate a mild temperature dependence in equilibrium absorption was noted.[3] The amount of unassociated water at saturation went from 0.8% at 23°C to 1.2% at 132°C. No clustered water was found in samples exposed below T_g (190°C). When the polymer was exposed to steam at 208°C for 1 minute and quenched to 23°C, cluster formation was noted. There was, however, only 0.04% water found in the clustered state. Five minute exposure to steam increased the clustered water content to 0.16%. Microscopic analysis of cross sections of dielectric specimens showed a non-uniformity of cluster size and distribution. This resulted in two DSC peaks (-34°C and -42°C) in cooling the sample at 20°C/min. and a broadened melting peak starting at 0°C on heating at the same rate.

Compression molded samples of poly(vinyl acetate) also showed a mild temperature dependence in equilibrium absorption. The amount of water went from 4% at 23°C to 6% at 70°C. This polymer was the only one we tested that formed clustered water while stored isothermally at room temperature. This clustering was obtained after 17h. as confirmed by DSC and could be seen visually as a whitening of the polymer.

Figure 2 shows the DSC cooling curve of a sample containing 6.3% total water, 2.1% of which was clustered. At a cooling rate of 20°C/min. the vitrification of the polymer was noted between 20 and 5°C. The next thermal event observed was the onset of freezing of clustered water at -5°C. The crystallization process proceeded sporadically until -35°C. At that temperature the major portion of the clustered water began to freeze and this process was completed by -38°C. The crystallization at smaller undercooling was believed to be due to freezing of large droplets.

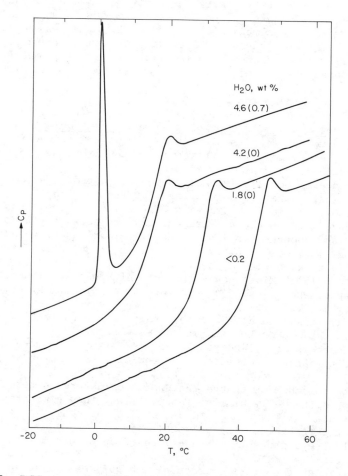

Figure 3. DSC curves for PVAc as a function of total percent water content (clustered percent in parenthesis)

Calorimetric T_g Behavior. The comparative C_p curves versus temperature of PVAc containing 0.2, 1.8, 4.2 and 4.6 weight percent water are plotted in Figure 3. These C_p curves in all cases were run after cooling each sample to $-70^\circ C$. The lowest curve represents a film that has been vacuum dried overnight at $41^\circ C$. The C_p increased linearly with temperature until the glass transition which was noted as a discontinuity in C_p between 35° and $45^\circ C$. T_g, which is defined in this work as the midpoint of the transition, is $43^\circ C$. The C_p curve for a PVAc film containing 1.8% water (all unclustered) has the T_g shifted $13^\circ C$ lower to $30^\circ C$. Increasing the level to 4.2% unclustered water lowered T_g to $19^\circ C$. In all the above cases the transition width of T_g was the same, $12^\circ C$.

The upper curve in Figure 3 represented 4.6% total water with 0.7% of that in the clustered state. It was noted that the clustered water melts near $0^\circ C$ followed by a T_g at $19^\circ C$ as in the 4.2% unclustered water sample. Thus clustered water has no plasticizing effect on T_g. This was shown dramatically in the curves of Figure 4 which show that when clustered water content was increased from 0.7 to 2.1% no shift in T_g behavior occurred.

Dielectric Behavior. The dielectric loss behavior of polysulfone samples was measured below $23^\circ C$ as a function of unassociated water content. The activation energy of the process was calculated to be 11.4 kcal/mole (1.1% H_2O) and was in agreement with Allen's prior determination (4). The areas under the loss curves of Figure 5 were directly proportional to the amount of unassociated water present. This dielectric loss increase of the polymer's β-mechanism can be understood in terms of the motion of the water dipoles correlated according to the dynamics of the polymer molecules. A calculation of the added dipolar contribution of the water was made and found to be one quarter the value that would be obtained if the water dipoles participated completely.

An enhanced dielectric loss maximum was observed at $-85^\circ C$ when a polysulfone sample which contained 0.76 wt. % unassociated water and no detectable level of clustered water (<0.01 wt. %) was run (Fig. 6, curve A). An apparent low temperature broadening of the dielectric loss dispersion was noted for another polysulfone specimen with 0.76 wt. % unassociated water and an additional 0.04 wt. % clustered water (Fig. 6, curve B). However, when a polysulfone sample which contained the same amount of unassociated water as the two prior samples but had 0.16 wt. % clustered water was analyzed, it had a significantly more intense loss peak centered near $-105^\circ C$ (Fig. 6, curve C). We believe that this shift in loss maximum and increase in loss intensity is caused by the development of an additional secondary loss peak about 20° below the β-transition (Figure 6). In earlier work we had observed the same phenomenon in polycarbonate where the new loss peak occurred about 40° below its β-transition as a separate loss peak.

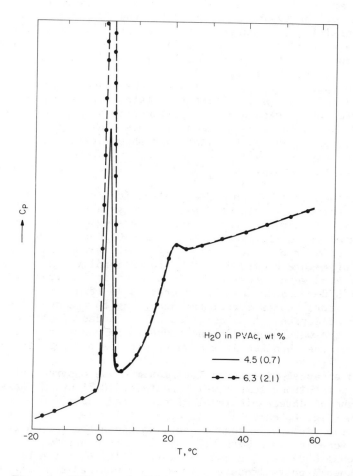

Figure 4.   DSC curves for PVAc (clustered percent in parenthesis)

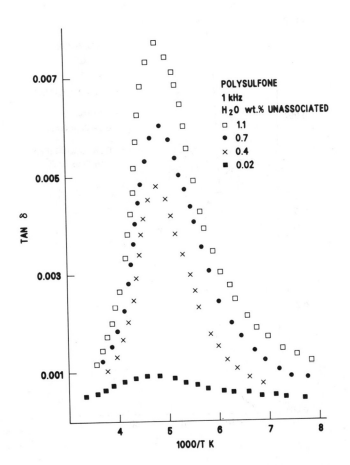

Figure 5. Polysulfone beta transition's dielectric loss as a function of unassociated water content

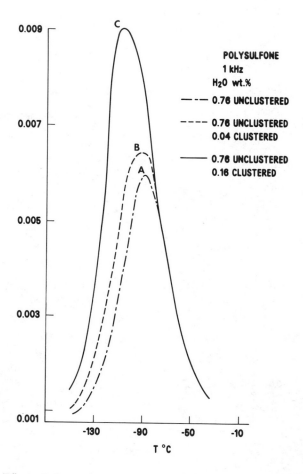

Figure 6.   Effect of clustered water on the polysulfone beta transition's dielectric response

Polyethylene samples were also exposed to conditions which created 0.4% clustered water and dielectric data taken at low temperatures on the samples. The same loss maximum noted in polycarbonate and polysulfone near -100°C at 1 kHz was also noted in polyethylene. A special polyethylene sample was molded around a PTFE sheet. The PTFE was removed and replaced with distilled water. This sample was equivalent to a thin water layer between polyethylene sheets. The dielectric behavior of this sample was quantitatively equivalent to that of the polyethylene containing spherical clusters of water if the difference in geometry of the water phase is taken into account. Figure 7 shows the logarithm of the frequency of loss maxima due to water clusters versus reciprocal temperature for polyethylene, polycarbonate, poly(vinyl acetate and polysulfone. The polysulfone data from Allen[4] are shown for comparison and it is seen that the data can be interpreted as a single mechanism with an activation energy of 7 kcal/mole.

The dielectric loss behavior of PVAc was similar to that of the other polymers. An increase in dielectric intensity of the polymer's β mechanism was directly proportional to the amount of unclustered water. In addition when clustered water was present two separate low temperature peaks occurred as shown in the frequency dependent data of Figure 8. The higher frequency peaks were the result of clustered water. This is confirmed by the similarity between poly(vinyl acetate) and the clustered water peaks of other polymers as plotted in Figure 7.

The frequency of ε'' maximum versus reciprocal temperature was plotted in Figure 9 for the poly(vinyl acetate)'s β mechanism. Samples with 0.2, 2.0 and 4.5% unclustered water are shown. All samples were held at room temperature and then quenched to -145°C to initiate the frequency dependent data collection at discrete temperatures. The apparent activation energy of the β mechanism decreases with increasing unclustered water content providing that the samples absorbed water at 23°C. A sample quenched from 70°C rather than 23°C showed an activation energy close to the dry (0.2% water) sample although 4% unclustered water was present.

With the new cell it was possible to measure dielectrically the α transition behavior of poly(vinyl acetate) while retaining water within the sample. Figure 10 shows the normalized loss for the α transition for 0.2, 2.0 and 4.5% unclustered water. The shape of the transition was independent of water content. This was in agreement with the DSC curves on similar samples which also showed no difference in T_g transition breadth as a function of unclustered water.

The plot of frequency of ε'' maximum versus reciprocal temperature for the α transition showed typical WLF behavior in Figure 11. All these data were fit to a single WLF equation with the fractional free volume at T_g equal to .0225 and the expansion coefficient of free volume above T_g equal to 5.3×10^{-4}.

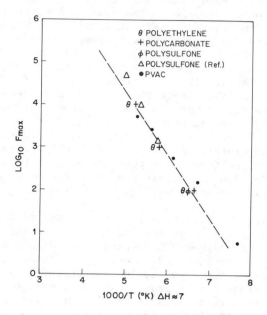

Figure 7. Dielectric loss maxima due to clustered water in various host polymers

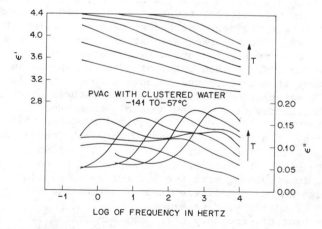

Figure 8. Fourier transform dielectric spectrometer data for polysulfone containing clustered and unassociated water

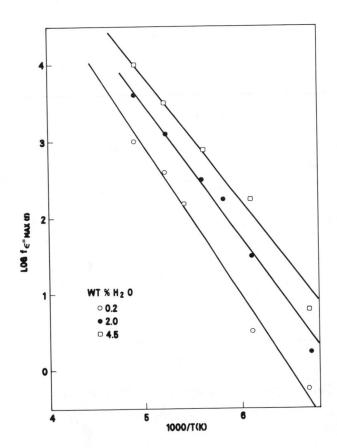

Figure 9. PVAc dielectric beta transition loss maxima vs 1/T as a function of weight percent water

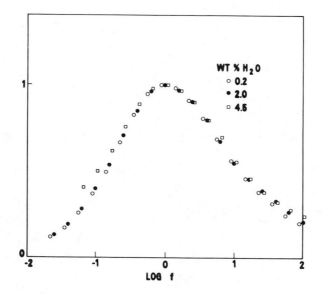

Figure 10.   Normalized dielectric loss for PVAc's alpha transition

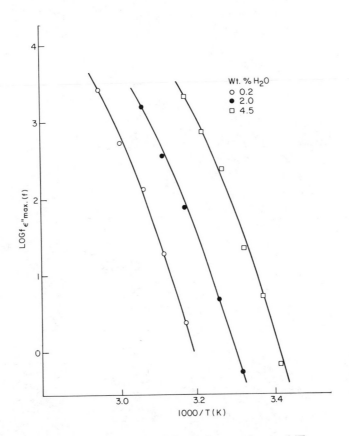

Figure 11. PVAc alpha dielectric loss maxima vs. 1/T

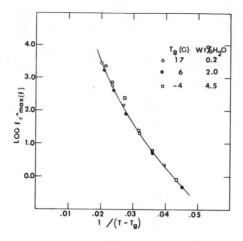

Figure 12.   WLF fit of log dielectric maximum vs. $1/(T - T_g)$

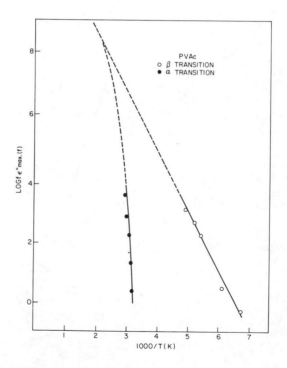

Figure 13.   PVAc dielectric maxima (alpha and beta) vs. $1/T$

$$\text{LOG } f_{\epsilon''max(f)} = \frac{19.6(T-T_g)}{42+(T-T_g)}$$

The T_g's of 17, 6 and -4°C for 0.2, 2.0 and 4.5% water in poly (vinyl acetate) were found dielectrically. The shifts of 11°C and 10°C were similar to the 13°C and 11°C shifts noted by DSC. When all the dielectric maxima were plotted versus $(T-T_g)^{-1}$ in Figure 12 an excellent fit to the above WLF equation was noted. The additional fractional free volume obtained when 1% water was absorbed was .003. This was in agreement with earlier polycarbonate data.

Figure 13 shows dielectric loss maxima versus 1/T for both the α and β transitions in poly(vinyl acetate) with 0.2% water. Using a WLF extrapolation for the α and an Arrenhius behavior for extrapolating β, the transitions merge at about 180°C. A plot of the dielectric strength, $\Delta\epsilon$, of the α transition versus reciprocal temperature indicated that the α transition would be inactive at 180°C. The data was consistent with the hypothesis that above the merging temperature only the β transition is active.

Conclusion

Water absorbed in a polymer can exist in an unassociated state or as a separate phase (cluster). In this investigation the DSC technique of water cluster analysis was used in conjunction with coulometric water content measurements to characterize the water sorption behavior of polysulfone and poly(vinyl acetate) The polysulfone had to be saturated above its T_g (190°C) and quenched to 23°C for cluster formation to occur while cluster formation occurred isothermally at 23°C in the poly(vinyl acetate)

Both polymers showed an enchancement of their low temperature β-loss transitions in proportion to the amount of unclustered water present. Frozen clustered water produced an additional low-temperature dielectric loss maximum in PVAc and polysulfone common to polyethylene and polycarbonate as well. Dielectric data obtained on a thin film of water between polyethylene sheets was in quantitative agreement with the clustered water data.

The dielectric data on poly(vinyl acetate)'s α transition showed a good WLF fit with a shift in T_g occurring with increasing unassociated water content but no change in the shape of the loss peak. This was in agreement with DSC data on the polymer. Both DSC and dielectric data showed clustered water had no effect on T_g.

Literature Cited

1. Porter, R. S.; Johnson, J. F., Ed. "Analytical Calorimetry, 4"; Plenum Press: N. Y., 1977; p219

2. Johnson, G. E.; Bair, H. E.; Anderson, E. W.; Daane, J. H. 1976 Annual Report on Electrical Insulation and Dielectric Phenomena, 1976, 510.

3. Bair, H. E.; Johnson, G. E.; Merriweather, R. J. Appl. Phys., 1978, 49, (10), 4976

4. Allen, G.; McAinslie, J.; Jeffs, G. M. Polymer, London, 1971, 12, 85

5. Jackson, J. B. Polymer, London, 1969, 10, 159

6. Johnson, G. E., Anderson, E. W.; Link, G. L.; McCall, D. W. Am. Chem. Soc. Organic Coatings and Plastics Preprints, 1975, 35, (1), 404

RECEIVED January 4, 1980.

Water Absorption in Acid Nafion Membranes

R. DUPLESSIX[1], M. ESCOUBES[2], B. RODMACQ, F. VOLINO, E. ROCHE,
A. EISENBERG[3], and M. PINERI

Centre d'Études Nucléaires de Grenoble, 38041 Grenoble Cédex, France

Nafion polymers have been developed recently by the du Pont Company. They are perfluorosulfonic acid membranes mainly used as separators in electrochemical applications. The backbone of the polymer chains consists of perfluoroethylene units whereas the side chains are of the form $- O - CF_2 - CF - O - CF_2 - CF_2 - SO_3M$. A large amount of work has been $\underset{CF_3}{\overset{|}{}}$ published on their commercial applications but little work has been done in terms of the molecular structure. Ion clustering has been proposed by Yeo and Eisenberg (1) from results of small angle X ray scattering and dynamic mechanical experiments. This phase separation has been confirmed both from experimental (2) and theoretical studies (3). Further support for this phase separation has also been provided by ^{23}Na NMR studies (4). A model has been proposed by T. Gierke (2) to explain the main features of these membranes. For the water soaked membrane, clustering of the ionic groups is proposed with an average cluster diameter of 40 Å and an average distance of 50 Å between cluster centers. The experimental support for this model comes from small angle X-ray scattering experiments, electron microscope analysis and water diffusion data In order to get more information about the various different phases we have performed different experiments : heat of water absorption measurements, small angle scattering of neutrons and X-rays, as well as NMR and quasi elastic scattering of neutrons on samples containing various amounts of water. We are therefore using the water molecules as a probe to obtain information about the Nafion structure. A summary of these results is given in this paper, which concerns only the interactions of the water molecules with the acid form of Nafion.

Current addresses:
[1] Institut Laue Langevin, BP 156, 38042 Grenoble Cédex, France.
[2] Université Claude Bernard, 69621 Villeurbanne, France.
[3] McGill University, Montreal, Canada.

0-8412-0559-0/80/47-127-469$05.00/0
© 1980 American Chemical Society

Experimental

The acid sample which has been studied here has an equivalent weight of 1200 (weight of acid polymer per SO_3H group).

In the water absorption study, a Setaran thermobalance B 60 and a Richard Eyraud isotherm differential microcalorimeter were used. The same experiment permits one to obtain both the sorption isotherm and the differential heat absorption values (5). The accuracies were better than 0.1 mg and 10^{-4} cal/sec, respectively. The relative humidity was obtained by changing the temperature of a water/ice bath (accuracy 0.1°C) which was connected to the sample. We measured both the amount of absorbed water and the heat of absorption after a change in the relative humidity level. We can therefore define the average heat of absorption per water molecule corresponding to the molecules which have been absorbed after this change in water relative humidity.

The D_{11} and D_{17} machines at Laue Langevin Institute (I.L.L.) in Grenoble have been used to perform the small angle neutron scattering experiments. The q values ($q = \frac{4\pi}{\lambda} \sin \theta$) which are accessible in these experiments are respectively, 5×10^{-3} to 20×10^{-3} Å^{-1} for D11 and 1×10^{-2} to 20×10^{-2} Å^{-1} for D17. Such experiments therefore permit a demonstration of the existence of clusters up to a few hundreds of angströms in size.

The NMR experiments were performed at 60 MHz using the Bruker WP 60 pulsed Fourier transform NMR spectrometer.

The incoherent neutron quasi elastic experiments were performed with the time of flight spectrometer IN5 of the I.L.L.. The scattering angle used corresponded to q values ranging from 0.2 to 1.2 Å^{-1}. The corresponding energy resolution was $\sim$ 18 μeV full width at half maximum.

Results

Figure 1 shows a plot of the water loss versus temperature for a sample which has been dried at room temperature for 24 hours under 10^{-4} torr. NMR experiments on the same starting material give evidence of the presence of residual water ; we also observed some neutron scattering at low angles which corresponds to some contrast because of the presence of water. After heating the sample at 220°C for 30 minutes we observed no further change in the weight loss, even after heating to higher temperatures. This difference in weight between the sample dried at room temperature and at 220°C corresponds to a water loss because of the reversibility of the absorption/desorption behaviour. We indeed cooled the sample which has been dried at 220°C down to room temperature, rehydrated it at 100 % R.H., then dehydrated it under vacuum at room temperature. Now, if we heat this sample up to 220°C, we find excactly the same weight loss curve as found previously. Such behaviour means that we have desorbed water during the heating treatment. In table I are given the amounts of absorbed water corresponding to different humidity levels and the corresponding numbers of water molecules per SO_3H groups.

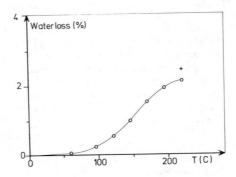

Figure 1.   Water loss on heating an acid Nafion sample (at 3°C/min) under 10⁻⁴
torr to 220°C (+, after 30 sec at 220°C)

TABLE I

Relative humidity	% of water	Number of water molecules per SO_3H
Room T. dry samples	2.5	1.7
50 %	7.8	5.2
90 %	13	8.7
soaked	20	13.3
boiled	30	20

Figure 2 shows the corresponding sorption isotherms. The reference is the sample which had been vacuum dried at 220°C. The absorption isotherms were taken at 25°C. We observe some hysteresis between the first sorption and desorption curves and we cannot come back to zero by vacuum drying, as was previously pointed out. The subsequent sorption and desorption curves are then quite identical. In figure 3, we have plotted the average heat of absorption versus the water content. The empty circles correspond to the samples dried at room temperature which contain around 2.3 % of water. By changing the humidity level we have absorbed water up to 4.5 % with an average energy of absorption of around − 12 kcal/mole. We obtain a plateau up to 7 % water, which corresponds to about 5 water molecules per SO_3H group. For higher amounts of water we then observe a continuous decrease of the exothermic value, down to − 4 kcal/mole. If we now start from the sample which has been dried at 220°C, we find that the heat of absorption of the first water molecules is exactly the same as we have found for the samples dried at room temperature. This result is important because it means that the water molecules which have not been desorbed at room temperature do not have a larger binding energy. We always observe a plateau, but the decrease in absorption energy occurs at a lower water content, in all other respects the shape of the curve is identical.

There is a relatively large contrast between the CF_2 and H_2O groups both for X-rays and for neutron scattering. The X-ray contrast is due to differences in electron density and the neutron contrast is due to differences in the coherent scattering length. Table II summarizes these different values. It is therefore very interesting to study water clustering using these two techniques. Preliminary experiments have been done and are reported in the literature (6). Figure 4 shows the neutron scattering curves corresponding to different amounts of absorbed water, while Table I gives the percentages of water and the relative numbers of water molecules per SO_3H for different humidity levels. In figure 4 we note a maximum corresponding to a Bragg spacing of the order of 180 Å. When the amount of water increases, we observe a change both in the position of this maximum and in the shape of the curve. To get more information about the phase separation phenomenon suggested by the shape of these curves we soaked some sam-

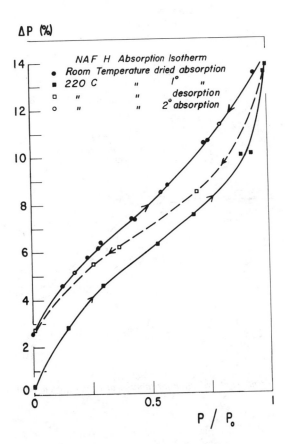

Figure 2. Sorption–desorption curves obtained after drying an acid Nafion sample at different temperatures vs. the humidity (P/P₀)

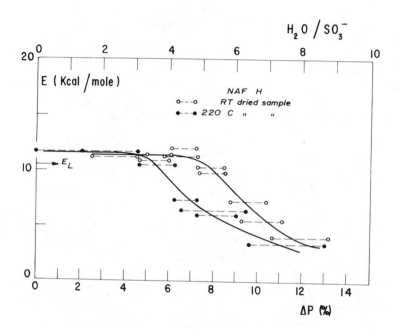

Figure 3.   *Energy of hydration for samples dried at different temperatures*

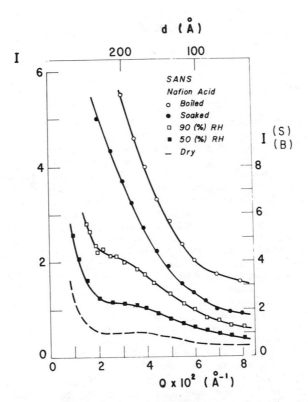

Figure 4.   Small-angle neutron scattering (SANS) curves for different amounts of absorbed water

TABLE II

Chemical unit	Diffusion length $b \times 10^{12}$ (cm)	Number of electrons
H^1	− 0.37	1
D^2	0.67	1
C^{12}	0.67	6
O^{16}	0.58	8
F^{19}	0.55	9
CF_2	1.68	24
H_2O	− 0.168	10
D_2O	1.95	10

ples with different H_2O/D_2O ratios. In Figure 5 we present the different scattering curves which have been obtained. If only two phases were present, then each curve would be identical with every other one, except for a multiplication of the intensity by a cons-tant factor. Since this is not the case, our results show that we have at least three different phases in our sample.

Further experiments have been performed at larger angles with the D 17 instrument. At these larger angles (Figure 6) a peak appears, its position and amplitude depending on the amount of absorbed water. This peak has been studied more extensively by small angle X-ray scattering (Figure 7). The upswing in scattering which is observed for 2θ values < 1° corresponds to the increase in scattering observed with neutrons at q values < 6.10^{-2} Å^{-1}. For large amounts of water we note the appearance of a peak which may correspond to an interference peak because of a large volume fraction of clusters. More information will be obtained about this behaviour by working on Nafions neutralized with Na ; the results of that study will be reported in the next paper.

Nuclear magnetic resonance is a powerful technique which can yield information on dynamic phenomena. Because of the lack of protons in the Nafions it is easier to study the water protons. A single line has always been observed for water contents between 2.7 and 20 %. Such behaviour can be explained in two ways : either we have only one kind of water molecule with a well defined envi-ronment or we have two or more different kinds of water molecules with an exchange rate larger than 10^{-3} to 10^{-4} sec. In Figure 8 are plotted the different room temperature chemical shifts and line widths corresponding to different water contents. We used a mixture of 95 % D_2O and 5 % H_2O as the reference for the chemical shifts. In Figure 9 are plotted the changes in line width versus temperature for the different samples and in Figure 10, we report the variations of the spin lattice relaxation time (T_1) versus the temperature. From these results we can make the following observa-tions. First of all there are two different regimes of water absorption. The first regime corresponds to the low water contents

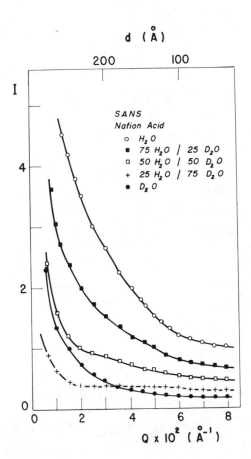

Figure 5. Small-angle neutron scattering (SANS) curves for an acid Nafion sample soaked in water with different H_2O/D_2O concentrations

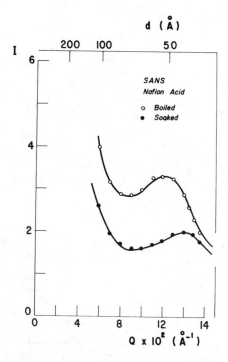

Figure 6. Small-angle neutron scattering curves at larger q values (D17) for high
water contents

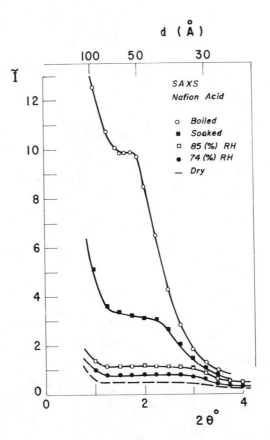

Figure 7. Small-angle X-ray scattering (SAXS) curves for different amounts of absored water

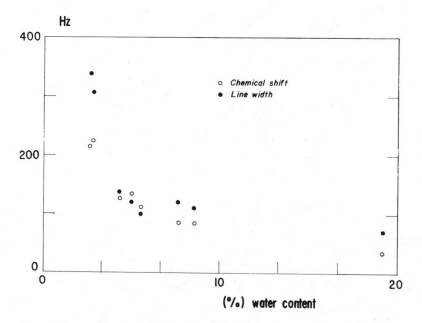

Figure 8.   Room temperature chemical shift and line width changes vs. water con-centration. The reference for the chemical shifts measured at 60 MHz is a mixture of 95% D_2O and 5% H_2O.

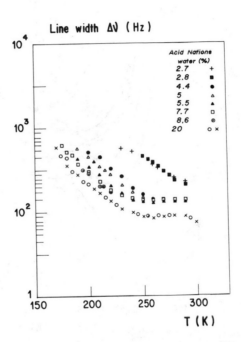

Figure 9.   Temperature dependence of the line width for samples with different amounts of water (NMR, 60 MHz)

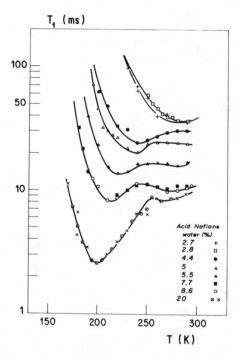

Figure 10. Proton spin lattice relaxation time dependence vs. temperature for different water contents (NMR, 60 MHz)

up to 4.4 %, and is characterized by significant decrease in both the chemical shift and the line width.

Only one kind of motion is detected with the samples, corresponding to this first regime of absorption (2.7 and 2.8 %). A continuous decrease of the line width is indeed observed when these samples are heated from low temperature up to room temperature, and a single minimum for T_1 appears around room temperature.

The second regime of water absorption extends from 4.4 to 20 % and no important change occurs either for the chemical shift or the line width. For all the samples corresponding to this second regime, we observe a similar behaviour of the line width versus temperature. When the samples are heated, the line width decreases up to 250 K and then remains constant. Two minima appear in the T_1 curves versus the temperature. Such behaviour suggests the presence of a more complex motion than is found in the first regime. At low temperatures we presumably observe a local diffusive motion. The characteristic time at the temperature corresponding to this low temperature T_1 minimum can be estimated to be $\sim 1.6 \times 10^{-9}$ sec. The activation energy associated with this minimum is 3-4 kcal/mole. When the amount of water decreases, the low temperature minimum moves to higher temperatures. The corresponding motion is therefore more and more difficult. The change in behaviour above 250 K for the line width probably corresponds to the onset of another kind of motion, presumably long range self-diffusion. The characteristic time for this second motion is always $1.6 \ 10^{-9}$ sec at the temperature corresponding to the high temperature T_1 minimum.

The incoherent neutron quasi-elastic scattering technique gives information about the motions of the hydrogen atoms on a shorter time scale than NMR. In figure 11 we observe a single peak at zero energy transfer and with a width corresponding to the resolution of the instrument ($\sim 18\mu eV$ in the present case for IN5). This spectrum has been obtained with a nafion sample dehydrated at high temperature. The spectrum of a soaked sample is different (Figure 12). We note here the presence of a broad line superimposed on the elastic line of Figure 11. Such a broadening proves the existence of a diffusive random motion of the water protons with a characteristic time $<10^{-10}$ sec. The line width dependence versus q is not proportional to q^2. Therefore the motion involved in this broadening is not only a long range self diffusion. In fact the broad line is not a simple Lorentzian line but seems to be the superposition of two or more Lorentzian lines. The broadest component has a width of about 100 µeV corresponding to a $\tau \sim 10^{-11}$ sec. This line can be associated with the low temperature motion detected by NMR spin lattice relaxation measurements. The value for this low temperature NMR motion extrapolated to room temperature is $\tau \sim 2.4 \ 10^{-11}$ sec. This value is consistent with the characteristic time obtained from neutron experiments. The high temperature motion detected by NMR would correspond to a neutron quasi-elastic linewidth of the order of magnitude of the instrumental resolution of the IN5 device. This second motion may therefore explain the presence of this complex line. More

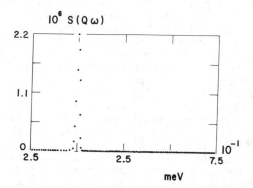

Figure 11.   Incoherent quasi-elastic neutron scattering spectrum obtained with a
220°C dehydrated acid Nafion sample (q = 0.59 Å⁻¹)

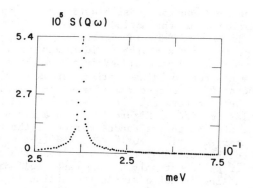

Figure 12.   Incoherent quasi-elastic neutron scattering spectrum obtained from a
water-soaked acid Nafion sample (q = 0.59 Å⁻¹)

experiments have already been performed and a more detailed analysis is in progress and will be published (7).

Conclusion

Let us summarize here the different results me have obtained. First of all, by vacuum drying at room temperature for 24 hours under 10^{-4} mmHg we do not desorb all the water. This undesorbed water has been found to have the same binding energy as the other water molecules. Such behaviour suggests that these undesorbed water molecules are trapped in the nafion and have to overcome a large energy barrier in order to diffuse out of the polymer. The water retension is thus a kinetic phenomenon. The drop in the exothermic value of the water absorption for larger amounts of water is probably the result of matrix swelling. The water-ion and the water-water interactions permit the elastic deformation of the polymer i.e. the motion of the hydrophobic chains out of the hydrated zone. This corresponds to an endothermic contribution which produces a drecrease in the total exothermic effect. By vacuum drying of the sample at 220°C we produce a change in the structure of the polymer and the decrease in the energy of absorption occurs for lower water contents because of a larger endothermic effect.

The main result from the small angle neutron and X-ray experiments is the demonstration of the presence of more than two phases in this sample. There are three regions of interest : an increase of the scattering curve at very low q values, a peak which appears in a q range corresponding to a few hundreds of Angströms in terms of Bragg distance, and another peak at larger angle which is seen both in X-ray and neutron experiments. The position and amplitude of these peaks depend on the amount of absorbed water.

Interesting results have been obtained by a combination of NMR and quasi-elastic incoherent neutron scattering. The presence of one single line in an NMR spectrum, for all the water concentrations, can be interpreted in two ways : either we have only one kind of water molecule with a very well defined environment or we have different kinds subject to a fast chemical exchange ($\tau < 10^{-3}$ sec.). Two regimes of absorption have been demonstrated and two different motions have been characterized both by NMR spin lattice relation time measurements and quasi-elastic incoherent neutron scattering. From these results and from results obtained on the Nafion salts (8) a structural model will be proposed (9).

Literature Cited

1. Yéo, S.C. ; Eisenberg, A. Journal Appl. Polym. Sci., 1977, 21; 875.
2. Gierke, T.D. Electrochemical Society Fall Meeting. Atlanta, Georgia. Oct. 1977.
3. Hopfinger, A.J. ; Mauritz, K.A. ; Hora, C.J. Electrochemical Society Fall Meeting. Atlanta, Georgia, Oct. 1977.

4. Komoroski, R. ; Mauritz, K.A. Submitted for publication.
5. Soulié, J.P. ; Escoubez, M. ; Douillard, A. ; Chabert, B.
   J. Chim. Phys., 1976, 4, 423-429.
6. Pineri, M. ; Duplessix, R. ; Gauthier, S. ; Eisenberg, A.
   Advances in Chemistry, to be published.
7. Volino, F. ; Dianoux, A.J. ; Pineri, M., to be published in
   J. Polym. Sci.
8. Rodmacq, B. ; Coey, J.M. ; Escoubez, M. ; Roche, E. ; Duplessix,
   R. ; Eisenberg, A. ; Pineri, M., following paper.
9. Pineri, M., et al, to be published in J. Polym. Sci.

RECEIVED January 4, 1980.

Water Absorption in Neutralized Nafion Membranes

B. RODMACQ, J. M. COEY[1], M. ESCOUBES[2], E. ROCHE, R. DUPLESSIX[3], A. EISENBERG[4], and M. PINERI

Centre d'Études Nucléaires de Grenoble, 38041 Grenoble Cédex, France

In the previous paper (1) we have studied the water-polymer interactions in acid Nafions $^-+$. In this paper we want to report some results we have obtained on the interactions between water and Nafion neutralized with different cations. The energy of water absorption has been measured over the whole range of relative humidities using the same technique as described previously (1). Mössbauer spectra were obtained in order to get information about the change of environment of the iron atoms during hydration. In addition, small angle neutron and X-ray scattering experiments have been performed to define a possible phase segregation. From these results, combined with those obtained in the preceeding (1) paper we propose a model of clustering in the Nafion membranes.

The neutralized Nafion samples have been obtained by soaking the acid samples in solutions containing iron chloride or sodium hydroxide.

Heat of Absorption Measurements

This experiment has been described in the previous paper (1). Figure 1 shows the water loss versus temperature for a heating rate of 3°C/min. The samples (.4mm thick) have first been dried for 24 hours under vacuum (10^{-4} torr) at room temperature. The weight loss obtained after such a heating procedure must correspond to a water loss because the behaviour observed after rehydration is completely reversible. It has to be noted that the amount

Current addresses:
[1] Trinity College, Dublin 2, Ireland.
[2] Université Claude Bernard, 69621 Villeurbanne, France.
[3] Institut Laue Langevin, BP 156, 38042 Grenoble Cedex, France.
[4] McGill University, Montreal, Canada.

0-8412-0559-0/80/47-127-487$05.00/0
© 1980 American Chemical Society

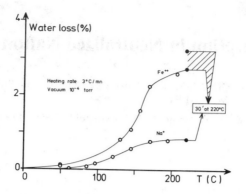

Figure 1. *Water loss corresponding to Na- and Fe-neutralized Nafion samples during a heating run*

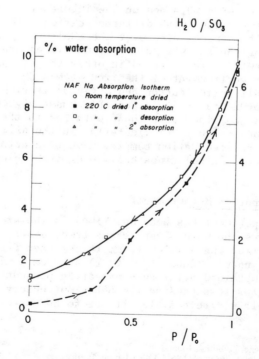

Figure 2. *Room temperature water absorption isotherms for the Na salt*

of undesorbed water at room temperature is three times larger for the iron salt than for the sodium salt and the equilibrium value is obtained after a 30 minutes annealing at 220°C for the iron salt. Figures 2 and 3 show the sorption-desorption isotherms for the Na⁺ and Fe⁺⁺ salts. The behaviour is very similar. For both samples we observe a different absorption curve for the sample dried at 220°C. For relative humidity values, only a few water molecules are absorbed. Then, beyond a relative pressure value of 0.25, there is a drastic increase in the amount of water absorbed. The corresponding desorption curves are then exactly the same as the sorption and desorption curves of the room temperature dried sample. In Figure 4 are plotted the average energy of absorption for the different water molecules absorbed in the iron sample. The empty circles correspond to the values obtained when starting with the sample dried at room temperature which contains about 3 % of water. The average energy for these first molecules is 13 Kcal/mole, it then decreases after a water content of about 8 %. The filled circles correspond to the sample which has been dried at 220°C. As was previously observed for the acid sample (1) the energy of the first absorbed water molecules is also 13 kcal. Therefore, as before, the water molecules which had not been desorbed at room temperature do not have a larger binding energy. For this sample the decrease in energy occurs for lower amounts of water (∿ 5 %) because more energy is needed to change the structure of the polymer. Corresponding curves are plotted in Figure 5 for the Na salts.

Mössbauer Spectroscopy

Experimental Methods. The acid form of the membrane with an equivalent weight (i.e. the weight of polymer per SO_3H group) of 1200 was neutralized to about 50 % by immersion of a thin foil (≈300 μm) in an aqueous solution of 57-ferric chloride. The samples were dried in vacuum at room temperature and then hydrated at different humidity levels. For a given humidity level, the quantity of absorbed water is largely dependent on the state of the polymer i.e. whether it is an acid or a salt. As the samples we have studied are neutralized to 50 %, it is not yet possible, from the weight increase, to estimate the amount of water molecules fixed by the neutralized groups which are the only ones observable by Mössbauer spectroscopy.

Mössbauer spectra have been recorded in a conventional transmission geometry in the constant acceleration mode. The temperature of the sample could be varied from 4.2 K up to room temperature by means of a liquid helium cryostat, the source (^{57}Co in Rh) being kept at room temperature. The Mössbauer experiments were carried out in the "Laboratoire d'Interactions Hyperfines" at the CEN-Grenoble.

Results. Figure 6 shows Mössbauer spectra of a sample dried in vacuum at room temperature. The main features are the following :

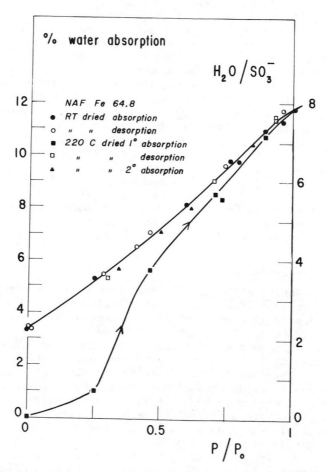

Figure 3. Room temperature water absorption isotherms for the Fe salt

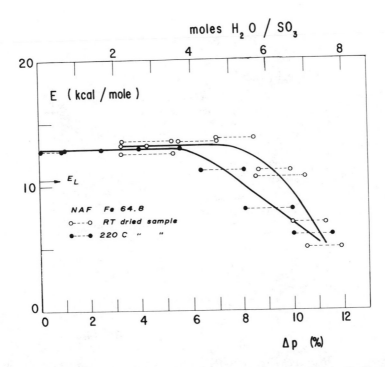

Figure 4. Heat of absorption vs. the amount of water absorbed for the Fe salt

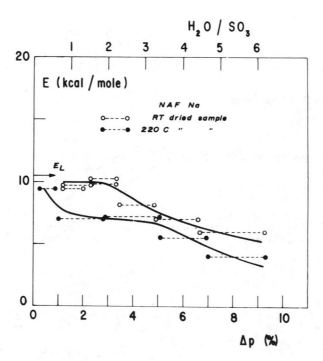

Figure 5.   Heat of absorption vs. the amount of water absorbed for the Na salt

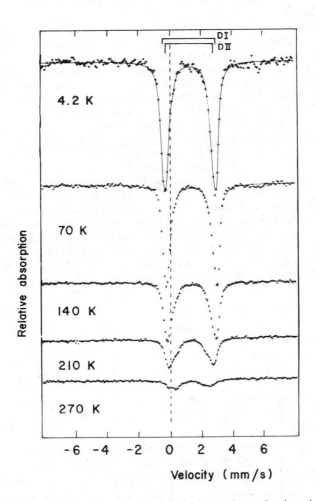

Figure 6. Mössbauer spectra of the Fe salt: (+) experimental points; (—) theoretical fit with two doublets.

- f - factor (i.e. the area of the absorption spectrum) decreases rapidly with increasing temperature and reaches zero at about 300 K.
- there is no evidence of a magnetic hyperfine structure at low temperature.
- the mean quadrupolar splitting is of the order of 3.2 mm/s, which is characteristic of ferrous ion, although the Nafion membranes have been neutralized with ferric chloride. The mean quadrupolar splitting decreases when the temperature increases.
- the shape of the peaks is asymmetric, and this asymmetry increases with increasing temperature. Moreover, the relative intensities of the peaks vary with increasing temperature.

The changes of the spectrum as a function of temperature showed us that it would not be possible to fit the experimental data by only one doublet. These doublets will be called DI and DII in the rest of this paper. Table I gives the values of the hyperfine parameters corresponding to the spectra of figure 6, δ is the chemical shift, Δ is the quadrupolar splitting, Γ is the line width and f are the recoil free fractions. We have performed other Mössbauer experiments on samples with different amounts of water. We have fitted the experimental spectra with the same components DI and DII. Such a decomposition gives a continuous change in the relative proportion of the iron atoms corresponding to the doublets DI and DII.

TABLE I

T_K		δ mm/s	Δ mm/s	Γ mm/s	f a.u	f_{tot} a.u
4.2	D I	1.34	3.46	0.33	0.32	0.97
	D II	1.35	3.03	0.47	0.65	
70	D I	1.35	3.40	0.33	0.32	0.72
	D II	1.35	2.98	0.48	0.40	
140	D I	1.31	3.25	0.39	0.24	0.47
	D II	1.30	2.73	0.53	0.23	
210	D I	1.26	2.92	0.49	0.18	0.26
	D II	1.20	2.15	0.52	0.08	
270	D I	1.16	2.70	0.67	0.06	0.09
	D II	1.11	1.78	0.60	0.03	

Discussion. The fit of the experimental spectra by means of two quadrupolar doublets corresponds to the existence of two different iron sites in water containing Nafion membranes. The proportion of the iron atoms connected with the doublet DI increases when the water content increases. In samples dried under vacuum at room temperature this proportion is 30 % and it increases up to 90 % for a water soaked sample. It seems therefore reasonable to identify this site with a $Fe(H_2O)_n^{++}$ complex. Indeed, we know both from NMR and heat of absorption measurements that there are some water molecules left in the sample dried at room temperature (1). We can also note, that the value of the quadrupolar splitting is not very different from that of frozen solutions of $FeCl_2$ and $FeSO_4$ as measured by Nozik and Kaplan (2). These authors showed that the dissolution of these salts in water led to the formation of $Fe(H_2O)_6^{++}$ complexes. The other doublet DII would represent a less hydrated ferrous iron, with the relative proportion of this species decreasing when the amount of water is increased.

Conclusions. From these results we can conclude the following :
- water is located close to the iron ions.
- the water molecules are not randomly distributed around the iron ions but form complexes with the cations with well defined structures.
- there are two different kinds of iron atoms with different environments, one with a high water content (DI) and the other with none or very few water molecules (DII).

A more detailed analysis of these results will be developped in another publication (3).

Small angle scattering results

The physical structure of Nafions salts has been explored by neutrons (SANS) and X rays (SAXS). The former method is sensitive to fluctuations in the coherent neutron scattering cross section while the latter detects fluctuations in the electron density. Such fluctuations arise in the Nafions from partial crystallization of the samples and from clustering of the ionic groups or water molecules in hydrated samples.

As discussed in the accompanying paper, three different scattering signals arise in Nafions, each occuring over a different range of values of the scattering vector Q. These signals are believed to arise from impurities (this is a tentative assignment), structures involving crystalline perfluoroethylene units, and structures involving clustered ionic groups and water molecules. These same three scattering signals have been observed in all acid, iron salt, and sodium salt samples which had not been quenched. Thus a basic similarity exists between the acid and the salt form in the overall structure, showing that the neutralization does not lead to a large scale reorganization as is observed in other ionomers.

In the accompanying paper it has been noted that the scattering arising from crystalline regions complicated the interpretation of the scattering arising from the ionic structures since
the two signals overlap in Q space. To eliminate this difficulty
the method of quenching from the melt has been used for Na^+ salts.
Such a procedure is not possible for the acid form since the
material degrades at temperature high enough to melt the crystals.
The quenching procedure involves maintaining a sample at 330°C for
one hour in the melt followed by quenching rapidly to room temperature by passing a cold gas over the film. Wide angle X ray scattering studies show the disappearance of the crystalline maximum
in quenched samples. The SANS signal arising from crystalline
superstructures is also seen to disappear in a quenched sample as
shown in Figure 7, leaving only the scattering component at very
low q, which is believed to arise from impurities.

The study of scattering from quenched Na^+ samples has been
used to analyze the structure arising from clustering of ions and
water molecules. Figure 8 shows SAXS curves for such samples at
different degrees of hydration. The curves for the samples dried
at room temperature and hydrated at 50 % or at 83 % relative humidity show no scattering maxima, as is characteristic of scattering
from widely separated particles. SAXS curves from soaked and boiled Na^+ samples show scattering maxima. The maxima are attributed
to interparticle interference effects which arise at higher particle concentrations. This maximum has also been observed by SANS.

The size of the scattering entities can be directly obtained
for scattering curves without maxima by means of a Guinier plot
of $\ln I$vs.Q^2. Such an analysis may also be made for the curves
exhibiting maxima but is only highly approximate due to the importance of the interparticle interference function which is not taken into account in the Guinier analysis. Results for the radius
of gyration obtained from the slope of such plots are listed in
table II. It is seen that radius of gyration of the particules is
constant at about 8 Å up to 83 % R.H. but then increases to 15 and
20 Å for the soaked and boiled samples. An analysis of the SANS
curve for a boiled sample has been made on the basis of a hard
sphere model. In this model interparticle interference is taken
into account allowing a fit of the scattering maximum. A radius
of gyration of 24 Å has been found from such a fit.

TABLE II

Humidity level	Dry R.T.	50% R.H.	83% R.H.	Soaked	Boiled
$Rg(\mathring{A})$	8	6	8	15	24
$\bar{n} \times 10^{-19}$ (number/cm^3)	0.6	1.5	3.1	2.2	1.9

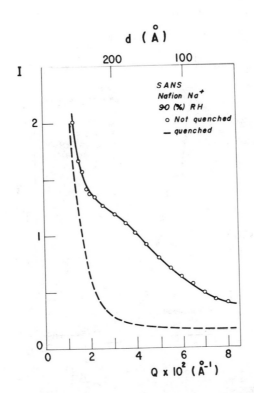

Figure 7. Small-angle neutron scattering curve of the quenched and unquenched Na salt. Measurements done at room temperature.

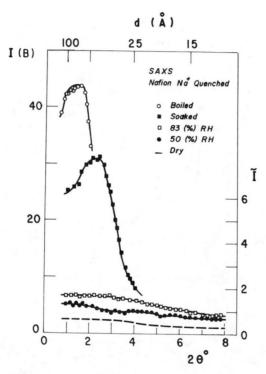

Figure 8. Small-angle X-ray scattering curve of the quenched Na salt with different water contents

The number of scattering particles per unit volume, $\bar{n}$, may also be calculated at different hydration levels from the equation:

$$\bar{n} = \frac{3 \, \emptyset \, (H_2O)}{4 \, \pi \, R_s^3} \tag{1}$$

where $R_s = (5/3)^{1/2} \, Rg$ and $\emptyset \, (H_2O)$ denotes the volume fraction of water which is determined from the weight fraction and overall density. It is assumed here that the scattering particles contain only water and that all of the water in the sample is clustered. The latter assumption has been verified within experimental error from an analysis of the total scattering invariant which has been calculated from the absolute intensity of scattering. The results for $\bar{n}$ listed in Table II show an apparent increase at low water contents and then a slight decrease at large water contents. It is noted that this decrease in $\bar{n}$ implying particle coalescence is in apparent contradiction to the hard sphere model used above.

A more detailed analysis of these results will be given in another publication (4).

Conclusion

In this conclusion we just want to propose a possible model for the structure of Nafion consistent with all the results we have obtained. It should first be pointed out that there is no large change in the macrostructure between the acid and salt forms of the Nafion membranes. Indeed, we observe exactly the same multiphase separation as seen from small angle X-ray and neutron scattering experiments. The changes in these curves with the amount of absorbed water are quite similar. We only note a small change in the ionic cluster sizes depending on the nature of the ion (H^+, Na^+ or Fe^{++}). The structure of the ionic phase is not changed by quenching the sample from 330°C. Such a procedure permits us to get rid of the crystalline phase. It is therefore possible to get information about the structure of these Nafion polymers from our experimental results obtained from the acid, as well as the quenched an unquenched salt forms.

The kind of model which can be proposed is summarized in Figure 9. It is a three-phase model with a crystalline phase, an ionic cluster phase and an intermediate "ionic" phase of lower ion content. For the sample dried at room temperature we have the microcrystallites, the diameters of which are a few hundreds of angströms. Such a result is in agreement with both electron microscope and X ray experiments. Dark field pictures obtained with these materials show the presence of small microcrystallites of the size mentionned above. By quenching the Na salt, the peak in the SANS curves corresponding to the microcrystallites disappears. An X ray pattern of such a sample shows that there is no crystalline phase left. The ionic clusters have a diameter of around 20 Å as measured by small angle X ray scattering experiments. Most of the water left in this sample is trapped inside these cluster. In this phase, we also have the iron atoms corresponding to the

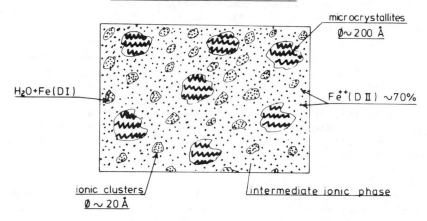

ROOM TEMPERATURE DRIED SAMPLE

microcrystallites
Ø ∼ 200 Å

H₂O+Fe(DI)

Fe⁺⁺(DⅡ) ∼70%

ionic clusters
Ø ∼ 20 Å

intermediate ionic phase

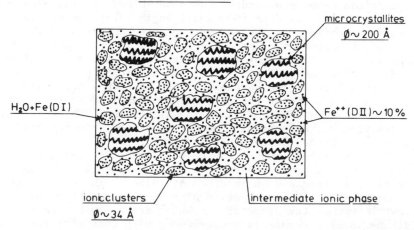

SOAKED SAMPLE

microcrystallites
Ø ∼ 200 Å

H₂O+Fe(DI)

Fe⁺⁺(DⅡ) ∼ 10%

ionic clusters
Ø ∼ 34 Å

intermediate ionic phase

Figure 9.   Model of the Nafion structure

doublet DI, the percentage of which is around 30 % as defined from Mössbauer measurements. The last phase is an intermediate phase which contains the iron atoms corresponding to the doublet DII (a possible structure of which may be $- SO_3- Fe^{++}- O_3S$). For the soaked sample we have an increase in both the average size and the volume fraction of the ionic clusters. The relative number of Fe^{++} ions with a water environment (DI) is increased up to 90 %.

The model presented above is in accord with all our experimental results. A more detailed analysis, taking into account the quantitative aspects of the water motion obtained from the quasi-elastic neutron scattering experiments, will be presented in the near future (5).

Literature Cited

1. Duplessix, R. ; Escoubes, M. ; Rodmacq, B. ; Volino, F. ; Roche, E. ; Eisenberg, A. ; Pineri, M.; This Conference, precceding paper.
2. Nozik, A.J. ; Kaplan, M. ; J. Chem. Phys, 1967, 47, 2960.
3. Rodmacq, B. ; Coey, J.M.D. ; Pineri, M., Revue de Physique Appliquée.
4. Roche, E. ; Duplessix, R. ; Levelut, A.M. ; Eisenberg, A. ; Pineri, M. ; Journal of Polymer Science. To be published.
5. Pineri, M. ; Dianoux, A.J. ; Volino, F. ; Journal of Polymer Science. To be published.

RECEIVED January 4, 1980.

SYNTHETIC POLYMERS:
WATER AND POLYMER PERFORMANCE

The Interactions of Water with Epoxy Resins

P. MOY and F. E. KARASZ

Polymer Science and Engineering, University of Massachusetts, Amherst, MA 01003

The structural properties of a synthetic polymer can often be modified by the service conditions to which it becomes exposed, an example being interaction with atmospheric moisture. Water induced plasticization is a common occurrence, yet depending upon the chemistry and morphology of the polymer encountered, the nature of the interaction process may take widely different routes. Some degree of specificity must be assumed in discussions relating the state of sorbed water in different polymeric systems.

The results presented in this paper are part of a continuing study dealing with the water-induced plasticization of a high performance epoxy resin. The resin, based on a tetrafunctional epoxy monomer cured with an aromatic diamine, is a highly cross-linked system with a high attainable T_g (>200°C). This system has utility as a matrix material for reinforced composites widely used in aerospace applications. Although the glass transition temperature of the dry resin is far above ambient temperatures, the moisture-depressed T_g of the material can fall within the use temperature of the resin. Instabilities in the mechanical properties brought about by the onset of the glass transition is sometimes sufficient to initiate failure processes in these materials.

In an earlier paper (1), the ambient sorption behavior of this resin was examined. These studies have now been extended to high temperatures. More relevant insight may be gained in this manner since these experiments will more closely simulate the actual conditions at which the plasticization process occurs. Knowledge of the equilibrium weight gain of the resin at various temperatures also permits a quantitative evaluation of the plasticization process in terms of theoretical expressions. The effect of sorbed water on the state of the hydrogen bonds in the epoxy system and its possible correlation to glass transition depression phenomena has also been examined in this study. Finally, it was found in the course of these studies that the observed discontinuity in the heat capacity of this network system at the glass transition is a very strong function of the degree of cross-linking. In fact, at high extents of cure, the transition becomes impercep-

0-8412-0559-0/80/47-127-505$05.00/0
© 1980 American Chemical Society

tible by scanning calorimetry. Further discussion will be given
to this finding and its relation to the observed plasticization
phenomenon.

Experimental

The principal resin investigated in these studies is formu-
lated from a 27 weight percent mixture of 4,4'-diaminodiphenyl
sulfone (DDS) and 73 weight percent tetraglycidyl-4,4'-diaminodi-
phenyl methane (TGDDM). Curing was effected thermally at 177°C
for 24 hours. Samples used in sorption measurements were molded
films, 10 to 15 mils thick.

The isothermal sorption behavior of the TGDDM-DDS system at
elevated temperatures was studied using a quartz spring micro-
balance (2). Samples suspended from calibrated springs of sensitivity 3 mm/mg were sealed in evacuated glass tubes together with
bulbs containing water. Measurements were initiated following the
release of the water by successive freeze-thaw cycles. The temperature of the measurement was regulated by a furnace enclosing
the spring assembly. The pressure within the system was varied by
changing the temperature of a heated bath into which the lower
portion of the sorption cell was immersed. Changes in the weight
of the sample were determined by monitoring the extension of the
spring assembly with a cathetometer.

The depressed glass transition temperatures of the resin at
various concentrations of sorbed water were determined using scanning calorimetry. Samples of the resin were first conditioned
with an excess of water in large volume stainless steel DSC pans
at predetermined temperatures then scanned at 20 deg/min. Heat
capacity measurements were also made with a Perkin-Elmer DSC-2
scanning calorimeter.

The infrared analysis was performed using samples cured between silver chloride plates. Conditioning of the resin was
achieved by immersion in water. High temperature scans were made
using a temperature regulated sample cell. Spectra were obtained
with a Perkin-Elmer 283 spectrophotometer.

Results and Discussion

Isothermal sorption plots of the TGDDM-DDS system are shown
in Figure 1. From these results two distinct regimes of sorption
behavior may be discerned. While the lower temperature plots are
sigmoidal curves of the type II isotherms in the BET classification (3), the high temperature isotherms approach a linear dependence at the higher partial pressures. The former is often identified with a multilayer sorption process in which initial sorbed
water molecules are able to interact with binding sites on the
polymer while subsequent molecules associate with the primary
layer in liquid water-like structures. The latter regime however,
approaches simple solution sorption. At elevated temperatures,

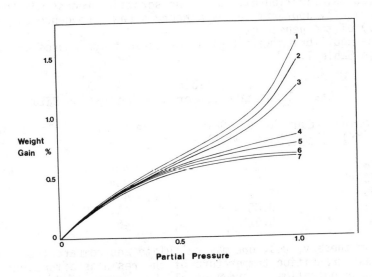

Figure 1.   Sorption isotherms of the TGDDM–DDS system: (1) 25°C; (2) 35°C; (3) 75°C; (4) 100°C; (5) 125°C; (6) 150°C; (7) 175°C

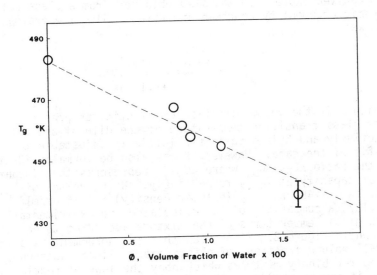

Figure 2.   Glass temperature depression data: (---) calculated line as predicted by Equation i

the average thermal energy of the system exceeds that due to any localized polarization effects. The sorption process may then be asymptotically approximated by a Henry's Law dependence on the activity of the penetrant.

The equilibrium weight gains at these temperatures are compiled in Table I.

Table I.
Glass Transition Temperature Depression Data

Conditioning Temp. (°K)	Equilib. Wt. Gain of Water (wt %)	T_g Experimental (°K)	T_g Calculated (°K)
348	1.27	438	443
373	0.85	454	455
398	0.77	457	460
423	0.68	461	461
448	0.67	467	462

From these values, one may calculate and compare the effective glass transition temperature of the resin as a function of diluent concentration. Gordon et al. have recently derived an expression relating the glass temperature of a polymer-plasticizer mixture to the glass temperatures of the components on the basis of the configurational entropy theory of glass formation (4). Their derived expression was also obtained from a classical thermodynamic argument by Couchman and Karasz (5). The expression was given as:

$$T_g(\phi) = \frac{\phi T_{g1} + (1-\phi)T_{g2}K}{\phi + (1-\phi)K} \tag{i}$$

in which ϕ is the volume fraction of plasticizer present, $T_{g1,2}$ is the glass transition temperature of the diluent and polymer respectively and K is a parameter initially adjustable to yield a best fit of the data. However, K may also be formally identified with the ratio $\Delta C_{p2}/\Delta C_{p1}$ where $\Delta C_{p1,2}$ represents the discontinuity in heat capacity at $T_{g1,2}$ respectively. Using values of T_{g1} = 134°K, T_{g2} = 483°K, ρ_2 (polymer density) = 1.26 gm/ml and K = 0.127, a comparison of the calculated and experimentally obtained glass temperatures of the mixture according to equation (i) is shown in Figure 2. It is seen that the agreement is quite good at small volume fractions of plasticizer. Since equation (i) was derived for binary mixtures which obey the laws of regular solutions, it may be assumed that under these experimental conditions, the epoxy-water system may also be treated as such. The implications of the comparatively low value of K are discussed below.

It has been shown that the majority of the polar moities in epoxy-diamine networks, such as hydroxyl groups generated by the cross-linking reaction and residual amino hydrogens, take part in inter- or intramolecular hydrogen bonds (6,7). Indeed, at ambient temperatures, the infrared hydroxyl stretch frequency of epoxy-diamine systems are characteristically shifted from its free value of 3600 cm^{-1} to a hydrogen bonded value of 3440 cm^{-1} (8). Williams and Delatycki have shown that for a homologue of the epoxy-diamine systems studied, the internal hydrogen bonds are only disrupted when the polymers are heated to above their glass temperatures (7). Conversely, one may expect that the state of the hydrogen bonds will have an effect on the relaxational behavior of the network. In Figure 3, the wave number of the hydroxyl stretch frequency of the TGDDM-DDS system is plotted as a function of temperature. It can be seen that with increasing temperature, the band displays a transition-like process in approaching its unassociated or free value. Noteworthy also is the fact that the transition process is initiated at correspondingly lower temperatures for the wet resin than the dry. The implications of this finding may be more fully explored on a molecular basis. As water molecules are introduced into the network, it is not unreasonable to assume that some degree of exchange will take place between the hydrogen bonded groups and the water molecules. Since these interactions are largely dipolar in nature, the hydrogen bonds formed between polymer segments and those between polymer segments and water molecules will be energetically similar. It is then expected that the degree of interchange will be dependent solely on the activity of the water. However, as the temperature of the system is raised, the relative difference in mobility between polymer segments and the sorbed water molecules become significant. The hydrogen bonds containing water bridges will be weaker in this regard and will be expected to dissociate at lower temperatures, due to the enhanced thermal mobility of the water, than those formed between physically constrained polymer segments. The onset of segmental flexing will introduce fluctuations in the free volume of the system. Such segmental mobility and changes in free volume will all have the effect of initiating the glass transition in the system. This is one mechanism by which plasticization in epoxy resins may proceed.

It was found that the heat capacity discontinuity at the glass transition of the TGDDM-DDS system is strongly dependent on the extent of cure. At high degrees of cross-linking, the glass transition (which concomitantly increases with increased cross-linking) becomes undetectable by scanning calorimetry as ΔC_p approaches zero. To investigate this point further a slightly modified resin system was employed. This overcame the problems associated with characterizing an incompletely cured system and the possibilities of further reaction when the resin is heated above its glass temperature. By reacting TGDDM with a varying stoichiometric mixture of aniline and p-phenylene diamine, a

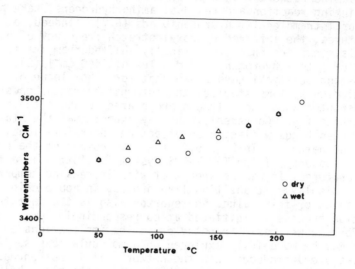

Figure 3.   Spectroscopic transition of the TGDDM–DDS system

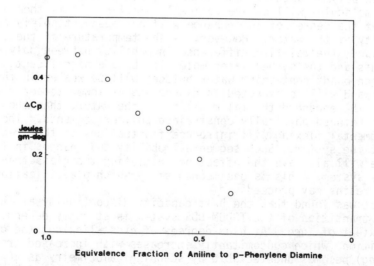

Figure 4.   Variation in ΔC_p at the glass transition for resins of TGDDM cured with
different equivalence fractions of aniline and p-phenylenediamine

series of completely cured compositions was obtained with values of ΔC_p dependent on the average cross-link density of the resin. ΔC_p as a function of the resin composition is thus shown in Figure 4 and results are also tabulated in Table II.

Table II.
Heat Capacity Data and Transition Temperatures
for Different Resin Compounds

Equivalence Fraction of Aniline to p-Phenylene Diamine in Stoichiometric Cure of TGDDM	T_g (°K)	$C_p(\frac{Joules}{gm\text{-}deg})$
1.0	389	0.472
0.9	376	0.485
0.8	408	0.414
0.7	438	0.326
0.6	445	0.284
0.5	450	0.201
0.4	483	0.109

It is, of course, commonly understood that the manifestation of the glass transition decreases in intensity as the cross-link density increases, but its effect does not seem to have been previously studied quantitatively. If the formalism of Wunderlich's rule of constant heat capacity increment per "bead" is used (9), the average "bead" size must increase with the average cross-link density in the resin so that at very high degrees of cross-linking the "bead" contributions of 11.3 joules per mole beads per degree becomes vanishingly small on a unit mass basis. However, the generality of this semi-empirical relationship is in part derived from its arbitrariness in assignment of the "bead" unit. A more rigorous treatment of the excess specific heat at the glass transition has recently been given by DiMarzio and Dowell (10). By incorporating lattice vibration arguments into the Gibbs-DiMarzio configurational entropy theory of glasses, an expression was derived for the specific heat discontinuity. The total ΔC_p is considered to arise from three contributions. There is a configurational term arising from shape changes of the molecules at the glass transition, a configurational term from the volume expansion and a vibrational contribution from the change of force constants with temperature. The value of K in equation (i) found above is thus understandable in these terms. The particular resin used was highly (though not completely) cross-linked and would correspondingly be expected to have small ΔC_p values. The size of the heat capacity discontinuity for water is not known exactly, but its estimated value of 35 to 37 joules per mole degree lies in the "normal" range of low molecular weight materials (11,12). The implications of K<<1 are important, for from equation (i) it is easily shown that the magnitude of the depression in T_g by plasticization is approximately proportional to K^{-1}. This fact probably accounts for the unusual effectiveness of water as a plasticizer for highly cross-linked epoxy systems often observed. Further quantitative evaluation of this heat capacity data is in progress and will be reported in a subsequent paper.

Conclusion

In a previous paper, it was shown that the diffusion process for water in the TGDDM-DDS epoxy-diamine system was non-Fickian at near ambient temperature (1). Furthermore, evidence from broadline NMR results was presented which showed that water is probably sorbed in the system in a multi-layer process. It was found that at low concentrations, the sorbed water exhibited greatly restricted mobility so that a distinct liquid phase was undetectable. This was in turn attributed to the localized binding of water molecules by adjacent polar groups in the epoxy network. Results presented in this paper may be used to further develop our understanding of the epoxy-water interaction process. While the sorption behavior at low temperatures deviate from Fickian diffusion, indication is given that ideal behavior is approached with increasing temperature. This trend is of course consistent with the expectations that the probability of any site interaction will be diminished by increasing the average energy of the system. Infrared analysis has indicated that a factor in the plasticization process may be the disruption of existing hydrogen bonds in the network. Furthermore, the facility with which water substituted hydrogen bonds are broken relative to those between polymer segments is important to the discussion of the state of sorbed water in macromolecular systems. It has been shown that strong perturbations are often observed for the thermodynamic properties of the sorbed water in polymeric systems containing polar groups, i.e., those from heat of fusion measurements (13, 14). The notion of bound water is one that is commonly invoked to account for these findings. However, the calorimetric data of Hoeve and Kakivaya as well as others show that the thermal stability of water bound to a polymer is not larger than that of bulk liquid water (15,16). This reinforces the concept that dipolar interactions are not significantly different for monomeric or polymeric species. Instead, one must consider the limiting spatial fluctuations that physically constrained polymer segments are able to undergo relative to those between two small molecules. At low temperatures, all polymer segments are effectively frozen and interacting small molecules will be expected to display a high degree of immobility. As the temperature of the system is raised, the greater thermal mobility of the sorbed water molecules will be reflected by an earlier onset of the dissociation of water containing hydrogen bonded segments. These expectations are in accordance with the NMR and infrared results.

Abstract

The high temperature sorption behavior of an epoxy-diamine system has been investigated. Correlation of the resulting data is made with a theoretical expression relating the depressed glass transition temperature of the system to the concentration of sorbed water. The effect of moisture on the hydrogen bonding in the network has also been examined. Discussion is given to the mechanisms through which plasticization may occur. It was found that the heat capacity discontinuity at the glass transition for this resin is a very strong function of the cross-link density. At high extents of cure, the ΔC_p per unit mass may become vanishingly small and undetectable by scanning calorimetry. This finding largely accounts for the unusually effective plasticization of cross-linked epoxies by water.

Acknowledgement

We are grateful for the support of this research by Grant AFOSR 76-2198.

Literature Cited

1. Moy, P.; Karasz, F.E., in press.
2. McBain, J.W.; Baker, A.M. J. Amer. Chem. Soc., 1926, 48, 690.
3. Brunauer, S. "The Absorption of Gases and Vapors"; Oxford University Press: London, 1945, p. 150.
4. Gordon, J.M.; Rouse, G.B.; Gibbs, J.H.; Risen, W.M. J. Chem. Phys., 1977, 66, 4971.
5. Couchman, P.R.; Karasz, F.E. Macromol., 1978, 11, 117.
6. Vladimirov, L.V.; Zelenetskii, A.M.; Oleinik, E.F. Vysokomol. Soed., 1977, A19(9), 2104.
7. Williams, J.G.; Delatycki, O. J. Polym. Sci., 1970, A-2(8), 295.
8. Silverstein, R.M.; Bassler, G.C.; Morrill, T.C. "Spectrometric Identification of Organic Compounds", John Wiley and Sons, Inc.: New York, 1974, p. 91.
9. Wunderlich, B. J. Phys. Chem., 1960, 64, 1052.
10. DiMarzio, E.A.; Dowell, F., in press.
11. Sugisaki, M.; Suga, H.; Seki, S. Bull. Chem. Soc. Jpn., 1968, 41, 2591.
12. Johari, G.P. Philosophical Mag., 1977, 35, 1077.
13. Lee, H.B.; Jhon, M.S.; Andrade, J.D. J. Colloid. and Interface Sci., 1975, 51(2), 225.
14. Illinger, J.; Karasz, F.E., in press.
15. Hoeve, C.A.J.; Kakivaya, S.R. J. Phys. Chem., 1976, 80, 745.
16. Pouchly, J.; Biroŝ, J.; Beneŝ, S. Makromol. Chem., 1979, 180, 745.

RECEIVED January 4, 1980.

Glass Transition Temperature of Wet Fibers

Its Measurement and Significance

JOHN F. FUZEK

Tennessee Eastman Co., Kingsport, TN 37662

When a polymer is heated to a temperature above its melting point (T_m), the crystallites of the polymer melt, and an abrupt change in volume occurs. Since melting involves a discontinuity in a primary thermodynamic variable (volume), melting can be considered a first-order transition. At temperatures above its melting point, the polymer is a liquid, and the slope of the line AB shown in Figure 1 is the coefficient of thermal expansion for the melt. When a molten amorphous polymer cools to a temperature below its melting point, the polymer behaves as a rubber (BE) until the glass transition temperature (T_g) is reached. Below this temperature, the polymer behaves as a glass (EF). If a polymer crystallizes, the path BCD is followed. The crystallization is not sharp, however, and both solid and liquid phases are present between B and C; so the melting point must be defined as a break in the curve. For a truly crystalline solid, the path ABGCD is followed. For practical polymers, the crystallization is usually far from complete, and a transition region BE'F' is observed as a temperature range similar to that for the amorphous polymer (BEF). The evident interpretation of this phenomenon is that a glass transition has occurred in the amorphous portion of a semicrystalline polymer (1).

The glass transition is a second-order transition caused by relaxation of the chain segments in the amorphous portion of the polymer. It is at that temperature, T_g, that a noncrystalline polymer changes from a glassy solid to a rubbery liquid. In terms of structure, T_g is generally considered to represent the beginning of motion in the major segments of the backbone molecular chain of a polymer. The temperature at which this transition occurs is of great practical importance, since it determines the specific utility of a polymer.

The wide use made of the term glass transition temperature to characterize polymer glasses implies that there is a well-defined temperature that relates to a freezing point and at which a substance transforms discontinuously from liquid to rubber to glass. Actually, the value found for T_g depends not only on the nature of the substance but also on the method of determination,

0-8412-0559-0/80/47-127-515$05.00/0
© 1980 American Chemical Society

since the measured value is rate dependent (2). For a typical dilatometric test, the time involved (hours) is long enough to give a virtually static value; that is, increasing the time scale by an order of magnitude has little effect on the measured T_g value. Other methods, such as calorimetric and dynamic methods, are carried out at much faster rates and can give T_g values significantly different from those obtained in a dilatometric test.

Many factors are known to influence T_g values: pressure, crystallinity, molecular weight, molecular branching, crosslinking, copolymerization, the presence of monomer, and the presence of a low-molecular-weight liquid or plasticizer (3-11). Almost all fibers are composed of polymeric materials. However, the crystallinity and the molecular orientation of the polymeric material in a fiber differ from those of the bulk polymer from which the fiber is made. Hence the T_g of a fiber may be significantly different from that of a bulk polymer. An increase in crystallinity generally increases T_g by 5 to 15°C, and an increase in molecular orientation generally increases T_g by 3 to 12°C. However, these increase are not directly additive. For example, a highly crystalline and oriented polyester fiber made from poly(ethylene terephthalate), PET, will have a T_g about 15°C higher than that of the amorphous bulk polymer.

In contrast to bulk polymers, which generally contain little or no water, almost all fibers absorb some water when conditioned at normal atmospheric temperature and humidity. The presence of this moisture can result in a substantial lowering of the T_g. For example, poly(vinyl alcohol) fiber has a bone-dry T_g of 85°C, but the T_g of the fiber at its equilibrium moisture regain of 6% is about 21°C (12). Hence, 6% water reduces the T_g by 64°C. Very little work has been reported on the T_g of wet fibers (13-18).

Experimental

Samples of several fibers were obtained for this investigation. Poly(ethylene terephthalate) and poly(cyclohexylenedimethylene terephthalate) (PCHDMT) were commercial Kodel polyester fibers; the PET fibers were filament polyester, whereas the PCHDMT fibers were taken from tow before it was cut into staple. The poly(1,4-butylene terephthalate) (PBT) fibers were experimentally spun. The nylon 6,6, Qiana nylon, and Orlon acrylic were Du Pont commercial fibers. The nylon 6 was Allied Chemical commercial fiber. The Verel modacrylic and Estron acetate fibers were Eastman commerical materials. The Rhovyl vinyl was Rhodiaceta commercial fiber. The silk was a commercial, degummed yarn.

For the determination of wet T_g's, the fibers were allowed to soak in distilled water for 24 hr before testing. Tensile testing was carried out on an Instron tester with the fibers totally immersed in water at the specified temperature. Testing was carried out at 0, 10, 23, 30, 40, 50, 60, 70, 80, 90, and 97°C for the wet determination and for the conditioned determination at -40°C in addition to these temperatures. Samples were conditioned for 24 hr at 70°F $\pm$ 2°F and 65% RH $\pm$ 2% after the samples had been oven dried for 1 hr at 110°C.

For determination of wet T_g by specific volume, densities were determined pycnometrically with water as the displacing liquid. The fibers were allowed to remain in the water for 24 hr before weighing. Measurements were made at 10°C intervals from 20 to 90°C and at 97°C. For determination of dry T_g's, a low-viscosity mineral oil was used as the displacing liquid.

For determination of T_g by differential scanning calorimetry (DSC), samples were weighed in the sample container, and the temperature was scanned at a rate of 20°C/min. For wet determination, the fibers were wet out for 24 hr before testing.

For thermomechanical analysis (TMA), applicable only to dry fibers, a load of 0.005 g/den. was applied to the fiber and a scan rate of 20°C/min was used. An X-Y recorder was used to record fiber length as a function of temperature.

For determination of T_g from shrinkage measurements, fibers were allowed to shrink at the specified temperature, and length measurements were made manually.

Discussion of Results

Some of the methods that have been used to measure T_g of polymers include:
1. Dilatometry (specific volume).
2. Thermal analysis.
      Specific heat (DTA, DSC)
      Thermal expansion (TMA)
      Thermal conductivity
3. Dynamic methods.
      Mechanical
          Free oscillation
          Resonance
          Forced oscillation
          Wave propagation
      Dielectric
      Nuclear magnetic resonance
4. Stress relaxation and creep.
5. Tensile behavior.
      Elastic modulus
      Stiffness
      Stress decay
      Resilience
6. Others.
      Penetration (TMA)
      Brittle point
      Compressibility
      Refractive index

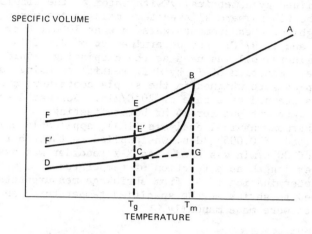

Contemporary Physics

Figure 1.   Melting and glass transitions in a polymer (1)

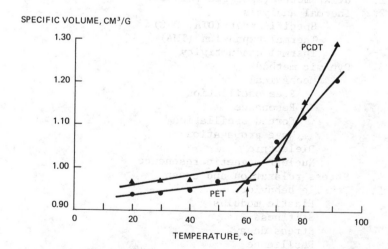

Figure 2.   Effect of temperature on the specific volume of polyesters

X-Ray diffraction
Small-molecule diffusion
β-Radiation absorption
Viscosity-activation energy

Many of these methods are not suitable for use with fibers because of their physical nature. For example, many of the dynamic methods involving loss modulus are not suitable for fiber T_g measurements because two of the fiber dimensions are very small. Monofilaments can be tested, but they are not representative of the majority of textile fibers in use today. Obviously, measurement of penetration, compressibility, small-molecule diffusion, and activation energy in viscosity methods is not possible for fibers. Of the methods that can be used for fibers, the most common involves specific volume. The use of specific volume for determining the T_g of two polyester fibers is shown in Figure 2. These data were obtained pycnometrically with water as the displacing liquid. The fibers were steeped in water for 24 hr before testing to obtain wet T_g's for these fibers. To obtain T_g's of dry fibers by this method, a low-viscosity mineral oil can be used as the displacing liquid.

Another common method used for measuring T_g is thermal analysis. The glass transition is associated with changes in specific heat, not with latent heat. Thus the transition occurs as a base-line shift rather than as distinct endotherms in DSC or differential thermal analysis (DTA). As shown in Figure 3, wet determinations are more difficult than dry determinations, since the wet fibers must be sealed in DSC capsules to prevent moisture vapor from escaping during the determination. Further, the base-line shift is usually very small, and an accurate measure is difficult to accomplish.

A thermal method that allows measurement of fiber-length change as a function of temperature can be used to determine the T_g of a fiber if the length change is great enough. This criterion is easily met by undrawn and partially oriented polyester fibers as well as by other fibers that exhibit moderate shrinkage at temperatures above their T_g's. The method used can involve simple shrinkage measurements (Figure 4) or the more elaborate differential technique, TMA (Figure 5). The T_g is taken as the abrupt length change from the base line. Because of experimental difficulties, the TMA method is applicable only to dry or conditioned fibers.

Tensile properties that are related to fiber stiffness can be used to measure the T_g of almost all fibers. The elastic modulus, that is, the slope of the Hookean region of the fiber stress-strain curve, is a measure of the fiber stiffness and can be used for T_g determination since, by definition, a glass is stiffer than a rubber (Figure 6). Since the transition from a glassy to a rubbery state involves a reduction in stiffness, the temperature at which the modulus is abruptly lowered is taken as

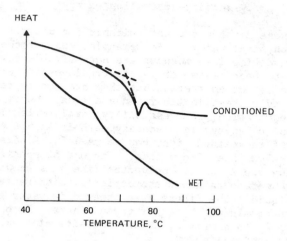

Figure 3.   Differential scanning calorimetry curves for undrawn PET fibers

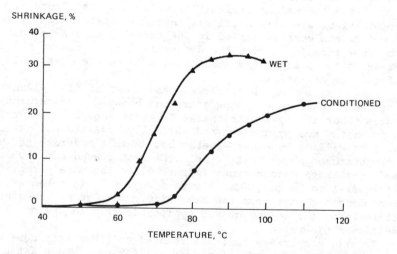

Figure 4.   Effect of temperature on shrinkage of undrawn PET fibers

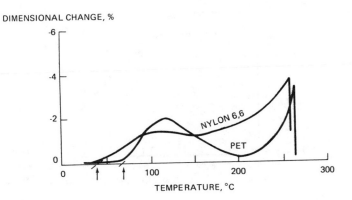

Figure 5. Thermomechanical analysis (TMA) of fibers

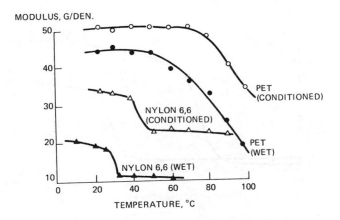

Figure 6. Effect of temperature on modulus of elasticity

WORK RECOVERY FROM 3 G/DEN. STRESS, %

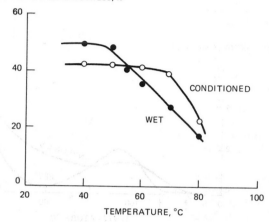

Figure 7.   Effect of temperature on work recovery of PET fibers

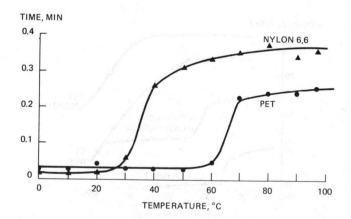

Figure 8.   Effect of temperature on stress decay time for wet fibers

the T_g. Other tensile properties, such as break tenacity and break elongation show no inflection in property-temperature curves. A few fibers, notably the modacrylics and the acrylics, show a rapid increase in elongation just above their T_g's. For determination of wet T_g by this method, actual tensile testing was carried out with the fiber totally immersed during the test.

Tensile work recovery can be considered a dynamic-loss modulus obtained at very low cycling rates. A fiber is stressed to 3 g/den. and immediately allowed to recover. The work recovered exhibits a marked reduction at the glass transition temperature (Figure 7).

Stress relaxation, or stress decay, occurs when a fiber is extended in length to a predetermined stress level and held at that length. We arbitrarily stressed fibers to 0.5 g/den., permitted the decay to occur, and measured the time required for the stress to decay to 0.4 g/den. This time is plotted against temperature (Figure 8). For wet nylon 6,6 and wet PET fibers, the time for the stress to decay increased rapidly when the T_g's had been reached.

As indicated earlier, T_g is not a fixed temperature for a particular substance; it is a value that is influenced by the method used for its determination. A comparison of the results from six methods used to measure the T_g of PET fibers is given in Table I.

TABLE I

Glass Transition Temperature (T_g) of Conditioned and Wet

PET Fibers Determined by Various Methods

Method	T_g (Conditioned), °C	T_g (Wet), °C
Specific volume	69	60
DSC	71	63
Shrinkage	73	59
Modulus	71	57
Work recovery	70	50
Stress-decay	--	58

For conditioned PET, the T_g ranges from 69 to 73°C and, for wet PET, from 50 to 63°C. The wider range obtained with the wet fibers probably results from increased experimental difficulty in obtaining some of the data on wet fibers. In all cases, however, the wet T_g was 10 to 20°C lower than the conditioned T_g for PET fibers. Of these methods, the use of fiber modulus is preferred because it is applicable to all fibers, that exhibit distinct

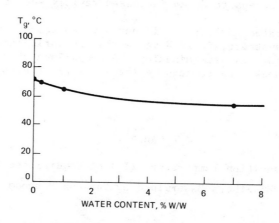

Figure 9.   Effect of moisture content on the T_g *of PET fibers*

glass transitions, and this procedure can be carried out more easily than the others, thus requiring less time than many of the
other methods. Conditioned and wet T_g's were determined from
modulus data for a number of other fibers (Table II). In each
case, the wet T_g was substantially lower than the dry T_g.

TABLE II

Glass Transition Temperatures of Various Fibers --
Wet and Conditioned

Fiber	T_g (Conditioned), °C	T_g (Wet), °C	ΔT_g
PET polyester	71	57	14
PCHDMT polyester	91	71	20
PBT polyester	47	38	9
Nylon 6,6	40	29	11
Nylon 6	41	28	13
"Qiana" nylon	150	90	60
"Orlon" acrylic	97	30	67
"Verel" modacrylic	59	38	21
Acetate	84	30	54
"Rhovyl" vinyl	75	55	20
Silk	108	30	78

The lowering of the T_g by organic liquids, plasticizers,
and monomers has been well documented, but the lowering of the T_g
of polymers by water is relatively obscure. Most polymers are
hydrophobic; hence they do not absorb much water. However,
enough water is taken up by most fibers to lower the T_g substantially. PET fiber regains about 0.4% moisture from an atmosphere
(21°C) and 65% RH. This amount of water lowers the T_g by about
2°C. Soaking the fiber in water can introduce as much as 7%
water into the fiber over a 24-hr period, and this amount of
water lowers the T_g by about 16°C. Further soaking in water does
not increase the water uptake. Intermediate levels of water uptake can be obtained by soaking for periods of time up to 24 hr.
The effect of water in PET fiber on the T_g is shown in Figure 9.
A similar curve for nylon 6, which is more hydrophilic than PET,
was previously published (16). For nylon 6, the wet fiber (after
24 hr in water) contains about 6% water and has a T_g of about
28°C, whereas the conditioned fiber (70°F and 65% RH) with 4%
water has a T_g of about 41°C. The bone-dry fiber has a T_g of over
100°C.

Equations that relate T_g to the presence of plasticizers,
organic liquids, and monomers in polymers have been proposed by a
number of investigators. Among the first was the equation proposed by Fox and Flory in 1954 (19):

$$\frac{1}{T_g} = \frac{W_p}{T_{g_p}} + \frac{W_s}{T_{g_s}} \qquad (i)$$

W_p = weight fraction of polymer.

W_s = weight fraction of solvent.

T_{g_p} = T_g of polymer.

T_{g_s} = T_g of solvent.

This equation fits the data well for small amounts of these additives. It also fits the data for water in most fibers, since the amount is usually quite small. A more refined equation was proposed by DiMarzio and Gibbs in 1963 (20).

$$\frac{\Delta T_g}{\Delta V_s} = \frac{M \cdot V_p \cdot T_g}{2V_s}$$

ΔT_g = T_g (polymer) -- T_g (plasticized polymer).

ΔV_s = Difference in solvent volume fraction.

V_p = Polymer molar volume.

V_s = Solvent molar volume.

T_g = T_g of polymer.

M   = Constant = 7.5 perfectly flexible solvent molecule.
            = 6.0 flexibility $\sim$ to polymer.
            = 4.8 rigid solvent molecule.

This equation fits fiber moisture-content data over a wider range than does Fox's equation.

The T_g's for conditioned and bone-dry fibers and fibers soaked in water for 24 hr as determined from modulus data are shown in Table III for a number of fibers with normal moisture regains of 0.4 to 9.4%.

The more-hydrophilic fibers show the greatest effect of water on T_g. Attempts have been made to correlate the reduction in T_g caused by water to moisture-related fiber properties such as moisture regain, advancing contact angle, and water imbibition (Table IV).

T_g reduction due to water content cannot be predicted on the basis of these properties; but, generally, a fiber with a high moisture regain, a low advancing contact angle, and a high water imbibition shows a large reduction in T_g.

TABLE III

Glass Transition Temperature -- Wet,
Conditioned, and Dry Fibers

| | $T_g, °C$ | | | Water in | Moisture |
Fiber	Continued[*]	Dry[**]	Wet[***]	Wet Fiber, %	Regain, %
PET polyester	71	73	57	6.8	0.4
Nylon 6,6	40	59	29	6.1	4.2
"Qiana" nylon	150	175	90	10.1	2.5
Silk	108	197	30	22.2	9.4
Acetate	84	118	30	15.3	5.0

[*]At normal equilibrium regain (70°F, 65% RH).

[**]Bone dry -- T_g calculated.

[***]Wet out by soaking 24 hr in water.

TABLE IV

Difference Between Wet T_g and Conditioned T_g for Various Fibers

Fiber	ΔT_g, Conditioned-Wet, °C	Moisture Regain, %	Contact Angle, °	Water of Imbibition, %
PET polyester	14	0.4	41.3	1.2
PCDT polyester	20	0.2	44.6	1.6
PBT polyester	9	0.3	--	3.0
Nylon 6,6	11	4.2	43.1	5.7
Nylon 6	13	4.0	47.7	7.8
"Qiana" nylon	60	2.5	41.3	2.0
"Orlon" acrylic	67	2.0	43.7	2.6
"Verel" modacrylic	21	2.1	36.5	7.9
Acetate	54	5.0	19.0	18.0
"Rhovyl" vinyl	20	0.2	--	2.2
Silk	78	9.4	22.0	35.8

T_g has great significance in the processing and use of textile fibers. For all fibers except the elastomerics and the polyolefins, the T_g is higher than ambient temperature. That is, the amorphous polymer of the fiber must exist in the glassy state. Since fibers are often processed under hot, wet conditions (as during dyeing and finishing), the wet T_g may be of even greater importance than the dry T_g. Of particular importance in the performance of wash-and-wear fabrics, the wet T_g of the fiber must be at least as high as the temperature the fabric will reach during laundering. If the wash water temperature exceeds the wet T_g of the fiber, molecular motion will permit changes that will result in dimensional instability and cause the fabric to have poor wash-and-wear performance. In other words, wrinkling will occur, and the fabric will require ironing. The wet T_g of a fiber that exhibits good wash-and-wear, or permanent-press, behavior should be at least 60 to 70°C, since the typical home laundry uses water at temperatures up to 65°C. The wet T_g of PET fiber is slightly below this level; consequently, this fiber exhibits only marginal wash-and-wear performance. The fiber is usually blended with cotton and the cotton is crosslinked to improve dimensional stability. The ability of PET fibers to function reasonably well in wash-and-wear garments is probably due to the relatively long time (compared with laundering time) required for the T_g to be lowered to the equilibrium wet T_g value. A high wet T_g should ensure adequate dimensional stability during laundering. Dimensional stability can also be obtained in a fiber with a low wet T_g if the fiber can be lightly crosslinked in the fabric or, preferably, in the garment configuration. Durable-press resins and permanent-press finishes on cellulosics, particularly cotton, function in this manner because the wet T_g of cotton is well below 0°C. Untreated cotton demonstrates very poor wash-and-wear performance, but cotton with its wet T_g increased by crosslinking is quite satisfactory.

Conclusions

Methods for measuring the glass transition temperature that are particularly adaptable to fibers, both wet and dry, have been proposed. In particular, the use of the elastic modulus has been shown to give reliable estimates of the T_g of fibers. The effect of the presence of water in the fiber, both at regain level and totally wet out, on the fiber T_g and the significance of the wet T_g of fibers on the stability of fabrics and garments made from them have been discussed.

Abstract

The glass transition temperature (T_g) of an amorphous polymer is the temperature at which motion occurs in the major segments of the backbone molecular chain of the polymer. However, the effect of water on the T_g of a hydrophobic polymer has generally not been recognized. Almost all fibers are only partially crystalline; hence they exhibit glass transition phenomena. And all fibers have an equilibrium moisture content at 70°F and 65% RH. This moisture content has been shown to result in a marked lowering of the T_g for many fibers. Further reductions in T_g occur when the fiber is allowed to wet out in water until equilibrium saturation is reached. These lowered T_g's are of significance in the processing and care of textile products. Several methods particularly useful for measuring the wet and dry T_g of fibers have been explored.

Literature Cited

1. Gee, G., Contemp. Phys., 1970, 11, 313-34.

2. Billmeyer, Jr., F. W., "Textbook of Polymer Science"; 2nd edition, Wiley-Interscience, New York, 1971, pp 207-211.

3. Miller, M. L. "The Structure of Polymers;" Reinhold Book Corp., New York, 1966, pp 284-94, 476-81, 540-42, 564-65.

4. Matsuoka, S., J. Polym. Sci., 1960, 42, 511.

5. Woods, D. W., Nature, 1954, 174, 753.

6. Newman, W., and Cox, W. P., J. Polym. Sci., 1960, 46, 29.

7. Hoffman, D., and Weeks, J. J., J. Res. Natl. Bus. Stds., 1958, 60, 465.

8. Griffish, H. and Ranby, B. G., J. Polym. Sci., 1960, 44, 369.

9. Vematsu, I., and Vematsu, Y., Kobunski Kagaku, 1960, 17, 222.

10. Jenckel, E. and Hensch, R., Kolloid Z, 1953, 130, 89.

11. Hata, N., Tobolsky, A. V., and Bondi, A., J. Appl. Polym. Sci., 12, 1968, 2597.

12. Pritchard, J. G., "Poly(vinylalcohol) -- Basic Properties and Uses," Gordon and Breach Science Publishers, New York, 1970, p 60.

13. Bryant, G. M. and Walker, A. T., Text. Res. J., 1959, 29, 211.

14. Hurley, R. B., Text. Res. J., 1967, 37, 746.

15. Tan, Y. Y., and Challa, G., _Polymer_, 1976, _17_, 739.

16. Kettle, G. J., _Polymer_, 1977, _18_, 742.

17. Buchanan, D. R. and Walters, J. P., _Text Res. J._, 1977, _47_, 398.

18. Reimschuessel, H. K., _J. Polym. Sci._, Chem. Ed., 1978, _16_, 1229.

19. Fox, T. G. and Flory, P. J., _J. Polym. Sci._, 1954, _14_, 315.

20. Dimarzio, E. A. and Gibbs, J. H., _J. Polym. Sci._, 1963, A1, 1417.

RECEIVED January 4, 1980.

Effect of Moisture on Fatigue Crack Propagation in Nylon 66

P. E. BRETZ, R. W. HERTZBERG, J. A. MANSON, and A. RAMIREZ

Materials Research Center, Lehigh University, Bethlehem, PA 18015

The fatigue phenomenon in polymers is of increasing fundamental and technological interest. From the standpoint of polymer
science, the kinetics and energetics of failure under cyclic
loading reflect an interesting balance between energy input due
to the applied stress and energy dissipated by dynamic viscoelastic processes. In turn, the position of this balance reflects
not only the loading conditions and environment but also polymer
structure and morphology. From the technological standpoint,
fatigue is important because many applications of engineering
plastics involve repetitive or cyclic loads. While the overall
fatigue process in a smooth specimen containing no significant
flaws includes both the initiation of an active flaw and its
growth, many (if not most) real specimens contain pre-existent
flaws that can, under appropriate loading conditions, develop
into catastrophic cracks (1,2) (often at inconvenient or dangerous
times and places). Thus the proper selection of materials and
design of parts requires particular attention to the ability of
a polymer to resist fatigue crack propagation (FCP).

For these reasons, an extensive program on engineering
plastics has been conducted in this laboratory to characterize
the FCP rates as a function of loading conditions, to identify
the micromechanisms of failure, and to elucidate the role of the
polymer chemistry and morphology (3-10). Interestingly, as a
group, crystalline polymers [notably nylon 66, polyacetal, and
poly(vinylidene fluoride)] have exhibited higher values of fracture toughness and lower values of FCP rates than amorphous or
poorly crystalline polymers. Following detailed studies of the
behavior of amorphous polymers, it was decided to explore the
effects of structural and morphological parameters on FCP in
nylon 66, polyacetal, and poly(vinylidene fluoride). This paper
updates earlier reports on nylon 66. (For other studies on
crystalline polymers see references 3, 7-12).

When relating polymer chemistry to mechanical behavior the
environment is, unfortunately, often neglected. However, the
effects of sorbed moisture on the general mechanical behavior of

0-8412-0559-0/80/47-127-531$05.75/0
© 1980 American Chemical Society

polyamides have long been documented. Such effects have been reviewed by McCrum et al. (13), Papir et al. (14), and Kohan (15), and described in the trade literature (16). Generally Young's modulus and yield strength are said to decrease, and impact toughness to increase, with increasing water content. Dynamic mechanical spectroscopy and other studies provide evidence that water exists in both tightly bound and relatively free states (13,17), and it has been shown that some properties exhibit a transition when the water content exceeds the ratio of one water molecule per two amide groups (corresponding to tightly bound water). [Use of the term "bound" does not imply the permanent binding of a particular molecule to a particular site; as with adhesion, the possibility of a dynamic interchange is assumed.]

Much less has been reported about fatigue in polyamides. The conditioning of nylon 66 at 50% relative humidity (RH) was said to reduce fatigue life by 30% (16), and long-term soaking in water was reported to increase FCP rates at a constant stress-intensity-factor range (3).

The aim of the present program is to establish the base-line behavior of FCP in nylon 66 as a function of water content, and to deduce the micromechanisms of failure. In the first progress report, significant effects of moisture content on both FCP rates and fracture surfaces were observed. In this continuation study, using a different series of nylon 66, some previously reported effects are confirmed, but some are not. Possible reasons are discussed, and future directions outlined.

Experimental

Two series of injection-molded plaques of nylon 66 were obtained through the courtesy of Dr. E. Flexman, E. I. duPont de Nemours and Co. The grade concerned, Zytel 101, had a nominal number-average molecular weight of 17,000. In the first study (A), the plaques were 8.6 mm thick, and were received in the dry, as-molded condition; these plaques had been sealed in plastic bags immediately after molding to prevent moisture pick-up and were stored in a desiccator following the opening of the bags. In the second study (B), the plaques were 6.4 mm thick, and were received containing approximately 2.2 wt % water. Equilibration to various moisture contents was performed according to the methods outlined in Table I.

Number-average molecular weights (M_n) were measured after various times of water immersion in order to check whether or not the water equilibration procedures induced a significant degree of hydrolysis. Thus values of M_n were determined viscometrically using solutions in formic acid and the following equation (18):

$$M_n = K^{-1/a} [\eta]^{1/a} \tag{1}$$

where K and a = 1.1×10^{-3} and 0.72, respectively, and $[\eta]$ is the

Table I. Moisture Conditioning Procedure for Nylon 66

Series	%RH[a]	wt % H_2O	Method
A-1	0	<0.2[b]	As—molded ("dry")
A-2	23	0.8-0.9[b]	Suspended for 24 hr above boiling saturated aqueous sodium bromide
B-0	44	2.2	As—received
A-3	50	2.6[b]	Boiled for 100 hr in a saturated aqueous potassium acetate[b]
B-1	51	2.7[c]	Boiled for 100 hr in saturated aqueous sodium nitrite
B-2	52	2.8[c]	Boiled for 100 hr in saturated aqueous sodium acetate
B-3	66	4.0[c]	Boiled for 100 hr in saturated aqueous sodium phosphate
B-4	72	4.5[c]	boiled for 100 hr in saturated aqueous potassium bromide
B-5	80	5.7[c]	Boiled for 100 hr in saturated aqueous potassium dichromate
A-4,B-6	100	8.5[b]	Boiled for 138 hr in water[b]

[a]Corresponding to the water contents (at equilibrium) of column 3

[b]Standard methods described in ref. 15

[c]Measured gravimetrically

intrinsic viscosity.

In preparation for thermal-history studies, specimens were
first dried (19) and melted at 290°C to form 0.6-mm-thick sheets
(20). The specimens were then subjected to the following treatments: 1) Quenching in mineral oil held at 0°C; 2) slow cooling
in mineral oil which was allowed to cool from 290°C to room
temperature over a period of 4 hours; 3) air cooling from 290°C
to room temperature; and 4) annealing at 235°C for 60 minutes.
Crystallinities were then estimated by two techniques: differential scanning calorimetry (DSC) and density. A Perkin-Elmer unit
(model DSC-1) was used for the former, and a density gradient
column (21) for the latter.

Fatigue crack propagation tests were performed on 73-mm x
73-mm (2.9 in. x 2.9 in.) compact-tension specimens that had been
machined from the plaques and then precracked. An electrohydrau-
lic closed-loop test machine was used to produce a constant-
amplitude, 10-Hz sinusoidal load. Most of the tests were
performed in duplicate at room temperature in laboratory air at
an average ambient RH of 40%. In view of the slow rate of
moisture equilibration in air at room temperature (15), the
difference between the ambient RH and that of the nylon being
tested was presumed to be unimportant within the time period of
the test (3-5 hr).

Crack growth rate data were analyzed to compare the incremental crack growth rate per cycle, da/dN, with the applied stress
intensity factor range at the crack tip, ΔK. The value of ΔK is
given by

$$\Delta K = Y \Delta \sigma \sqrt{a} \quad [\text{MPa}\sqrt{m}] \tag{2}$$

where $\Delta \sigma$ = applied remote stress range [MPa]; a = crack length
[m]; Y = geometric correction factor. Test results revealed that
da/dN depended on ΔK with a relationship of the form (22):

$$\frac{da}{dN} = A(\Delta K)^m \tag{3}$$

Crack-length readings were recorded with the aid of a calibrated traveling optical microscope equipped with a 50X eyepiece.
Examination of the fracture surfaces revealed significant crack
front curvature for specimens with moisture contents ≥ 2.2 wt %.
Reference 11 describes a procedure by which such non-linearity in
the crack front was treated.

Fractographic studies were conducted on an ETEC Autoscan
scanning electron microscope (SEM) at an accelerating potential of
20 kV. Each fracture surface was coated with vacuum deposited
layers of gold and carbon prior to examination to prevent specimen charging and minimize the degradation of the fracture surface
under the electron beam.

Results and Discussion

Viscometric determination of M_n yielded a value of 1.78×10^4, confirming the nominal value. On boiling for 100 hr, the value of M_n was reduced by ~20%. Since such a reduction would increase the absolute values of FCP rate by an amount not greater than the maximum experimental error, and would not affect the general shape of the curve of FCP rate vs water content, a correction for degradation was not applied to the fatigue data.

Values estimated for the percent crystallinity varied not only with thermal treatment, but also with the measurement techniques. As shown in Table II, the DSC technique gave consistently lower values of crystallinity than the density-gradient column. In general, more reproducible results were found by using the DSC than was the case with the density method.

Table II. Crystallinity in Nylon 66 by DSC and Density Methods

History	Cryst. by DSC,[a] %	Cryst. by Density, %
Quenched in oil	24	37.5
Air cooled	27	38.3
Slow cooled in oil	30	44.9
Annealed at 235°C (50 min.)	35	43.1

[a]Assuming a heat of fusion of 46.8 cal/g (196 J/g) (23).

In the case of slow cooling the water content corresponding to complete bridging of amide groups by water molecules ($\equiv$ 1 water molecule per 2 amide groups) would range between 4.4 wt % and 5.6 wt % (for % crystallinities of 30% and 45%, respectively).

Fatigue Crack Propagation. Results obtained with Series B specimens confirm the earlier observations that water content in nylon 66 not only significantly affects FCP rates but also may decrease or increase FCP rates, depending on the percent water (3,11). Figures 1-3 show the FCP results for one specimen of each of the moisture contents investigated. For those moisture contents for which duplicate tests were performed, the maximum difference in growth rate between the two specimens was a factor of 2; generally, much better agreement was observed. As noted previously (11), the range of response from "best" to "worst" is broad; at ΔK~3MPa$\sqrt{m}$, the FCP rate for the worst specimen is about 25 times that of the best specimen. When FCP rates (at constant ΔK) are plotted against water content (Figure 4), a pronounced

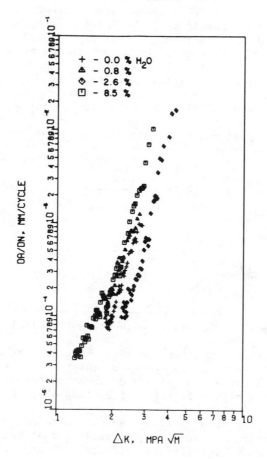

Figure 1.   Typical logarithmic plots of FCP rate per cycle (da/dN) *vs. the stress intensity factor range* (ΔK) *for nylon 66 (Series A) containing various levels of moisture*

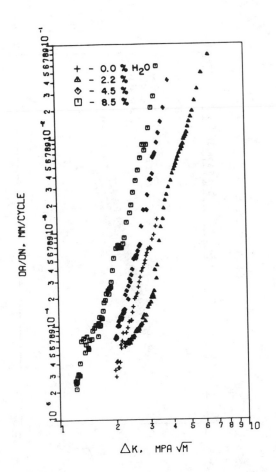

Figure 2. Crack growth rate vs. stress intensity factor range for nylon 66 (Series B) containing various levels of moisture

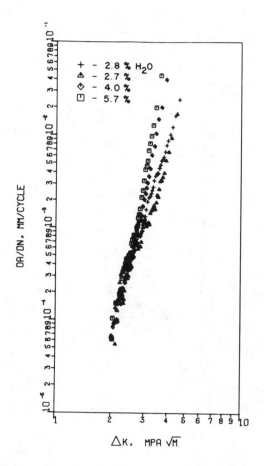

Figure 3. Crack growth rate vs. stress intensity factor range for the remaining Series B nylon 66 specimens

minimum is noted at about 2.5% water. Thus, as water is imbibed by an essentially dry polymer, the FCP rate first decreases to a minimum value and then rapidly increases, reaching a maximum value on saturation. This behavior certainly suggests the existence of competitive effects (and hence mechanisms), the balance depending on the amount of water present.

While a complete rationalization of this interesting behavior must await completion of dynamic mechanical characterization, an explanation must surely take into account the hydrogen-bonding and plasticizing nature of water in polyamides (12,13,14, 15). The sorption of increasing proportions of water has long been known to shift both the glass-to-rubber and secondary transitions to increasingly lower temperatures (13); for example, T_g decreases from 80°C to −15°C for this nylon as the moisture content of nylon 66 increases from ∿ 0 wt % to 8.5 wt % water (15). It is important to note, however, that the rate of decrease of T_g is greatest below about 2 wt % of water (14,24,26). This decrease in T_g has been attributed to the breaking of hydrogen bonds between amide groups and the bridging of amide groups by water (13,14). Paradoxically while T_g is being decreased by the sorption of water, water molecules are apparently able to pack well into the free volume, as shown by a decrease in the specific volume of the tightly bound water (26). At the same time, the small-strain modulus does not decrease (and may even increase slightly) when water is added to a concentration of up to ∿2.6% (15,25).

It is interesting to consider the water/amide group stoichiometry associated with water contents of about 2.6 wt %. Assuming a degree of crystallinity of 40%, and that the water remains in the amorphous phase, 2.6 wt % water would correspond to one water molecule per 3.7 amide groups. [This value is lower than that reported earlier (11) due to correction of the degree of crystallinity.] It is also interesting that a slight DSC peak begins to occur at ∿100°C when the water content reaches ∿2.6%; this suggests that some water begins to behave in a relatively "free" manner at about this concentration.

Returning to the question of FCP rates, it seems likely that an increased segmental mobility must be associated with the presence of small (<2 wt %) proportions of water. If the damping is low enough to minimize gross heating of the specimen (as may well be the case for tightly bound water), the localized deformation at the crack tip may be expected to result in blunting of the crack (6). With a more blunt crack, the effective stress concentration will be reduced, thus lowering the rate of crack propagation.

At higher concentrations of water, clustering of water molecules, rather than bridging, is believed to take place and become dominant. Such clustering causes increases in damping and decreases in the room temperature modulus, the latter remaining nearly constant for moisture contents ≤ 2.6 wt % (15). While

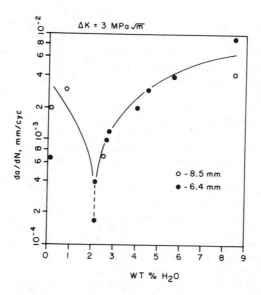

*Figure 4.   Crack growth rate vs. moisture level at a constant stress intensity factor
for both series of nylon 66 specimens*

this dominance would have been expected to be seen at water concentrations above about 5 wt % (equivalent to completion of tight packing and bridging of water molecules), the weakening effect of free water apparently comes into play when only about one-half the amide groups are bridged by water molecules. [If the water contents and crystallinities are in fact correct, it is possible that the combination of cyclic stressing with water content may induce the effects typical of loosely bound water at lower concentrations of water than would be the case for static loading (i.e., 4-5%). At the same time, as mentioned above, an endothermic DSC peak began to appear at about 2.6% wt.]

In any case, the effect of loosely bound water must be to weaken the resistance to fatigue crack growth. First, the lowering of the modulus for moisture contents >2.6 wt % water will increase the extent of plastic strains experienced by the bulk material ahead of the crack tip, and thus tend to increase the FCP rate at a given applied load range. In other words, the strain per loading cycle, $\Delta\varepsilon$, must then increase, since the test is performed under constant load range $\Delta\sigma$, and $\Delta\varepsilon = \Delta\sigma/E$. Thus, the nylon that contains loosely bound water accumulates more damage per loading cycle than would be the case with the drier specimens; a higher FCP rate will then be expected at a given value of $\Delta\sigma$ (and hence ΔK). This behavior indicates that the effect of a decreasing modulus overwhelms the beneficial effect of high localized segmental mobility (a question of the relative scale of motions). Second, this weakening effect will be accelerated by any hysteretic heating that occurs (see discussion below).

If we accept the conclusion that a minimum in the FCP rate occurs when the water content is between 2 wt % and 3 wt %, it is interesting to see if any other properties exhibit discontinuities in a similar range of water content. Let us consider in turn, fracture toughness, the development of hysteretic heating, and the nature of the deformation as revealed by microscopic examination of the fracture surface (fractography).

Fracture Toughness. With typical glassy materials subjected to an FCP test of the kind performed in this study, crack growth proceeds at an ever-increasing rate as cycling proceeds, until the specimen fractures catastrophically. Also, FCP resistance tends to be positively correlated with static fracture toughness (3). In such cases, it is often possible to estimate a critical value of ΔK for fast fracture from the last value of ΔK prior to fracture, ΔK_{max}. While this value does not correspond to a true value of fracture toughness (or energy) it does give a relative ranking of toughness (6). With more ductile materials, catastrophic fracture may not occur, and definition of an ultimate failure point in terms of ΔK becomes impossible.

As summarized in Table III, terminal fast fracture was obtained only at water contents lower than 4.0 wt %. Thus

TABLE III

FRACTOGRAPHIC OBSERVATIONS OF

MOISTURE-BEARING NYLON 66

MW = 17,000
ALL TESTS - 10 HZ

MOISTURE LEVEL	0%	0.8	2.2	2.6	2.8	4.0	4.5	5.7	8.5
FRACTURE SURFACE WHITENING	NO	NO	SOME	SOME	YES	YES	YES	YES	YES
TERMINAL FAST FRACTURE	YES	YES	YES	YES	YES	YES	NO	NO	NO
STRIATIONS	NO	NO	YES	YES	YES	YES	YES	YES	NO
D G BANDS	NO	NO	NO	NO	NO	NO	NO	NO	NO
SPHERULITIC FAILURE	YES	YES	NO	NO	NO	NO	NO	NO	NO

relative values of ΔK_{max} cannot be obtained at higher water contents. However, while the differences in ΔK_{max} for specimens containing <4.0 wt % water are small, there does seem to be a tendency to first increase from $\sim$3MPa$\sqrt{m}$ for "dry" specimens to $\sim$4-6 MPa$\sqrt{m}$ between 2 and 3 wt % water, and then to decrease to $\sim$3-4 MPa$\sqrt{m}$ at concentrations of water between $\sim$3 to 4 wt %. In other words, the pronounced minimum in da/dN (Figure 4) seems to be associated with a peak in relative toughness as defined by ΔK_{max}.

Onset of Hysteretic Heating. In our preliminary report on Series A specimens (11), it was noted that samples A-1 and A-2 (% water $\leq$$\sim$1) exhibited no significant temperature rise at the crack tip, but that all other specimens exhibited increases at least 15°C to 20°C. A more careful temperature measurement with the Series B specimens indicated that, while temperature increases at the crack tip occurred only in the moisture range previously observed, the temperature rise was only 8-10°C.

Heating of this kind reflects the effects of a combination of a high value of damping at a given frequency with the low thermal conductivity characteristic of polymers; heat is generated faster than it can be dissipated to the surroundings. The occurrence of hysteretic heating implies the occurrence of a corresponding irreversible deformation process such as large-scale chain motion at the crack tip or beyond. Thus if crack-tip heating is localized, permitting localized deformation and crack blunting, FCP rates may be decreased; however, if the heating involves more than a small volume element at the crack tip, the consequent softening is believed to result in enhanced specimen compliance and higher crack growth rates. [The latter condition has been encountered in the case of impact-modified nylon 66 (27) and poly(vinyl chloride) (28).] Hence, the effects of hysteretic heating on FCP behavior is dependent on the magnitude of the heated volume relative to the overall specimen dimensions (see especially Reference 27).

With this subtle behavior in mind, we may now address this question: do not the present findings disagree with the well-known fact (15,16) that the fatigue life of nylon 66 is monotonically decreased by increasing the water content within the range of this study? First, it must be noted that the fatigue tests quoted in references 15 and 16 are conducted on smooth, un-notched specimens at a frequency of 30 Hz and between fixed-load limits (so-called S-N tests, in which maximum stress is plotted against the number of cycles to failure). These conditions are precisely those that will enhance hysteretic heating on a large scale: the frequency is high, and the load is borne by the entire specimen rather than confined to the crack tip region. In fact, even "dry" nylon 66 experiences a significant temperature elevation during S-N tests because of its relatively large loss tangent (damping). If, as expected, the damping increases with water

content, the specimens will experience increasingly severe
strains and damage (see above) as the water content increases.
In other words, hysteretic heating on a gross scale, and its
ability to lower the modulus of the whole specimen undoubtedly
dominates the fatigue response, and progressively weakens the
material as the water content increased from near zero.

In contrast, the specimens of this study were notched (to
simulate the realistic expectation that adventitious flaws will
be present in "real" specimens) and the frequency was lower than
in the fatigue tests discussed. Under these circumstances, it
is reasonable to suppose that gross hysteretic heating of large
volume elements might not be noted at low water contents (that
are high enough to induce significant heating in the S-N tests).
As mentioned above, heating of only a small volume element at
the crack tip may induce crack-tip blunting. [It is also
possible (see discussion above) that there may be a higher
degree of segmental mobility per se at low water contents, as
the lowering of T_g would imply.

Experiments at different frequencies and with different wave
forms are planned to further elucidate the role of hysteretic
heating (see Ref. 27 for an example of how changing frequency
changes the relative importance of crack-tip vs large-volume
heating). In any case, the contrasting behavior of nylon 66
in S-N and notched-specimen tests as a function of water content
should be taken into account in decisions on component design and
materials selection.

Fractographic Examination. In the preliminary study of
Series A specimens (11) low- and high-magnification examinations
(10X and 380X, respectively) of fracture surfaces of nylon 66
(water contents: $\sim$0, 0.8, 2.6, and 8.5 wt %) revealed several
features whose existence or characteristics depended on water
content: (1) Stress whitening; (2) Evidence of trans-spherulitic
failure; (3) Surface markings (fatigue striations).

Visual and microscopic examination of the fracture surfaces
of Series B specimens has confirmed the earlier observations, and
extended them throughout the range of water contents up to
saturation (8.5 wt %). Results are summarized in Table III and
discussed in more detail below. It will be seen that, in at
least some cases, observations vary as a function of water content.
Not surprisingly, discontinuities in behavior do not necessarily
occur at water contents precisely corresponding to the minimum
in da/dN; however, there are several clear differences between
behavior at low and at high water contents.

Macroscopic Appearance of Fracture Surfaces. The most
striking observations (Table III) of fracture surface topography
are the occurrences of stress-whitening and extensive plastic
deformation at the higher water contents, and the occurrence
of arrest lines at high values of ΔK.

Thus, while not observed at water contents ≤0.8–0.9 wt %, stress–whitening to some degree was observed at water contents ≥2.6 wt %. The expected association of the stress–whitening with a significant degree of plastic deformation is confirmed by observation of the surface topography. For low water contents (∿0 and 0.8 wt %), the surfaces of all specimens (Series A and B) were flat and relatively featureless (Figure 5a). In contrast, for water contents ≥2.2 wt %, the fracture surfaces exhibited a much more rugged appearance, suggestive of extensive drawings (Figures 5b and 5c).

In addition (see Figure 6), the samples containing ≥2.2 wt % water showed large crack arrest lines at high values of ΔK ($\Delta K > 3.7$ MPa√m). These arrest lines, which were associated with the periodic interruption of the FCP test to read the crack-tip position, imply the occurrence of creep. (If creep is closely confined to the crack-tip, it may well be associated with crack blunting.) Some finer lines were observed as well (for discussion see the following section).

As mentioned above, it was observed that specimens (Series A and B) containing ≤4.0 wt % water failed by rapid, unstable (catastrophic) crack propagation in the final load cycle (see Figures 5a and 5b). At higher water contents, however, the specimens failed by stable but very rapid crack propagation (da/dN > 1 mm/cycle) at values of $\Delta K > 4$ MPa√m (see Figure 5c). The fast-fracture surfaces (Figures 5a and 5b) were characterized by a grouping of crisp, curved lines that emanated from some central point along the boundary of the crack front just prior to instability. Interestingly, this fast fracture region exhibited no evidence of stress–whitening (though such whitening was often seen in the stable crack growth region); this was so regardless of the water level in any specimen. The distinguishing feature noted on the terminal fracture surfaces in the specimens containing 8.5% water was the existence of widely separated arrest lines corresponding to the extent of tearing in each load cycle (Figure 5c). (These arrest lines will be discussed below.) Other fracture markings were observed and are described in the preliminary report (11).

Microscopic Appearance of Fracture Markings. As was the case with macroscopic observations, major differences were noted in the microscopic (380X) fracture surface micromorphology of specimens equilibrated to different moisture levels. At this higher magnification (cf. Figure 5), the "dry" specimens revealed crisp, facet-like markings over the entire fatigue fracture surface [see Figure 7a]. The average size of these small facets is approximately 10 μm, close to the spherulite size of 6.5 μm (determined by etching in xylene). Such faceting suggests that limited crack tip plastic deformation had occurred, an observation which is consistent with the fact that the surfaces observed at low magnification were flat. A generally similar appearance was noted

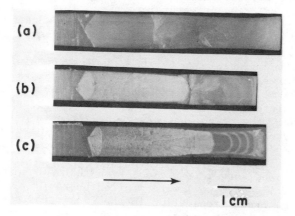

Figure 5.   Appearance of fatigue fracture surfaces of nylon 66 (Series A) as a
function of moisture content: (a) ⌐0 wt %, (b) 2.6 wt %, (c) 8.5 wt %. Series B
specimens behaved in a similar fashion, with evidence for ductility at water contents
≥2.7%. Terminal fracture regions also visible. Arrow shows direction of fatigue
crack growth.

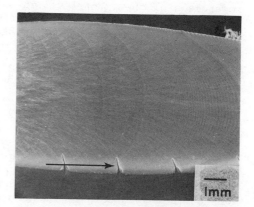

Figure 6.   Coarse and fine crack arrest
lines found on the fatigue fracture surface
of nylon 66 equilibrated to contain 5.7 wt
% water. Arrow shows direction of fatigue crack growth.

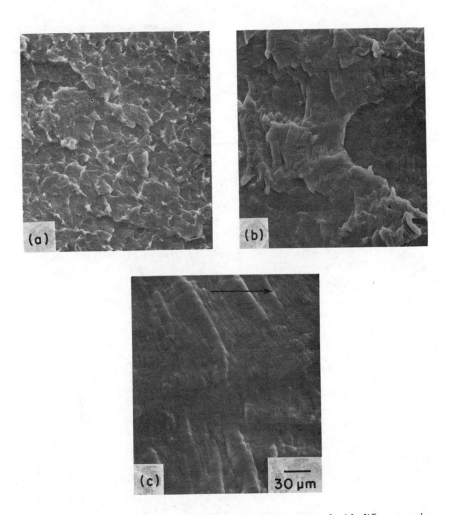

Figure 7. Fracture surface appearance of nylon 66 associated with different moisture contents: (a) ~0 wt % water, (b) 5.7 wt % water, (c) 8.5% water. $\Delta K = 2.6$ MPa√m̄. Arrow shows direction of fatigue crack growth.

Figure 8.   Fatigue striations in nylon 66 equilibrated to contain 5.7 wt % water. Arrow shows direction of fatigue crack growth.

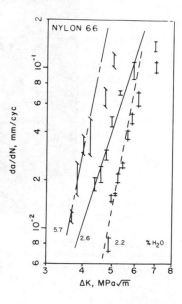

Figure 9.   Correlation between macroscopic growth increments per loading cycle and striation spacings as a function of ΔK in nylon 66 equilibrated to several moisture levels

with specimen A-2 (0.8 wt % water). However, at higher moisture
contents [see for example, Figures 7b and 7c], severely perturbed
fracture surfaces were observed, consistent with the occurrence
of extensive plastic deformation and drawing at the crack tip.

The use of a higher magnification also permits a more detailed assessment of other surface markings. Recent studies (5,
29) of the fractography of polymers subjected to FCP tests have
identified two types of linear fracture markings oriented parallel
to the advancing crack front: discontinuous growth bands (DGB)
and classical fatigue striations. The DGB markings, usually found
at low values of ΔK, represent discrete increments of crack growth
corresponding to the dimension of the plastic zone associated with
the crack tip. In this region, the macroscopic crack advances
in a discontinuous fashion, with the crack remaining periodically
dormant for as many as 10^5 cycles. True fatigue striations are
found at higher ΔK levels, and reflect the extent of crack
advance in a single load cycle.

Regardless of the water content or ΔK level, no DGB's have
yet been observed in any nylon specimens (either Series A or
Series B). By contrast, the discontinuous growth of fatigue
cracks has been described in studies of both crystalline and
amorphous polymers [for example, in polyacetal (29,30), poly-
ethylene(31), and a variety of poorly crystalline or amorphous
polymers (5)]. While the process of DGB formation in poorly
crystalline or amorphous polymers is fairly well understood (5,
12,29,32) the role of well developed crystallinity is not clear.
Hence the reason why DGB bands are not observed in nylon 66 is as
yet unknown.

Markings believed to be classical fatigue striations
(Figure 8) were, however, observed in specimens having water
contents in the range between 2.2 and 5.7 wt %, inclusive.
Similar markings have been noted by others for several crystalline polymers (31,33,34,35,36,37), and interpreted as fatigue
striations (though without firm evidence). The critical test of
whether or not such markings are true fatigue striations is to
determine by measurement whether the spacings of the markings
agree with the corresponding increments of macroscopic crack
advance. In fact, examination of our specimens suggests that the
markings are in fact true fatigue striations, for the spacings
do correspond to the increments of crack advance (see, for
example, Figure 9). The findings confirm the data obtained in
the preliminary study (11), which constitute the first unequivocal
evidence for the formation of true fatigue striations in crystalline polymers. [As previously noted, the measurements of crack
length received a special averaging technique because of the para-
bolic nature of the crack front (11)].

Summary and Conclusions

Several major conclusions (tentatively drawn in the earlier

study) are confirmed (or revised) by this more extensive study
of the effects of water on fatigue crack propagation in nylon 66:

1. Fatigue crack growth rates in nylon 66 are very sensitive
to moisture content in the range from $\leq$0.2 wt % ("dry") to 8.5 wt
% (saturated). A 25-fold variation exists between the fastest and
slowest rate (at a constant ΔK of 3 MPa$\sqrt{m}$).

2. In contrast to the case of fatigue in un-notched specimens, the fatigue crack growth rates exhibit a sharp minimum at
water contents in the range of 2 wt % to 3 wt % — a range
corresponding to about 1 water molecule per 4 amide groups.

3. The existence of the minimum must surely reflect the
presence of competing fatigue mechanisms. It is proposed that
the decrease of crack growth rates at low water contents reflects
the ability of tightly bound water to increase fracture energy
or to increase crack blunting due to hysteretic heating localized
at the crack tip. At higher concentrations of water, the relative
increase of crack growth rate is attributed to the rapid decrease
in the room temperature modulus of nylon 66, causing a lowering
of the resistance exerted by the bulk material to the advance of
the crack and an increase in crack growth rate. In addition, the
greater hysteretic heating noted at higher moisture contents may
contribute to a further decrease in the modulus.

4. These observations are supported by the onset of per-
ceptible heating at water contents above about 2 wt % water, and
by the onset of stress-whitening and a significant degree of
plastic deformation in the same concentration range. While plastic deformation is often associated with high fracture energies,
it is likely that the increase in fracture energy is overwhelmed
by a decrease in modulus due to plasticization by free water and
to hysteretic heating (see conclusion number 2). There is also
some evidence that trans-spherulitic failure occurs at low water
contents and reorganization of the spherulitic structure at high
water contents.

5. Classical fatigue striations are observed at intermediate
water contents, confirming earlier findings. However, no evidence
was found (at any water content or ΔK level) for discontinuous
growth bands, which are found in many other polymers.

6. From a technological viewpoint, it is important to
recognize that the fatigue response of nylon 66 as a function
of water content is strongly dependent on the test method — in
particular on the presence or absence of notches and, presumably,
the frequency.

Acknowledgement

This work was supported in part by the Office of Naval
Research.

Literature Cited

1. Andrews, E. H., "Fracture in Polymers", American Elsevier, New York, 1968.

2. Hertzberg, R. W., "Deformation and Fracture Mechanics of Engineering Materials", Wiley, New York, 1976.

3. Manson, J. A., Hertzberg, R. W., Crit. Rev. Macromol. Sci., 1973, 1, 443.

4. Hertzberg, R. W., Manson, J. A. and Skibo, M. D., Polym. Eng. Sci., 1975, 15, 252.

5. Skibo, M. D., Hertzberg, R. W., Manson, J. A., and Kim, S. L., J. Mater. Sci., 1977, 12, 531.

6. Kim, S. L., Skibo, M., Manson, J. A., and Hertzberg, R. W., Polym. Eng. Sci., 1977, 17(3), 194.

7. Hertzberg, R. W., Skibo, M. D., and Manson, J. A., J. Mater. Sci., 1978, 13, 1038.

8. Skibo, M. D., Janiszewski, J., Hertzberg, R. W., and Manson, J. A., Proceedings, International Conference on Toughening of Plastics, July 1978, the Plastics and Rubber Institute, London, paper 25.

9. Skibo, M. D., Manson J. A., Hertzberg, R. W., and Collins, E. A., J. Macromol. Sci.-Phys., 1977, B14(4), 525.

10. Andrews, E. H. and Walker, B. J., Proc. R. Soc. London A, 1971, 325, 57.

11. Bretz, P. E., Hertzberg, R. W., and Manson, J. A., J. Mater. Sci., 1979, 14, 2482.

12. Hertzberg, R. W. and Manson, J. A., "Fatigue in Engineering Plastics", Academic Press, New York, 1979 (in press).

13. McCrum, N. G., Read, B. F., and Williams, G., "Anelastic and Dielectric Effects in Polymeric Solids", Wiley, New York, 1967.

14. Papir, Y. S., Kapur, S., Rogers, C. E. and Baer, E., J. Polym. Sci., 1972, Part A-2, 10, 1305.

15. Kohan, M. I., "Nylon Plastics", John Wiley and Sons, Inc. New York, 1973.

16. "Zytel" Design Handbook, E. I. duPont de Nemours and Co., 1972.

17. Puffr, R. and Sebenda, J. in Macromolecular Chemistry, 1965, Prague, (J. Polym. Sci. C, 16), Witcherle, O. and Sedlacek, B. Eds., 1967, Interscience, New York, 79.

18. Taylor, G. B., J. Am. Chem. Soc., 1947, 69, 635.

19. ASTM Specification D789.

20. Starkweather, H. W., Jr., and Moynihan, R. E., J. Polym. Sci., 1956, 22, 363.

21. Boyer, R. F., Spencer, R. S. and Willey, R. M., J. Polym. Sci., 1946, 1, 249.

22. Paris, P. C. and Erdogan, F., Trans. ASME D, 1963, 85, 528.

23. Inoue, M., J. Polym. Sci., A-1, 1963, 2697.

24. Prevorsek, D. C., Butler, R. H., and Reimschussel, H., J. Polym. Sci., 1971, A2, 9, 867.

25. Illers, K. H., Makromol. Chem., 1960, 38, 168.

26. Starkweather, H. W., J. Macromol. Sci., 1969, B3, 727.

27. Skibo, M. D., Hertzberg, R. W., and Manson, J. A., Proceedings, Conference on Deformation and Fracture, April 1979, Cambridge, the Plastics and Rubber Institute, London, paper 4.1.

28. Skibo, M. D., Manson, J. A., Webler, S. M., Hertzberg, R. W., and Collins, E. A., ACS Symposium Series, No. 95, 1979, 311.

29. Hertzberg, R. W., Skibo, M. D., and Manson, J. A., ASTM Symposium on Fatigue Mechanisms, 1978, Kansas City, to be published.

30. Hertzberg, R. W., Skibo, M. D., and Manson, J. A., J. Mater. Sci., 1978, 13, 1038.

31. Laghouati, A. F., Thesis, 3rd Cycle, 1977, Université de Technologie de Compiègne.

32. Skibo, M. D., Hertzberg, R. W., and Manson, J. A., J. Mater. Sci., 1976, 11, 479.

33. Tomkins, B. and Biggs, W. D., J. Mater. Sci., 1969, 4, 544.

34. Broutman, L. J. and Gaggar, S. K., <u>Intern. J. Polym. Mat.</u>, 1972, <u>1</u>, 295.

35. Plumbridge, W. J., <u>J. Mater. Sci.</u>, 1972, <u>7</u>, 939.

36. Yamada, K. and Suzuki, M., <u>Kobunshi Kagaku</u>, 1973, <u>30</u>, 206.

37. McEvily, A. J., Jr., Boettner, R. C., and Johnston, T. L., in <u>Fatigue - An Interdisciplinary Approach</u>, 1964, Syracuse University Press, Syracuse, NY.

RECEIVED January 4, 1980.

The Influence of Water Concentration on the Mechanical and Rheo-Optical Properties of Poly(methyl methacrylate)

R. S. MOORE and J. R. FLICK

Eastman Kodak Co., Rochester, NY 14650

Although several studies have been done of the influence of H_2O on the dynamic mechanical properties of methacrylate polymers (1, 2, 3), relatively little is generally known about the influence of H_2O content on the rheo-optical and tensile properties of such polymers. This study of poly(methyl methacrylate), PMMA, was undertaken to obtain further understanding of the effect of H_2O on these properties and to serve as a model system for examining experimental methods for use in future studies of the mechanical and rheo-optical properties of other methacrylate polymers. In addition, we hope to gain insight into the potential importance of a variation in H_2O content in general studies on other polymers, how it influences the manner and degree of orientation of the polymer in response to stress, and how it influences the correlation between mechanical and optical properties in terms of the stress-optical law. Study of H_2O in PMMA is also of interest as a model system for studying the effect of a diluent at low concentrations.

Experimental

Material Characterization and Sample Preparation. Polymer. The PMMA was a commercial material, Plexiglas RV 811, made by Rohm and Haas Company. This polymer was especially chosen because of its widespread use as an acrylic thermoplastic. It was analyzed by gel permeation chromatography. Molecular weights, determined from the respective polystyrene-equivalent values with appropriate conversion factors, indicate that the polymer has a weight-average molecular weight, $\bar{M}_w$, of 1.5×10^5 and a weight-to-number average molecular weight ratio, $\bar{M}_w/\bar{M}_n$, of 2.3. Hence, the polymer has a reasonably narrow molecular weight distribution.

Specimen Preparation. Specimens for tensile and birefringence studies were prepared from sheets molded at about 200°C between stainless steel platens to the appropriate thickness.

0-8412-0559-0/80/47-127-555$05.00/0
© 1980 American Chemical Society

The initial pellets of the polymer were milled to a granular powder. The polymer was then dried at 110°C to remove residual moisture, a potential source of bubbles during molding. No residual birefringence was observed in the molded sheets as judged by viewing them between crossed polarizers. Specimens for tensile studies were typically 3 x 1 x 0.03 in., and specimens for birefringence studies were 3 x 0.5 x 0.01 in. The thinner specimens were cut on a Thwing-Albert apparatus; thicker specimens were prepared on an electrically driven saw designed for cutting glass. Samples were stored at room temperature in various environments, as discussed below.

Specimen Conditioning. Specimens were conditioned in three different environments as follows: samples placed in a wet environment were stirred in distilled H_2O in capped bottles at room temperature for as long as 29 days before use; samples placed in a dry environment were stored in a desiccator over P_2O_5 for a like period at room temperature; samples stored in a regular environment were kept at ambient conditions for a similar period such that the usual equilibrium H_2O content (see below) obtained.

Figure 1 shows a plot of the percent gain in weight as a function of time for the thin (birefringence) and thick (tensile) specimens after immersion in H_2O. The influence of specimen thickness on the rate of equilibration is evident, thin specimens imbibing H_2O somewhat faster than thick ones. (This effect of thickness is rather common and has its origin in the more rapid attainment of a finite H_2O concentration in the central core in the thin specimens (4).) The equilibrium value is nearly reached after 8-10 days. Results of desorption studies indicate that a like period is required to remove excess H_2O during specimen storage in a normal room environment. Since specimens were immersed until just before testing, no H_2O loss is considered to occur.

Figure 1 also indicates the percent weight loss as a function of time for samples under desiccation. The effect of sample thickness is less pronounced. Again, a period of 10 days appears to be sufficient to attain nearly complete equilibrium. Other studies on previously dried samples indicate that at least 4 h must elapse before a significant increase in H_2O content is observed in desiccated samples exposed to room air. Since these specimens were desiccated until just before testing, we consider that no moisture was present. Results of Figure 1 can be summarized as follows: normal samples contained /0.6% H_2O, dried samples had /0% H_2O, and saturated samples had 2.2% H_2O. Analysis by gas chromatography indicated that a small amount of residual monomer ($\underline{F}$ 0.6%) was present in these specimens. The analysis also indicated that this monomer was <u>not</u> removed by desiccation. Hence, any weight loss reflects loss of H_2O and not monomer.

Apparatus

Tensile Tests. Tensile tests were done on an Instron tensile tester under ambient conditions at strain rates, $\dot{\varepsilon}$, ranging from 1.0×10^{-4} to 5.5×10^{-3} sec^{-1}. Experiments were done in triplicate at each strain rate. Stress, σ_y, and strain, ε_y, at yield (not shown) were determined using the 0.2% offset method (5). Stress, σ_b, and strain, ε_b, at break and the work to break, W_b, (the area under the stress-strain curve) were also calculated. The latter was evaluated via a computer program using a Simpson's Rule method.

Birefringence. A separate apparatus using a laser light source was constructed for measuring the absolute relative birefringence $|\Delta|$, the absolute value of the difference between the refractive indices parallel and perpendicular to the deformation axis, $|[n_{\parallel} - n_{\perp}]|$, of specimens subjected to constant ε deformation in tension. (The absolute values of Δ would equal $(n_{\parallel} - n_0)$ and $(n_{\perp} - n_0)$, where n_0 is the value of n at zero stress.) This apparatus is similar to one described previously (6) and is pictured schematically in Figure 2.

The birefringence is determined from a comparison of changes in transmitted intensity, I, using crossed polarizers, with the transmitted intensity using parallel polarizers, I_0, through the relationship

$$I/I_0 = \sin^2(\delta/2) \quad . \tag{1}$$

Here δ is the retardation, defined as

$$\delta \equiv 2\pi d\Delta/\lambda \quad , \tag{2}$$

where d is the specimen thickness and λ is the wavelength of the incident light. The appropriate derived functions of Δ were calculated by computer. Graphical computer outputs of composite graphs of measurements at several ε were also prepared using specially developed programs. Nominal values of ε were corrected for effects of specimen load on the drive-motor response for the thicker specimens. Values of ε_b were considered to be small enough to not warrant modifying the values of ε from their initial values for both kinds of measurements.

The absolute sign of the relative birefringence was determined as described previously (6). The PMMA studied has negative birefringence at room temperature, in agreement with results of other workers (7).

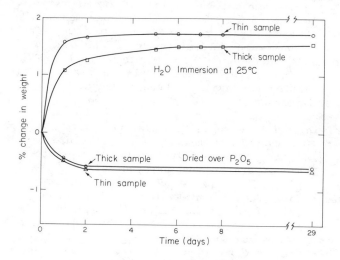

Figure 1. Percent gain or loss in weight for PMMA as a function of time

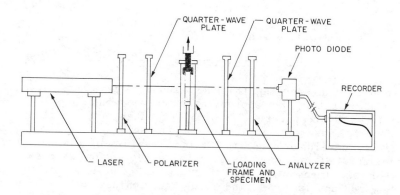

Figure 2. Schematic of the birefringence apparatus

Results

Ultimate Response. A plot of σ_b as a function of $\dot{\varepsilon}$ is given in Figure 3 for specimens conditioned in the three environments. Within the experimental uncertainty, no appreciable dependence on $\dot{\varepsilon}$ or on the environment is apparent at the two highest rates. The rather large scatter in the data reflects in part the sensitivity of ultimate properties of PMMA to adventitious flaws, especially in the materials of lower H_2O content.

Figure 4 is a plot of W_b as a function of $\dot{\varepsilon}$ for materials conditioned in the three environments. No pronounced dependence on $\dot{\varepsilon}$ is observed for the two materials of lower water content, although the maximum in W_b is probably a real effect. The sample with 2.2% H_2O shows a rather dramatic increase in W_b with increasing $\dot{\varepsilon}$. This is primarily due to an increase in ε_b, about 11% at the highest rate, and reflects the enhanced ability of the material to deform. This suggests that the added H_2O acts as a plasticizer, increasing the material's ductility.

The plot in Figure 5 shows the effect of H_2O content on σ_b and ε_b at constant $\dot{\varepsilon}$. With increasing H_2O content, there is a decrease in σ_b while ε_b increases, again suggesting that the H_2O is functioning as a plasticizer. The range in measured values decreases markedly as H_2O content increases. Thus, the effect of adventitious flaws diminishes as the system gains mobility.

Intermediate Response. Figure 6 is a double logarithmic plot of σ/ε vs. time in seconds at three different strain rates for the samples as a function of H_2O content. To extend the time scale and to correlate results at various $\dot{\varepsilon}$, we have used the reduced-variables procedure shown to be applicable in describing the viscoelastic response of rubbery materials (8) as well as of several glassy polymers (6). (To compensate for the effect of different $\dot{\varepsilon}$ we plot σ/ε vs. $\varepsilon/\dot{\varepsilon}$; the latter is simply the time, t.) Superposition over the entire time scale for 0% H_2O (upper curve) is excellent except for times close to the fracture times of the materials tested at the higher strain rates. For example, a deviation occurs at $10^{1.18}$ sec for the material at $\dot{\varepsilon}$ = 3.3 x 10^{-3} sec^{-1}.

Results for samples containing 0.6% H_2O, the middle curve in Figure 6, are generally similar to those having 0% water. However, yielding is more pronounced at 0.6% H_2O, deviations occurring at $10^{1.22}$ and $10^{1.58}$ sec. More curvature (and hence a smaller slope) than is seen in the upper curve is evident at the lowest rate at long times. Since the magnitude of σ/ε is also less, the slope of the corresponding linear plot of σ/ε vs. t (at the same t) would also be lower. Because $d(\sigma/\varepsilon)/dt$ is equal to $E(t)$, the tensile relaxation modulus, the modulus thus decreases with increasing H_2O content.

A similar effect is seen in the lowest curve in Figure 6 at the lowest $\dot{\varepsilon}$ for samples of highest H_2O content. Generally,

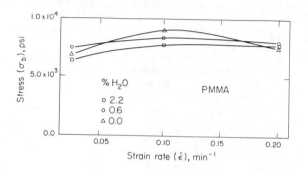

Figure 3. *Plot of stress at break for PMMA vs. strain rate for specimens with the H_2O content indicated*

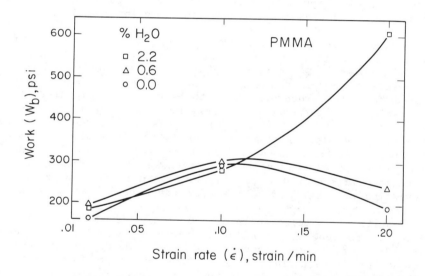

Figure 4. *Plot of work to break for PMMA vs. strain rate for specimens with the H_2O content indicated*

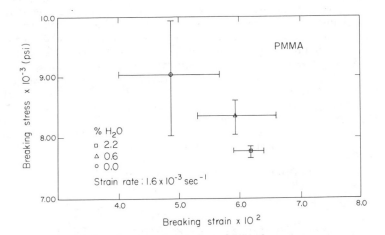

Figure 5. Plot of breaking stress vs. breaking strain for materials at the H₂O content indicated at a strain rate of 1.67 × 10⁻³ sec⁻¹

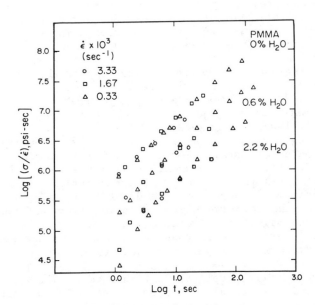

Figure 6. Double logarithmic plot of stress/strain rate vs. time (sec) for PMMA at three strain rates for samples of: (upper curve) ⌐0% H₂O; (middle curve) 0.6% H₂O; (lower curve) 2.2% H₂O. Upper curve displaced up by 0.5 log unit; lower curve displaced down by 0.5 log unit.

however, these effects of H_2O content are rather minor in comparing changes in curvature. Although still minor, the extent of
departure from the general curve <u>does</u> appear to be a function of
the H_2O content, owing to its influence on the extent of yielding, the deviation being greatest at $10^{1.6}$ sec for the sample of
highest H_2O content, as indicated in Figure 6 at 2.2% H_2O. The
dependence of σ/ε on $\dot{\varepsilon}$ at the shortest times in this case is
somewhat uncertain owing to difficulties of measurement. The
results suggest, however, that if the effect is real, a lower
reduced stress level ($\sigma/\dot{\varepsilon}$) obtains for a lower $\dot{\varepsilon}$ so that the
material appears to be slightly less rigid than at the higher
rates. This is consistent with orientation effects determined by
birefringence, as discussed below. Except for the deviations
noted above, superposition of all the data (not shown) is generally good, indicating that the influence of H_2O is relatively
minor except at the shortest times and at times after yielding
just before failure.

    <u>Rheo-Optical Properties. Intermediate Response.</u>   Plots
using reduced variables similar to those for mechanical response
were prepared from birefringence measurements. These are shown
in Figure 7 as double logarithmic plots of the absolute value of
birefringence divided by strain rate vs. time. This type of plot
was discussed previously for other glassy polymers ([6]). Relatively little influence of $\dot{\varepsilon}$ on the reduced variables response is
observed in the upper curve, superposition being generally excellent for the polymer specimens of ~0% H_2O content.
    The middle curve shows the corresponding results for the
samples having 0.6% H_2O content. Here considerable dependence on
$\dot{\varepsilon}$ is observed at times below 10 sec. The reduced absolute bire-
fringence, $|\Delta|/\dot{\varepsilon}$, becomes larger as $\dot{\varepsilon}$ decreases for constant t.
Thus, the material acts proportionately as though it were more
mobile at the lower strain rates.
    A strong dependence on $\dot{\varepsilon}$ is observed for the specimens of
2.2% H_2O content at short times, as indicated in the lower curve
of Figure 7. Here a difference of almost a factor of $10^{0.6}$ is
found between results at the two extreme strain rates. Clearly,
the influence of changes in $\dot{\varepsilon}$ is largest in samples of highest
H_2O content, the relative degree of orientation and mobility, as
judged by the magnitude of $\Delta/\dot{\varepsilon}$, being greater the lower the rate
of strain. This trend is consistent with the decrease in $\sigma/\dot{\varepsilon}$
observed with decreasing $\dot{\varepsilon}$ as discussed previously.

    <u>Strain and Stress Optical Coefficients.</u>   The relationship
between Δ and ε is usually defined in terms of results obtained
in the customary stress relaxation experiment in which a specimen
is deformed to a constant strain, and σ (or in this case Δ) is
measured as a function of time. The strain optical coefficient,
$K(t)$, is defined as

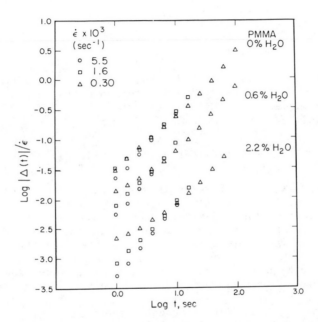

Figure 7. Double logarithmic plot of the absolute value of birefringence divided by strain rate vs. time for PMMA at the three strain rates indicated for samples of: (upper curve) ~0% H_2O; (middle curve) 0.6% H_2O; (lower curve) 2.2% H_2O. Upper curve displaced up by 0.5 log unit; lower curve displaced down by 1.0 log unit.

$$K(t) \equiv \Delta(t)/\varepsilon \; , \tag{3}$$

and the tensile stress relaxation modulus is given by

$$E(t) = \sigma(t)/\varepsilon \; . \tag{4}$$

Just as $E(t) = d(\sigma/\dot{\varepsilon})/dt$ at time (t) at constant strain rate ([6], [8]), so $K(t) = d(\Delta/\dot{\varepsilon})/dt$ at time (t) at constant $\dot{\varepsilon}$ ([6], [9], [10]).

The stress optical coefficient, C, is defined for the stress relaxation experiment as

$$C \equiv \Delta(t)/\sigma(t) \tag{5}$$

so that

$$C(t) = K(t)/E(t) \; . \tag{6}$$

Experimentally, C is generally a constant ([9]), as predicted by theory ([6], [9]). Previously ([6]) it was shown that a more sensitive test than Equation 6 of the constancy of C is to consider $[\Delta(t)/\dot{\varepsilon}]/[\sigma(t)/\dot{\varepsilon}]$ as a function of t, since

$$\frac{\Delta(t)/\dot{\varepsilon}}{\sigma(t)/\dot{\varepsilon}} = \int_0^t K(t)dt / \int_0^t E(t)dt \; . \tag{7}$$

These integrals, which involve integration over the corresponding relaxation spectra ([6], [9]), will yield a constant value for C only if the corresponding relaxation spectra are equivalent over the whole time interval of integration ([6]). This test of the constancy of the value of C is then to compare logarithmic plots of $\Delta(t)/\varepsilon$ and $\sigma(t)/\varepsilon$ with one arbitrary vertical shift, as shown in the plots in Figure 8. Here superposition is excellent over the entire range of values for the upper curve. The value of C so obtained is 5.3×10^{-13} cm^2/dyne. Although difficult to estimate, the uncertainty in C is probably about 5%.

Results for samples with 0.6% H_2O are shown in the middle curve in Figure 8. Here generally good superposition is obtained at longer times except for regions in which mechanical results are dominated by yielding. The lower plot in Figure 8 shows results for specimens with 2.2% H_2O. Within the limitations just cited for the middle curve, superposition is also very good. The value of C obtained for the plots of Figure 8 is virtually the same, 5.3×10^{-13} cm^2/dyne, suggesting that the stress optical coefficient is independent of time, of H_2O content, and of $\dot{\varepsilon}$ at sufficiently long times, t > 10 sec, but before pronounced yielding. Below 10 sec the results suggest that C increases with decreasing $\dot{\varepsilon}$, the effect being more pronounced the higher the H_2O content.

The magnitude of C is about half that for polystyrene and for a typical polystyrene-acrylonitrile copolymer (6), 1.0 x 10^{-12} cm^2/dyne. Further interpretation of this difference must await measurements of the absolute stress optical coefficients parallel and perpendicular to the deformation axis, $[(n_{||} - n_0)/\sigma$ and $(n_{\perp} - n_0)/\sigma]$, since the sign of C is positive for these two polymers (6), whereas it is negative for PMMA. Our studies of the dynamic mechanical response of these materials provide a more detailed understanding of the mechanism of enhanced mobility in the specimens of higher H_2O content. Studies of fracture surface morphology also indicate differences in ultimate response as a function of H_2O concentration.

Indirect Observation of Deformation Bands.   Although birefringence results shown in the various plots to this point imply a smoothly increasing value of the intensity, I, such was not always the case. An interesting and entirely unexpected behavior was observed in the response of many specimens. In some cases, though not in all, this phenomenon occurred only when the laser beam traversed the sample at the subsequent fracture point. A typical example, in which the relative photodiode response (see Equation 1) is plotted as a function of time, is seen in Figure 9. It is evident that a sinusoidal oscillation occurs about the steadily increasing transient value; the plots previously shown were based on a smoothed transient curve for calculations of Δ. The period of oscillation is nearly independent of time. We have also observed constant periodicity for other methacrylate polymers. The period appears to be proportional to the strain rate, as judged from several experiments, so that the calculated spacings (see below) are independent of $\dot{e}$. The magnitude of the fluctuations is typically 2-4% of the transient value in terms of Δ. (The amplitude increases with time, but the degree of modulation remains nearly constant because of the increase in Δ or in I.) Thus, fluctuations of this magnitude in σ, if they did occur, would not be likely to be observed in typical tensile tests.

In our birefringence apparatus the upper grip moves, but the lower one does not, so that the velocity of the center of the specimen relative to the light beam position is about half that of the upper grip. Calculations based on time, velocity, and period of oscillation indicate that if we are observing a periodic-banded structure, its width ranges from 8.5 to 18.3 µm, depending on the polymer, if we assume the structure bands lie at 45° to the tensile axis. (This assumption causes a decrease of a factor of only $\sqrt{2}/2$ from the width which would obtain for a structure at right angles to the deformation axis.) It is unlikely that sample slippage is the origin of such periodicity; the regularity of modulation argues against this. Similarly, mechanical oscillation of the drive mechanism would be expected to cause oscillations in all experiments. Such was not the case.

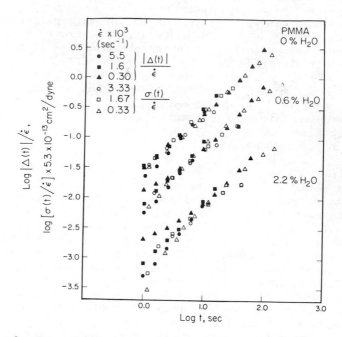

Figure 8.   *Composite logarithmic plots of stress divided by strain rate and birefrin-*
gence divided by strain rate vs. time at the strain rates indicated for samples of:
(upper curve) ⁓0% H₂O; (middle curve) 0.6% H₂O; (lower curve) 2.2% H₂O.
Upper curve displaced up by 0.5 log unit; lower curve displaced down by 1.0 log
unit.

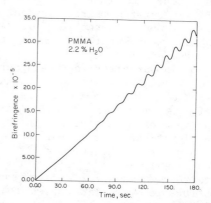

Figure 9.   *Recorder output of relative*
photodiode response as a function of time
for a birefringence experiment on PMMA
containing 2.2% H₂O at a strain rate of
1.0 × 10⁻⁴ sec⁻¹

Periodicities in laser intensity can be ruled out based on monitoring of the intensity before the start of the experiment.

Although our evidence is indirect, we feel the observed periodicity is due to the formation of incipient deformation bands, or Lüder's bands, in the specimen. Such bands have been observed in polystyrene (11,12) in compression and in preoriented polystyrene specimens in tension and in torsion (12). Such bands have also been observed in poly(vinyl chloride) (13). In polystyrene the angles between the band axis and the axis band of deformation were 38° in compression and 52° in tension (12). (Our results are calculated arbitrarily for the intermediate value of 45°.) The measured widths in polystyrene are 1-10 μm (11,12). Our results are in reasonable agreement with this band size.

In contrast to observations in polystyrene, we do not observe permanent bands; our specimens exhibit no residual birefringence upon release from stress. Neither do we observe crazing before failure. However, the specimens do whiten just before failure when viewed edge-on, and this whitening disappears within a few seconds after fracture occurs. We think the oscillations in intensity we observe are likely to be due to incipient shear deformation which disappears after specimen failure. Unpublished results of other workers are reported (see References 11 and 14 in the present Reference 12) to be consistent with the idea that such bands should be difficult to observe in PMMA and in polycarbonate because of their lower draw ratios compared to polystyrene. Studies of an unfilled epoxy polymer (14) in cyclic tensile deformation indicate that shear bands do not remain after removal of stress until a threshold amplitude of deformation is exceeded.

Results for polystyrene (12) indicate that the bands start to occur at a tensile yield stress defined (12) as the point at which stress is no longer proportional to strain. For our specimen, the onset of oscillation occurs at a stress of 3.9×10^7 dynes/cm^2, and its tensile yield stress as defined above is 4.3×10^7 dynes/cm^2, in excellent agreement. If our interpretation is correct, these results suggest that after tensile yielding an additional mode of deformation arises such that spatially periodic incipient yielding in shear occurs at ~45° to the deformation axis. After failure, this shear yielding is recoverable, leading to no residual birefringence in the specimen.

Summary

Results of studies of tensile deformation at constant $\dot{\varepsilon}$ in samples of PMMA containing from 0 to 2.2% H_2O indicate that mechanical response is relatively insensitive to $\dot{\varepsilon}$ when considered in terms of reduced variables at intermediate strains. At short times or just before failure, the specimens with higher H_2O

content have slightly enhanced ductility. Birefringence measurements indicate a strong enhancement of orientation at short times
when response is considered in terms of reduced variables. The
stress optical coefficient is, with these exceptions, independent
of H_2O content, $\dot{\varepsilon}$, and time, $|SOC|$ being about half that for
polystyrene or for a typical poly(styrene acrylonitrile) copoly-
mer (6). The present results provide a starting point for
studies of the absolute birefringence and the degree of orientation in deformed polymers. The results also show that the concentration of H_2O has a minor effect on the longer time response
of commercial PMMA.

Results in many specimens indicate that the gradually increasing intensity in birefringence is accompanied by periodic
oscillation. Although the evidence is indirect, we feel that the
latter observations are consistent with the formation of incipient deformation bands at ~45° to the tensile stress axis after
the onset of tensile yielding. These nonpermanent bands then
disappear upon sample failure.

Abstract

As part of a program directed toward understanding brittle-
ness in glassy polymers, this paper reports results of studies of
birefringence (Δ) and tensile stress measurements (σ) at constant
strain rates $(\dot{\varepsilon})$ ranging from 1.0×10^{-4} to 5.5×10^{-3} sec^{-1} on a
commercial poly(methyl methacrylate). The polymer of $\bar{M}$ 1.5 x
10^5 ($\bar{M}_w/\bar{M}_n = 2.3$) contained between 0 and 2.2% H_2O. All measurements were made at room temperature. Results of tensile
measurements indicate that plots of $(\sigma/\dot{\varepsilon})$ versus time (t) were
relatively insensitive to $\dot{\varepsilon}$ and H_2O content except at values of
strain (ε) greater than the yield values. Below 10 sec, the
dependence of $(\Delta/\dot{\varepsilon})$ on $\dot{\varepsilon}$ became larger the higher the H_2O content. The absolute value of the average stress optical coeffi-
cient, 5.3×10^{-3} cm^2/dyne as determined from these logarithmic
plots, was independent of H_2O concentration, of $\dot{\varepsilon}$, and of t, for
$t > 10$ sec. The influence of H_2O concentration on Δ, greatest at
short times, suggests that it contributes to enhanced mobility of
the system, functioning in many respects as a plasticizer. At
longer times a periodic oscillation of $|\Delta|$ about its gradually
increasing transient value was observed. Although the evidence
is indirect, this periodic variation in Δ is consistent with the
formation of incipient deformation bands in shear at ~45° to the
tensile stress axis after the onset of tensile yielding. The
periodic variation in birefringence disappears after specimen
failure, suggesting that in this case it is associated with a
recoverable elastic deformation.

Literature Cited

1. Gall, W. G.; McCrum, N. G. J. Polym. Sci. 1961, 50, 489.
2. Shen, M. C.; Strong, J. D. J. Appl. Phys. 1967, 38, 4197.
3. Janáček, J. J. Macromol. Sci. Phys. 1968, B2, 497.
4. Crank, J. S.; Park, G. S. "Diffusion in Polymers"; Academic Press: New York, 1968.
5. See, for example, American Society for Testing Materials procedure D-638-68.
6. Moore, R. S.; Gieniewski, C. J. Appl. Phys. 1970, 41, 4367.
7. Andrews, R. D.; Hammack, T. J. J. Polymer Sci. 1964, 5, 101.
8. Ferry, J. D. "Viscoelastic Properties of Polymers"; Wiley: New York, 1961; Chapter 13.
9. Stein, R. S.; Onogi, S.; Sasaguri, K.; Keedy, D. A. J. Appl. Phys. 1963, 34, 80.
10. Yamada, R.; Stein, R. S. ibid. 1965, 36, 3005.
11. Whitney, W. J. Appl. Phys. 1963, 34, 3633.
12. Argon, A. S.; Andrews, R. D.; Godrick, J. A.; Whitney, W. ibid. 1968, 39, 1899.
13. Bauwens, J. C. J. Polym. Sci. Part A-2 1970, 8, 893.
14. Ishai, O.; Bodner, S. R. Trans. Soc. Rheol. 1970, 14, 253.

RECEIVED January 4, 1980.

Water–Epoxy Interactions in Three Epoxy Resins and Their Composites

JOYCE L. ILLINGER and NATHANIEL S. SCHNEIDER

Army Materials and Mechanics Research Center, Watertown, MA 02172

Glass fiber reinforced epoxy resin composites are finding increasing use for critical structural applications in Army materiel. Of necessity these systems are exposed to varying environments, both in storage and in operation. Given the ubiquity of water in the environment, concern has arisen about the nature of the interaction of water with these systems, and the effects of water on the material properties of the composites. Of the many epoxy resin systems available, we have studied three in some detail. Resin III was part of a long term project on determining mechanisms of deterioration during environmental exposure. Resins I and II are materials which are under serious consideration for use in Army materiel. The results of these studies are reported here.

Experimental

Resins. The structural formulae for the major components of these systems are shown in Figure 1.

Resin I contains roughly equal amounts of diglycidyl ether of bisphenol A (DGEBA) and an epoxy cresol novolac. Sufficient dicyandiamide (DICY) as a curing agent is present such that the amine/epoxy ratio is 0.85. Monuron is present as an accelerator. The supplier's recommend standard cure is two hours at 127°C. Previous work (1) has shown that this cycle produces a fully cured system, as indicated by the disappearance of the epoxide absorbance band in the infrared spectrum.

Resin II contains tetraglycidyl methylene dianiline (TGMDA) as the principal epoxide with small amounts of a cresol novolac and DGEBA. DICY as curing agent is present in an amount such that the amine epoxy ratio is 0.25. Diuron is the accelerator. The supplier's recommended standard cure is the same as for SP250. However, in this system approximately 20 percent of the epoxide remains unreacted following the cure cycle (1). Selected resin plates were subjected to additional cure at 170°C and 220°C. The samples post cured at 220°C for forty minutes showed no residual epoxide absorbance in the IR spectrum.

This chapter not subject to U.S. copyright.
Published 1980 American Chemical Society

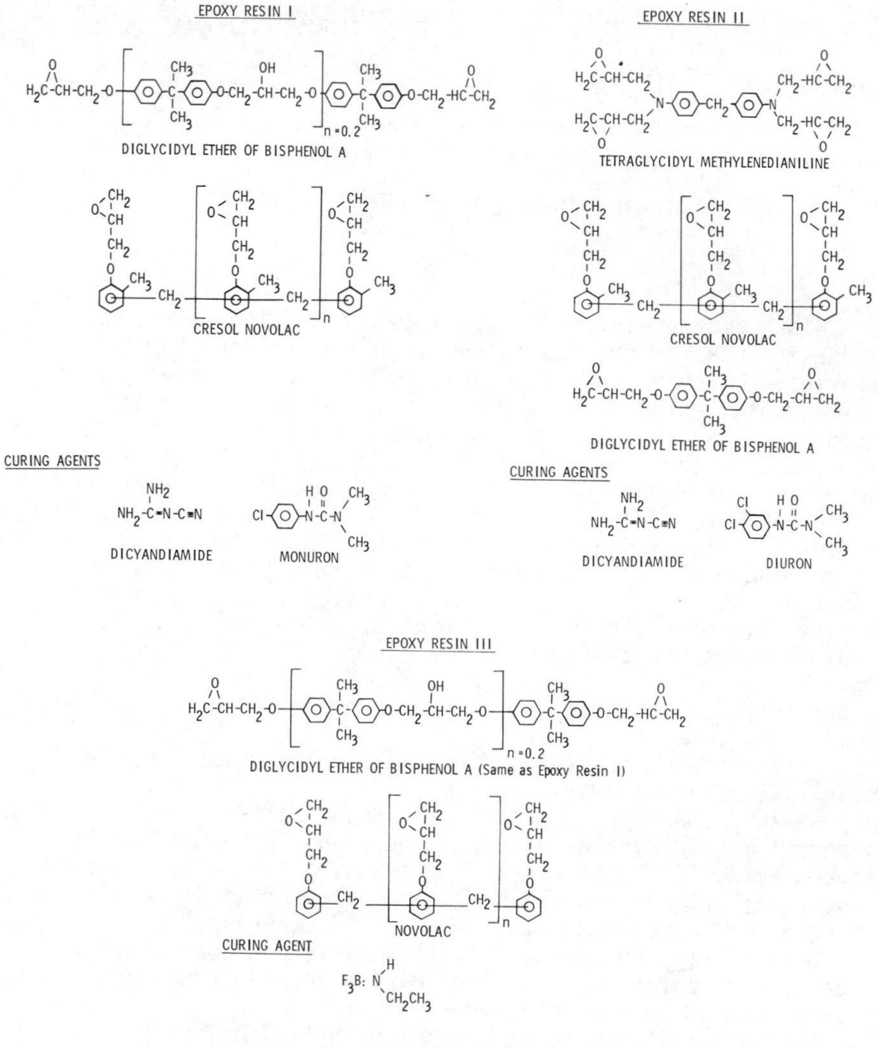

Figure 1. Structural formulas for resin components

Resin III is based upon DGEBA and an epoxy novolac in a one-to-two weight ratio. The curing agent is 3 percent boron trifluoride monoethylamine complex. The supplier's recommend cure cycle is 0.75 hour at 165°C followed by 4 hours post cure at 177°C. Some samples were cured for 24 hours at 80°C before post cure, while other samples were cured for the standard 0.75 hour at 165°C and subjected to different post cure times (2). There is still residual epoxide absorbance following initial 165°C cure. This absorbance decreases during post cure reaching zero after 4 hours.

Composites. Panels from prepregs of Resins I and II with S2 glass fiber were fabricated in a 2-ply unidirectional lay-up using the following cure cycle (3). After lay-up the panel was bagged in plastic, evacuated, and placed in an autoclave. Pressure was raised to 20 ± 5 psig. Temperature was raised to 70 ± 6°C after which pressure was increased to 90 ± 10 psig. The temperature was then raised at 3°C per minute to 127 ± 3°C and held for 2 hours. Following this, temperature was reduced to 60°C, pressure released and the panel removed from the autoclave.

Specimens of dimensions 165mm X 19mm were cut from the panels with the fibers at 10° off axis. When tested in tension, this configuration is sensitive to changes in interfacial bonding (4). Following conditioning (described below) they were tested in tension in an Instron. The samples contained 33 ± 3% resin as determined by resin burn off.

Crossply panels of Resin III and E glass in a six ply configuration (0/0/90/90/0/0) were supplied by the 3M Company with nominal 26% resin content. The cure cycle was reported as follows. The panels were bagged, evacuated, and placed in an autoclave. Pressure was applied for consolidation below the gel point. The temperature was then raised to 165°C at 3° per minute. The temperature was held at 165°C for 45 minutes, the panels were cooled to 60°C, removed from the autoclave and transferred to an oven. They were post cured for four hours at 177°C.

Tensile dumbbells were cut at 0 and 90 degrees to the outer plies producing samples designated 0/90/0 or 90/0/90. These samples were tested in an Instron following conditioning.

Water Measurements and Conditioning. The samples were suspended from glass T's above saturated salt solutions or distilled water which produced different humidities at each temperature (30,45,60°C). The humidity chambers were immersed in circulating water baths or forced air ovens to maintain constant temperature. Samples were also immersed in distilled water at each temperature. Periodic weighings were made and results recorded as detailed elsewhere (5,3). Desorption of water from selected samples was performed by holding the samples over dessicant at the same temperature as sorption had been performed.

Results and Discussion

Resins. Samples of Resin II (25mm x 25mm x 1.5mm) were monitored at 60°C and 80, 95, 100% RH and under immersion conditions. Equlibrium sorption increases with relative humidity as is generally the case in polymer systems, with an isotherm which is slightly concave upward. The 100% RH and immersion data were virtually identical. At these thicknesses, equilibration time was disturbingly slow. Figure 2A shows the results of these measurements plotted conventionally as a function of the square root of time. If the sorption followed a mechanism described by Fick's equations, the percent uptake would show a linear plot to up to 60 or 70 percent of equilibrium uptake, with a zero intercept. This is clearly not the case with this resin. There is evidence of early continuing curvature at all three humidities and there appears to be two or more breaks in the curves.

Another approach to the interpretation of sorption kinetics has been suggested by Jacques (5) et al following Alfrey (6) where one assumes that uptake can be represented by

$$M_t = kt^n$$

where M_t is sorption at time t, and k includes geometric constants and the diffusion coefficient. Plots of lot M_t vs log t should reveal changes in sorption mechanisms. A slope of 1/2 would indicate Fickian behavior, that of 1 would indicate control of sorption by a relaxation mechanism. Eventually, the slopes should decrease to zero as equilibrium is attained. Figure 2B is a plot of the data in this form. The slope at any portion of these curves is less than 1/2 so that the sorption mechanism cannot be identified within the above stated limits. These deviations occur well below 30% of equilibrium uptake.

This system shows additional anomalies, the total uptake is high and also some temperature dependence of equilibrium sorption occurs, contrary to the behavior of other systems (6,7,8) with the same epoxide but different curing agent.

Samples which had undergone the standard cure were then subjected to post curing at two elevated temperatures selected from DSC traces of the curing behavior (1). The fully cured system (no residual epoxide) did show different sorption kinetics as shown in Figure 3A. The early slopes are greater than 1/2 indicating possible combination Fickian and relaxation behavior. The approach to equilibrium sorption is faster and the final level is lower than that of the standard cure samples. As Figure 3B shows, it is only by going to the extreme cure that this behavior is seen. Intermediate cure at lower temperature produced some modification of early time kinetics but little change in the long time approach to very high equilibrium values.

Resin III (Figure 4) cured at the 80°C for 24 hours shows slopes at any point in the curve of less than 1/2 (similar to

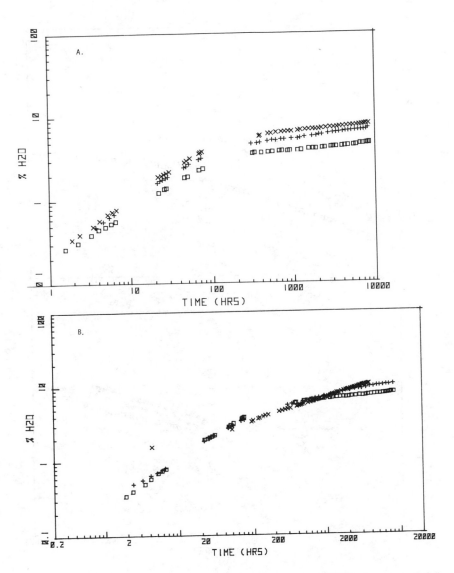

Figure 2.   Sorption kinetics for resin II, standard cure at 60°C; A. conventional plot; B. log M_t vs. log t: (×) immersion, (+) 95% RH, (□) 80% RH.

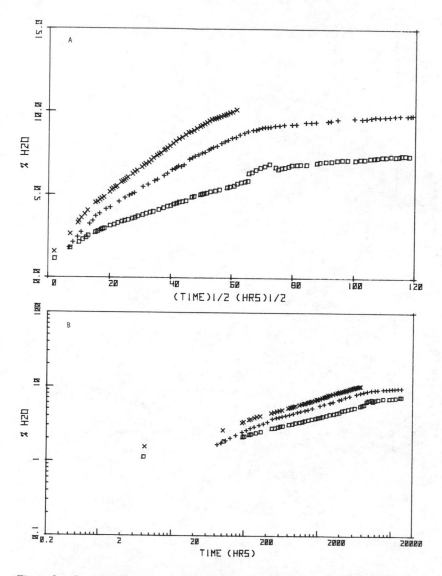

Figure 3.   Sorption kinetics for resin II at 60°C effects of post cure: A. cured at 220°C for 42 min, no residual epoxide: (×) immerision, (+) 95% RH, (□) 80% RH; B. immersion conditions, different degree of post cure: (×) no post cure, (+) 170°C for 20 min, (□) 220°C for 42 min.

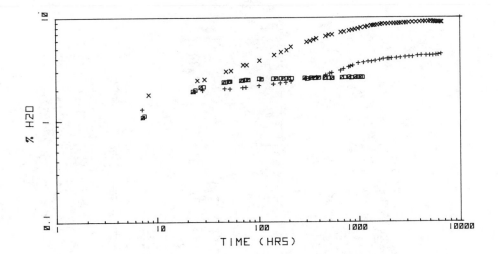

*Figure 4.   Sorption kinetics for resin III, immersion at 60°C (effects of post cure):
(×) 80°C for 24 hr, (+) 165°C for 1 hr, (□) 80°C for 24 hr followed by
177°C for 4 hr, (⧄) 165°C for 1 hr followed by 177°C for 4 hr.*

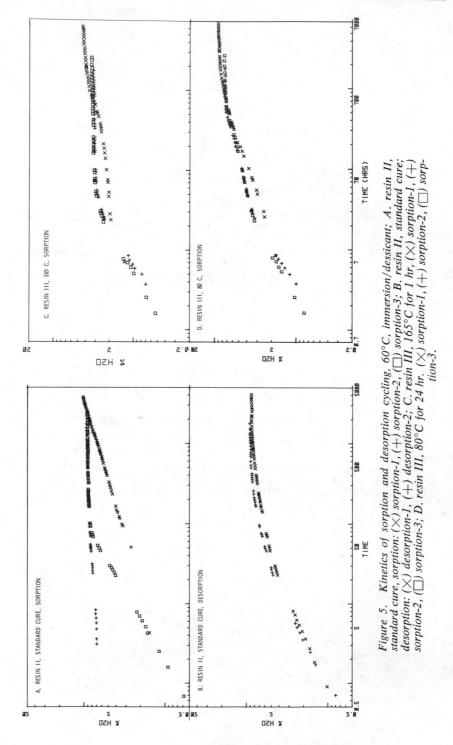

Figure 5. Kinetics of sorption and desorption cycling, 60°C, immersion/dessicant; A. resin II, standard cure, sorption: (×) sorption-1, (+) sorption-2, (□) sorption-3; B. resin II, standard cure, desorption: (×) desorption-1, (+) desorption-2; C. resin III, 165°C for 1 hr, (×) sorption-1, (+) sorption-2, (□) sorption-3; D. resin III, 80°C for 24 hr. (×) sorption-1, (+) sorption-2, (□) sorption-3.

standard cure Resin II), while the 165°C one hour cured material shows initial behavior which is apparently Fickian followed by another mechanism. Subjecting both types of samples to four hour post cure at 177°C removes the differences and yields a material whose sorption kinetics are well behaved.

The standard cure Resin II and both non post cured Resin III samples were subjected to immersion/drying cycles. The results are seen in Figure 5, A-D. The first cycle for Resin II has already been discussed. The second cycle shows extremely rapid short term sorption and then slow levelling off to equilibrium. The third cycle shows Fickian behavior. Obviously previous moisture cycling causes changes in the interactions and in the controlling sorption mechanism in subsequent cycles. The initial desorption (Figure 5B) does not remove all of the previously sorbed water, further indicating a change in the material due to moisture sorption. The mechanism appears to follow Fickian kinetics, as does the second desorption. In this case all of the water sorbed in the second cycle is removed. The 80°C 24 hour cured Resin III (Figure 5D) shows even greater differences in the sorption cycles and no indication that cycling effects are levelling off. The 165°C one hour cured Resin III (Figure 5C) material, however, shows the second and third cycles to be virtually identical indicating that the changes are limited to the first sorption.

Composites. Sorption plots for the composite systems Resin I/S-glass and Resin II/S-glass, are shown in Figure 6A,B. In general, the kinetics appear to be Fickian, i.e., in most cases, initial slopes of these plots are close to 0.5. The total uptakes are much lower than the corresponding pure resins and in the range expected if all sorption occurs in the resin of the samples. This rules out the possibility that any significant amount of water is accumulating along the fiber/resin interface in pockets or voids. Two of the Resin I/S-glass samples show a downturn in the amount of water uptake starting at about 2% water suggesting that leaching might be occurring (the samples after drying were lighter than initially.) The downturn appears at earlier times under immersion than for the 100% RH condition. The 95% RH curve is just levelling off while the 80% RH samples are still sorbing (the experiment was terminated by oven failure.) The Resin II/S-glass composites show parallel sorption behavior over the full range of relative humidity and are free of the complications exhibited by the Resin I/S-glass samples. The Resin III/E glass composites (not shown) also show straightforward sorption behavior similar to that of Resin II/S, achieving lower total uptake.

Resin II/S-glass and Resin I/S-glass composites were tested in tension on unidirectional samples at 10° off axis. This test is considered to be sensitive to changes in interfacial bonding and is useful as a measure of moisture induced degradation. The data scatter for both systems is quite broad. Statistical

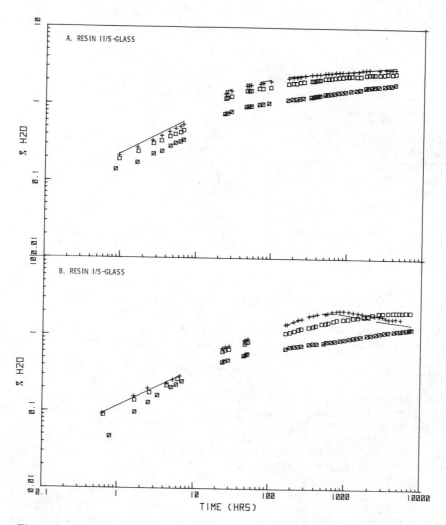

Figure 6.   Kinetics for composite sorption. A. resin II/S-glass, (—) immersion, (+) 100% RH, (☐) 95% RH, (☑) 80% RH; B. resin I/S-glass, (—) immersion, (+) 100% RH, (☐) 95% RH, (☑) 80% RH.

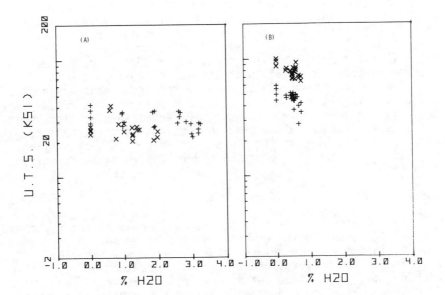

Figure 7.   Strength properties of composites vs. water content. A. 10° off axis test, (✕) resin I, (+) resin II, B. cross-ply specimens of resin III, (✕) 0/90/0 configuration, (+) 90/0/90 configuration.

analysis shows the variances to be so large that quantitative predictions are not possible from these limited data. However, some trends are apparent. Of the two sets of composites, Resin II/S shows higher dry ultimate tensile strength (U.T.S) and perhaps 20% degradation as a function of moisture content as opposed to Resin I/S which shows a significantly lower dry U.T.S. and a tensile strength reduction of perhaps 15%. However, it should be noted that the highest water uptake of Resin I is only about 60% of that for the Resin II/S glass composites. Initial modulus response to water is similar to U.T.S. response.

The Resin III/E glass composites (Figure 7B) showed even higher dry U.T.S. (although this may be due to crossply configuration rather than higher resin strength) and the 0/90/0 configuration showed higher strength than the 90/0/90. Temperature of moisture conditioning appears to have little or no effect except that the 60°C conditioned samples sorbed slightly more water than did those conditioned at 45 or 30°C. Again, there is a downward trend in U.T.S. with sorbed moisture (log plot).

The comparison of all configurations shows that Resin III/E dry U.T.S. is higher and its strength drops off more rapidly than Resin I/S or Resin II/S. However, the total sorption for Resin III/E is lower than in the other composites so that the strength remains higher, even at its highest water content, signifying superior retention of properties under 100% relative humidity conditions.

Conclusions

The results show that different bulk epoxy resins may exhibit large differences in the equilibrium amount of sorbed moisture as well as in the kinetics of sorption. Marked nonreversible effects are observed in at least one system which can not be reconciled with present guide lines for discriminating between such causes as matrix relaxation and other types of non-Fickian diffusion. However, the changes in sorption behavior on repeated cycling argues strongly for a change in the structure of the resin caused by the initial sorption cycle. Examination by optical microscopy or scanning electron microscopy should be helpful in identifying changes in the resin involving, possibly, surface or bulk microcracking and void formation. With Resin II the failure to remove all sorbed water in the first desorption cycle also argues strongly for changes in the resin chemistry.

Considering the complexity of the sorption behavior of the bulk resin, the almost classical Fickian behavior of the composites comes as a surprise and represents a welcome simplification. At the present time the only explanation that can be made is by analogy with the size dependence of relaxation controlled diffusion effects reported by Berens (13) for diffusion of vinyl chloride in poly(vinylchloride). It was noted that as the particle size and, therefore, the path length diminished, the contribution of the relaxation controlled behavior disappeared. Perhaps the small matrix sections between the fibers are analagous to the small particle condition in the above work. In drawing such an analogy, the constraint to matrix swelling introduced by bonding to the glass fibers is a problem. However, since in all the cases studied here diffusion occurs transverse to the fiber layers, these constraints might have little effect on the character of the time dependent sorption.

Literature Cited.

1. Schneider, N.S.; Sprouse, J.F.; Hagnauer, G.L.; Gillham, J.K. Polym. Eng. & Sci., 1979 19, 304.

2. Illinger, J.L; Sprouse, J.F. Org. Coat. & Plast. Preprints, 1978,338, 497.

3. Sprouse, J.F; Halpin, B.M. Jr.; Sacher, R.E. AMMRC TR 78-45, 1978.

4. Chamis, C.C.; Sinclair, J.H. NASA TN D-8215, April 1976.

5. Illinger, J.L.; Schneider, N.S. Org. Coat. & Plast. Preprint, 1978, 39, 462 and Polym. Eng. & Sci. (in press 1979).

6. Jacques, C.H.M.; Hopfenberg, H.B.; Stannett, V. Polym Sci. & Tech., 1974, 6, 73.

7. Alfrey,,T. Jr.; Gurnee, E.F.; Lloyd, W.D. J. Polym. Sci., 1966, C12, 249.

8. Angl, J.M.; Berger, A.E. NSWC-WOL-TR 76-7, 1976.

9. Augl, J.M.; Berger, A.E. NSWC-WOL-TR 77-13, 1977.

10. Browning, C.E., Polym. Eng. & Sci., 1978, 18, 16.

11. Browing, C.E. AFML TR 72-94, 1972.

12. Shirrell, C.D. Reported at ASTM meeting, Dayton, OH, Sept 1978 (in press).

13. Berens, A.R. Polymer, 1977, 18 697.

RECEIVED January 4, 1980.

INDEX